Water-Lubricated Bearings

Water-Lubricated Bearings

Editors

Zhongliang Xie
Yanfeng Han

MDPI

Basel • Beijing • Wuhan • Barcelona • Belgrade • Novi Sad • Cluj • Manchester

Editors
Zhongliang Xie
Northwestern Polytechnical University
Xi'an
China

Yanfeng Han
Chongqing University
Chongqing
China

Editorial Office
MDPI AG
Grosspeteranlage 5
4052 Basel, Switzerland

This is a reprint of articles from the Special Issue published online in the open access journal *Lubricants* (ISSN 2075-4442) (available at: https://www.mdpi.com/journal/lubricants/special_issues/4W6Y0Y0EQV).

For citation purposes, cite each article independently as indicated on the article page online and as indicated below:

Lastname, A.A.; Lastname, B.B. Article Title. *Journal Name* **Year**, *Volume Number*, Page Range.

ISBN 978-3-7258-2125-9 (Hbk)
ISBN 978-3-7258-2126-6 (PDF)
doi.org/10.3390/books978-3-7258-2126-6

Contents

lubricants

MDPI

Article

Theoretical Study on the Dynamic Characteristics of Marine Stern Bearing Considering Cavitation and Bending Deformation Effects of the Shaft

Tao He [1,*], Zhongliang Xie [2,*], Zhiwu Ke [1], Lu Dai [1], Yong Liu [1], Can Ma [1] and Jian Jiao [3]

1 Wuhan Second Ship Design and Research Institute, Wuhan 430205, China
2 Department of Engineering Mechanics, Northwestern Polytechnical University, Xi'an 710072, China
3 School of Mechanical-Electric Engineering, Xidian University, Xi'an 710071, China
* Correspondence: hetao05031213@163.com (T.H.); zlxie@nwpu.edu.cn (Z.X.)

Abstract: When the ship runs, owing to the superposition of the gravity of the shaft and resistance of water, with the increment in rotational speeds, the shaft will produce different degrees of bending deformation, which immensely reduces the power transmission efficiency. Based on the aforementioned problem, the present study focuses on the influences of bending deformation of the shaft with a cavitation effect on the dynamic characteristics of the stern bearing. The mixed lubrication model with bending deformation and cavitation effect is established. At present, the deflection curve equation is employed, the finite perturbation method is applied to calculate the dynamic coefficient, and the cavitation pressure is determined by the numerical method. According to the analysis, the variation laws of equivalent stiffness and natural frequency are exhibited. It is shown that the equivalent stiffness is more affected by the speeds, especially at low speeds; There is a critical speed between 130 rpm and 150 rpm, which makes the natural frequency strike the maximum value. Finally, the research results provide a theoretical basis for the ships to avoid large vibration during navigation.

Keywords: water lubricated bearing; dynamic coefficient; cavitation effects; bending deformation

Citation: He, T.; Xie, Z.; Ke, Z.; Dai, L.; Liu, Y.; Ma, C.; Jiao, J. Theoretical Study on the Dynamic Characteristics of Marine Stern Bearing Considering Cavitation and Bending Deformation Effects of the Shaft. *Lubricants* **2022**, *10*, 242. https://doi.org/10.3390/lubricants10100242

Received: 25 August 2022
Accepted: 17 September 2022
Published: 29 September 2022

Publisher's Note: MDPI stays neutral with regard to jurisdictional claims in published maps and institutional affiliations.

1. Introduction

The lubricants can be employed to form the hydrodynamic, decrease the friction and wear in bearings, intensify load–carry capacity and reduce the mechanical vibration. For the journal bearing, the lubricant forms the hydrodynamic film to support the external loads when the three conditions (the wedge gap, relative velocity and the viscous fluid) are met. In addition, the lubrication mode of journal bearings in ships is mainly water lubrication. Therefore, the present research focuses on water-lubricated bearings.

The studies on water-lubricated bearings have been implemented by plenty of scholars, and the relevant vibration performances and lubrication behaviors are acquired and analyzed. As the crucial part of the submarine or ship, the stiffness and damping properties have a crucial effect on the critical velocity and the stability of ships (as can be seen in Figure 1). For simplifying the analysis on stability, the bearing forces are linearized and adopted to be the linear function for velocity and displacement. According to the method mentioned above, there are eight dynamic coefficients, including four damping coefficients and four stiffness coefficients [1]. For these coefficients, plenty of works are implemented by scholars, and some typical conclusions and results are proposed. Sternlicht [2] expressed the bearing force vector as the linear function of the displacements and static bearing force found the equilibrium position. Sternlicht [2] carried out the relevant investigation. The force vector is adopted as the linear function of static bearing force and displacement to locate the equilibrium point. In his study, by using the Finite Difference Method (FDM), four stiffness coefficients and two direct damping coefficients were given. For the dynamic characteristics, a perturbation solution method of the Reynolds equation is proposed by Lund [3,4]. As noted, the method decreases the demand for numerical differentiation.

Figure 1. Location of the stern bearings.

The first order perturbation solution is employed in Reinhardt's study [5]. For the dynamic characteristics of the journal bearing with grooves, two perturbation methods (infinitesimal perturbation method, IFPM and finite perturbation method, FPM) are applied in Someya's investigation [6]. Nowadays, IFPM and FPM are frequently employed for analyzing dynamic characteristics [7–10]. Due to the shortcomings of conventional methods in studying the fluid–solid interaction of the bearing–rotor system, Lin [11] explored the two-way fluid–solid interaction and hydrodynamic lubrication performances with thermal effect and cavitation effect by computational fluid dynamics (CFD). In addition, the variation laws of dynamic coefficients under the misalignment are also explored [12–15]. The influences of the misalignment on the dynamic characteristics are shown in the research of Bou-Said [16] and San Andres [17]. In Qiu's research [18], for the static loads, the static and dynamic characteristics (attitude angle, friction force, moment and dynamic coefficients, etc) are analyzed under different misalignments and eccentricity ratios. For the water-lubricated bearings, the static and dynamic characteristics are investigated by Feng [19] and Hanawa [20]. Thermohydrodynamic, turbulent and inclined effects were taken into account in the revised models. Lin [21] and Liang [22] performed research on the dynamic characteristics of water-lubricated bearings with grooves. Cavitation effects were considered for the lubrication model. Research indicated that cavitation had more effects on the stiffness coefficients than the damping coefficients.

Therefore, based on the mentioned above, the studies on the influences of cavitation effect on journal bearing are abundant. However, for the coupling effect of bending deformation and cavitation effect, there is little literature. Consequently, a related investigation is carried out.

When the shaft runs, the journal deformation is caused by the force acting on it. Subsequently, the deformation has significant effects on the pressure profile and the asperity contact force under a mixed lubrication condition. Meanwhile, the hydrodynamic force and the asperity contact force jointly act on the journal and affect the deformation. In addition, when the bearing width is large, the deformation of the shaft cannot only be described by two misalignment angles (β and γ) [14]. With the exception of the influences of pressure, asperity contact force and bending deformation of the shaft, the shaft also supports the concentrated load from the propeller. In this study, the interaction between lubrication properties and deformation journals has been taken into consideration. The study applies the deflection curve equation to express the bending deformation. The finite perturbation method is selected to simplify the mathematical model. Moreover, the problem of the mixed lubrication with JFO cavitation condition on a marine stern tube bearing is presented. The effects of bending deformation of the propeller, as well as lubrication boundary conditions, on the dynamic coefficients, are analyzed.

2. Theory Modelling

2.1. Characteristics of Journal Bearing

A typical lubricated cylindrical journal bearing is displayed in Figure 2. According to the references describing the lubrication properties of journal bearing, it is obvious that the hydrodynamic pressure rises to a peak in the converging film region and drops to the atmospheric pressure level at both ends. However, in some special areas where the film thickness partially raises, the pressure descends to an ambient level on the contrary. The water evaporation is caused by the released gas from the lubricant or the ambient pressure below the vapor pressure. The rupture of water film is a crucial feature for the stability of the bearing, and the phenomenon is called lubricant cavitation. The research [23,24] explored the effects of cavitation on stability and bearing performances.

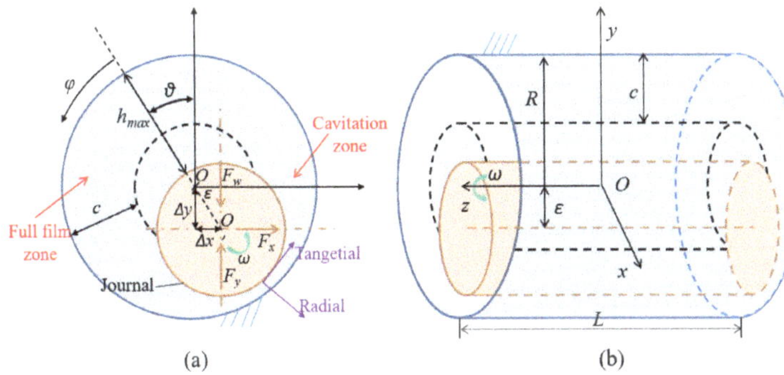

Figure 2. Cylindrical journal bearing with cavitation. (**a**) XOY view; (**b**) Three-dimensional view.

Because cavitation collapse can cause severe surface material damage, the cavitation phenomenon should be tried as a very important study direction. Moreover, the beginning and degree can change the load-carrying capacity. The cavitation has a significant effect on the stability of the bearing-rotor system and the maximum amplitude of vibration and whirl [25].

In common, the main function of hydrodynamic pressure is that it provides the wedge space and squeeze effect. The JFO model applies to the dynamic loads that the surface is at the squeeze motion of film [26]. The time variation parameters of rupture film are included in the continuity equation but the bubble dynamics are not considered. With the fluctuation of surface squeeze velocity, the shape of cavitation changes.

In the JFO model, however, since the cavitation zone is unknown during the operating condition, is hard to solve the Reynolds equation numerically. A current cavitation algorithm is developed by Elrod [27,28]. In his research, JFO model has included the Reynolds equation. For the model, the fluid bulk modulus (κ) is used to relate lubricant pressure and density. The boundary conditions at the cavitation area are automatically met by the switch function. Moreover, under cavitation area and integrity film, the feature of continuity equation is transformed from ellipse to parabola by the switch function. Then a variable (θ in [29,30]) is introduced to indicate different lubrication regions (full film region and cavitation).

For the compressible liquid, the relationship between density and pressure is shown as follows:

$$\kappa = \rho \frac{\partial P}{\partial \rho} \tag{1}$$

And define the density ratio as:

$$\alpha = \frac{\rho}{\rho_{cav}} \tag{2}$$

Then Equation (2) can be rewritten as:

$$g\kappa = \frac{\rho}{\rho_{cav}} \frac{\partial P}{\partial(\rho/\rho_{cav})} = \alpha \frac{\partial P}{\partial \alpha} \tag{3}$$

where g is defined as a switch function in [29,30] and values of g in different zones are described as:

$$g = \begin{cases} 1 & whole\ film\ zone \\ 0 & cavitation\ zone \end{cases}$$

For Equation (3), the direct integration is implemented, and then the following equation can be obtained:

$$P = P_{cav} + g\kappa \ln(\alpha) \quad (4)$$

It should be indicated that for the part of whole film, $g = 1$ and $P > P_{cav}$, since $\alpha = \frac{\rho}{\rho_{cav}} > 1$; and for cavitation area $P = P_{cav}$ because of $g = 0$. As noted, pressure gradient is a crucial factor for the thin film flow. Consequently, for P_1 and P_2, they are larger than P_{cav},

$$P_1 - P_2 = \kappa[\ln(\alpha_1) - \ln(\alpha_2)] \cong \kappa(\alpha_1 - \alpha_2) \quad (5)$$

Owing to the large magnitude of κ, the slight difference in density ratio can result in a tremendous difference in pressure. In addition, for the numerical model, there are a few difficulties because of the different considerations. For the variable α, under the whole film area and cavitation area, the explanation is different [31]. In cavitation area, due to the vapor material within the cavity, the density of water is even (ρ_{cav}) and the wedge gap is not stuffed. Therefore, α is recognized as the small film content and the void fraction can be expressed by the $(1 - \alpha)$.

Based on the mass flow conversation of thin film:

$$\frac{\partial(\rho h)}{\partial t} + \frac{\partial(M_x)}{\partial x} + \frac{\partial(M_z)}{\partial z} = 0 \quad (6)$$

Under the full film area, for the mass flow rate of fluid, it can be shown as:

$$M_x = -\frac{\rho h^3}{12\mu}\frac{\partial P}{\partial x} + \frac{\rho h U}{2}; \quad M_z = -\frac{\rho h^3}{12\mu}\frac{\partial P}{\partial z} \quad (7)$$

However, for the cavitation position, $\frac{\partial P}{\partial x} = \frac{\partial P}{\partial z} = 0$, the mass flow rate is:

$$M_x = \alpha \rho_{cav} h \frac{\Omega R}{2}; \quad M_z = 0 \quad (8)$$

Due to $\partial P = \frac{\kappa}{\rho}\partial\rho = g\frac{\kappa}{\alpha}\partial\alpha$, Equation (7) can be derived as:

$$M_x = -\frac{\rho_{cav} h^3}{12\mu} g\kappa \frac{\partial \alpha}{\partial x} + \alpha\frac{\rho_{cav} h U}{2}; \quad M_z = -\frac{\rho_{cav} h^3}{12\mu} g\kappa \frac{\partial \alpha}{\partial z} \quad (9)$$

As noted, under the full film area ($g = 1$ and $\alpha = \rho/\rho_{cav}$) and cavitation position, $g = 0$ with α is recognized as the small film content.

Equation (6) can be solved by the global mass conservation as:

$$\frac{\partial}{\partial x}\left(\frac{\rho_{cav} h^3}{12\mu} g\kappa \frac{\partial \alpha}{\partial x}\right) + \frac{\partial}{\partial z}\left(\frac{\rho_{cav} h^3}{12\mu} g\kappa \frac{\partial \alpha}{\partial z}\right) = \rho_{cav}\frac{\partial(\alpha h)}{\partial t} + \rho_{cav}\frac{U}{2}\frac{\partial(\alpha h)}{\partial x} \quad (10)$$

And in the cavitation zone, Equation (10) can be reduced to:

$$\rho_{cav}\frac{\partial(\alpha h)}{\partial t} + \rho_{cav}\frac{U}{2}\frac{\partial(\alpha h)}{\partial x} = 0 \quad (11)$$

This establishes a dynamic flow balance in the cavitation area. The mixed success is available in the present research by the common cavitation algorithm. However, compared with the rough technology, the common cavitation algorithm is obsolete.

2.2. Bearing Forces

There is a crucial relationship between rotational speed and journal position. The relation equation can be expressed by the bearing force.

$$F_x = F_x(x, y, \dot{x}, \dot{y}), \quad F_y = F_y(x, y, \dot{x}, \dot{y}) \tag{12}$$

By numerical procedure, as long as the film pressure is determined, the pressure in a horizontal and vertical direction is integrated to acquire the bearing force:

$$\begin{bmatrix} F_x \\ F_y \end{bmatrix} = -\int_{\phi_0}^{\phi_2}\int_0^L \begin{bmatrix} \sin\phi \\ \cos\phi \end{bmatrix}(Rd\phi)dz = -C_g\int_{\phi_0}^{\phi_2}\int_0^{L/R} \begin{bmatrix} \sin\phi \\ \cos\phi \end{bmatrix} d\phi dz \tag{13}$$

2.3. Attitude Angle

The angle between the load direction and the central line that connects the bearing and journal is the attitude angle. Under stable condition, the bearing forces are employed to balance the external loads. According to the external load W, the bearing force vector is shown by the load-carrying capacity vector:

$$\vec{W} = \vec{F}_x + \vec{F}_y \tag{14}$$

If the W acts on the vertical direction (such as the weight of the screw propeller in this study), the horizontal force is $\vec{F}_x = 0$. Therefore:

$$\tan^{-1}\frac{F_x}{F_y} = 0 \tag{15}$$

The error function is defined as:

$$d\phi_0^k = \tan^{-1}\frac{F_x^k}{F_y^k} \tag{16}$$

If $d\phi_0^k = 0$, the condition of Equation (15) is met automatically. In order to calculate $d\phi_0^k = 0$, the Newton-Raphson iteration method is adopted:

$$\phi_0^{k+1} = \phi_0^k - \frac{d\phi_0^k}{\left(d\phi_0^k - d\phi_0^{k-1}\right)/\left(\phi_0^k - \phi_0^{k-1}\right)} \tag{17}$$

The ϕ_0^k is considered the attitude angle ϕ_0. In the numerical procedure, if

$$d\phi_0^k = \left|\tan^{-1}\frac{F_x^k}{F_y^k}\right| \le \delta_\phi \tag{18}$$

Once ϕ_0 is converged, $P_{i,j}$ and water film boundaries can be determined.

2.4. Film Thickness Considering Bending Deformation of Journal

When the relative rotating speed exists between the journal and bearing, the eccentricity amount adjusts automatically until the film pressures calculated by the average Reynolds equation and the asperities contact pressures calculated by the contact model balance the external loads. From the description of lubrication properties, the film pressures and the contact pressures depend on the local film thickness. In addition, the misalignment parameters give the expression of film thickness of misaligned bearing (β and γ) [14], as Equation (19) shows:

$$h_\theta = c + e_0\cos(\theta - \varphi) + z\tan\gamma\cos(\theta - \beta - \varphi) \tag{19}$$

When it comes to the large width of the bearing, the deflect angle in each journal section is different as a result. After considering the bending deformation of the shaft, the film thickness can be briefly expressed as:

$$h_{i\theta} = c + e_i \cos(\theta - \phi) \tag{20}$$

Define $e_i = \sqrt{(v_{ix} + e_{ix})^2 + (v_{iy} + e_{iy})^2}$, when the shaft is bending deformation, for the ith section, the eccentricity ratio is e_i; e_{ix} and e_{iy}. These are the components of the eccentricity ratio without bending in horizontal (x) and vertical (y) directions (as shown in Figure 3). The deformation of this section is $v_i = \sqrt{v_{ix}^2 + v_{iy}^2}$, and in order to determine v_i accurately, the displacement superposition method is applied to calculate the bending deformation in this study (as shown in Figure 4).

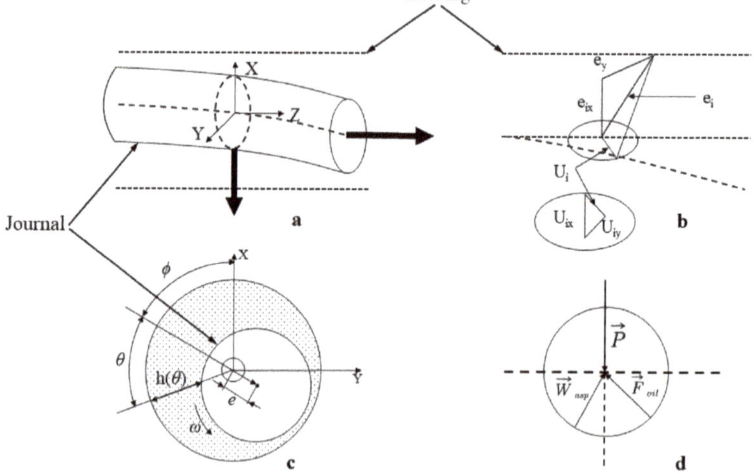

Figure 3. Basic geometry of journal bearing. (**a**) Overall drawing; (**b**) Horizontal amplification; (**c**) Cross-section amplification; (**d**) Load diagram of journal.

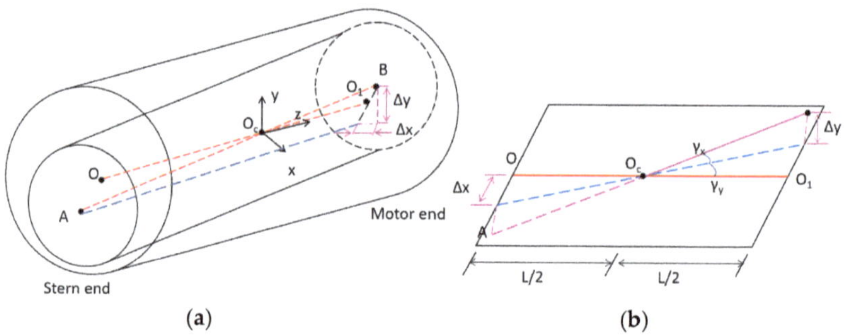

Figure 4. Basic geometry of journal bearing. (**a**) Three-dimensional view; (**b**) Cross-section view.

2.5. Determining the Bending Deformation of Journal

Obviously, the force that acts on the shaft results in the bending deformation of the shaft. Due to the reaction relationship, the deformation will change the profile of film pressure. If the ratio of film thickness to surface roughness is less than 4 [30], the asperity contact force of the end is seriously affected. In [14], Sun used γ to describe the misaligned angle of the shaft in bearing, and it can be expressed as:

$$\gamma = \frac{F_{ex}l^2}{16E_jI} \tag{21}$$

However, as per the reason mentioned in Section 2.4, it should be noted that the misaligned angle in each section of the shaft is various in the very wide bearing. During this study, in the axial direction, the whole bearing is divided into 20 sections. For the sake of accurate determination of the bending deformation, the differential equation of the deflection curve (given as following) is used:

$$\frac{d^2v}{dx^2} = \frac{M(x)}{EI} \tag{22}$$

When it comes to the loaded stern shaft shown in Figure 3. Under sophisticated load conditions, owing to the small bending deformation of the shaft, the shaft material is recognized as elastic, based on the conclusion of [31]. The displacement superposition method can be applied in this study to solve bending deformation. From superposition theory, in the systems subject to multiple loads, resultant stress or strain is the algebraic sum of its effects when the load is applied alone. Taking a cantilever beam as an example, the solution of this beam under typically single-load conditions is shown in Table 1. The sum of these solutions can be also employed to express the deformation caused by external loads and asperity contact force.

Table 1. The cases exhibition of in cantilever beam.

Loading Condition	Deflection (v)	
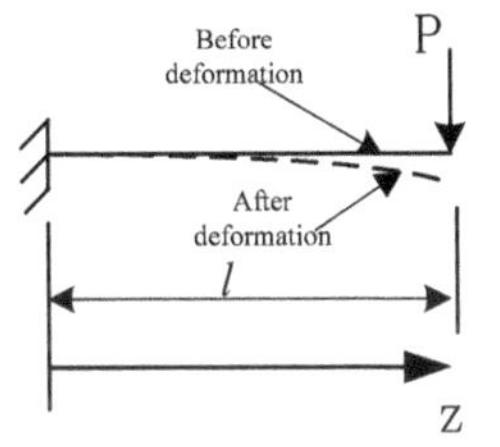 Concentrated load P at end	$v = -\frac{Pz^2}{6EI}(3l - z)$	(23)
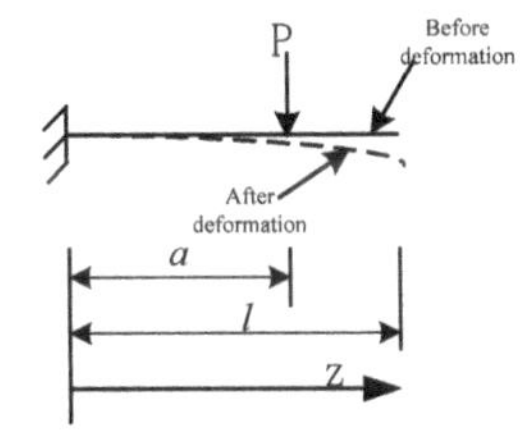 Concentrated load P at point a	$v = -\frac{Pz^2}{6EI}(3a - z)\ 0 \le z \le a$ $v = -\frac{Pa^2}{6EI}(3z - a)\ a \le z \le l$	(24)

As described previously, in the axis direction, the journal is divided into 20 rigid nodes, and in z direction, for the ith node, the coordinate is Z_i. For the bending deformation of the shaft, the ith section is v_{ix} and v_{iy}. For the hydrodynamic force and asperity contact force, at the ith section center, they are expressed as F_{ix} and F_{iy}, W_{ix} and W_{iy}. As noted, the subscript of x indicates the x direction, for the subscript of y, the meaning is identical. In addition, the resultant forces of both are F_i and W_{asp}, respectively.

According to Equations (15) and (16), for all nodes of the journal, the bending deformation is displayed as:

when $z = z_i\ (1,\ k)$,

$$
\begin{aligned}
v_{zx} &= -\frac{Pz_k^2}{6EI}(3l - z_k) - \sum_{i=1}^{k} \frac{(F_{ix}+W_{ix})z_i^2}{6EI}(3z_k - z_i) - \sum_{i=k+1}^{n} \frac{(F_{ix}+W_{ix})z_k^2}{6EI}(3z_i - z_k) \\
v_{zy} &= -\sum_{i=1}^{3} \frac{(F_{iy}+W_{iy})z_i^2}{6EI}(3z_k - z_i) - \sum_{i=4}^{n} \frac{(F_{iy}+W_{iy})z_k^2}{6EI}(3z_i - z_k)
\end{aligned}
\tag{25}
$$

when $z = z_n$,

$$
\begin{aligned}
v_{zx} &= -\frac{Pz_n^2}{6EI}(3l - z_n) - \sum_{i=1}^{n} \frac{(F_{ix}+W_{ix})z_i^2}{6EI}(3z_n - z_i) \\
v_{zy} &= -\sum_{i=1}^{n} \frac{(F_{iy}+W_{iy})z_i^2}{6EI}(3z_n - z_i)
\end{aligned}
\tag{26}
$$

According to the above expressions, the bending deformation at mixture load conditions, including the concentrated loads of the propeller, the hydrodynamic pressure and asperity pressures, can be determined accurately.

2.6. The Deformation of Bearing

As an external load is applied to a journal bearing, the bending deformation of the journal has happened, and as Figure 5 shows, the water film thickness is reduced at the sections close to the bearing end; as a result, the film pressure will increase rapidly at those sections. In this situation, in order to analyze the consequences of the pressure distribution and wear phenomena better, the method that the elastic deformation of bearing is included in the elastohydrodynamic model is crucial. At present, the elastic deformation is incorporated into the lubrication characteristics of dynamic bearing that analyzes the various external loads. By calculating the deformation of bearing ΔH in different conditions and adding it to the film thickness value (as shown in Equation (20)), the influences of elastic deformation on lubrication performances are explored.

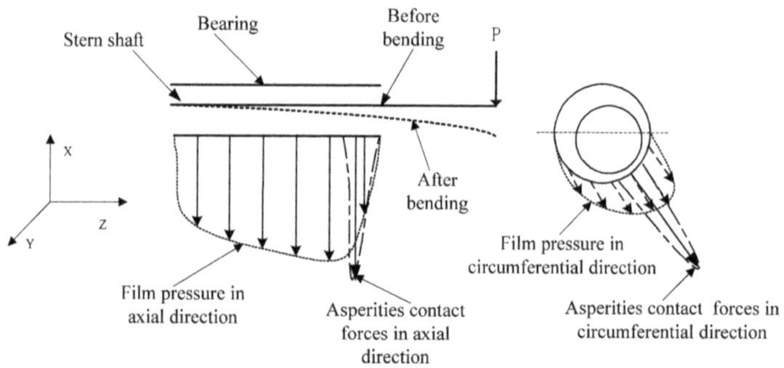

Figure 5. Bending deformation of propeller under sophisticated load conditions.

In this study, the equivalent model is employed to describe the deformation. From Figure 6, the whole elastic deformation of journal and bearing is expressed by ΔH, and E_1, E_2 is Young's modulus of journal and bearing, respectively. The composite of Young's modulus of bearing and journal is E'. The deformation corresponds to that of the bearing surface just with composite Young's modulus. According to the equivalent model, the changing length of the object just with Young's modulus E' and initial length L can be expressed by ΔH. According to the above analysis, the expression [32] is obtained as:

$$
P \cong E'\varepsilon \cong E'\frac{\Delta H}{L}
\tag{27}
$$

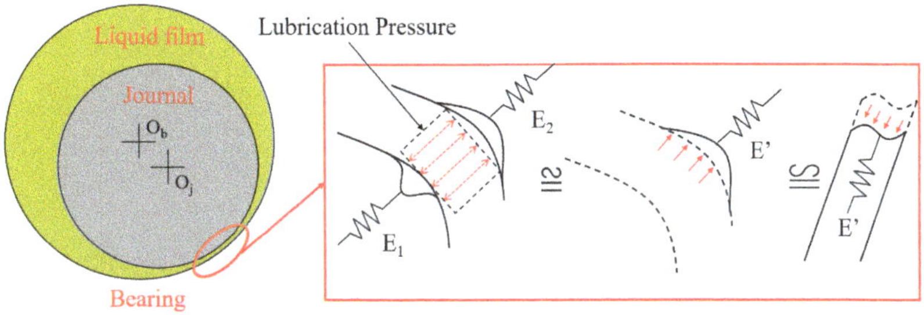

Figure 6. Equivalent model of surface deformation.

The normal pressure acting on the journal can be acquired by:

$$\Delta H \cong \frac{L}{E'}P \tag{28}$$

2.7. The Pressure in Cavitation Zone

To determine the pressure in the cavitation zone, the equations listed in reference [33] are used in this study, which is given as:

$$P_{cav} = P_{ml} \tag{29}$$

$$P_{ml} = \rho_v {C_v}^2 - N\log\left(\frac{{\rho_v}^2{C_v}^2}{{\rho_l}^2{C_l}^2}\right) \tag{30}$$

$$N = \frac{\rho_v {C_v}^2 \rho_l {C_l}^2 (\rho_v - \rho_l)}{{\rho_v}^2{C_v}^2 - {\rho_l}^2{C_l}^2} \tag{31}$$

As it is summarized in reference [33], Equation (29) could give a better correlation in the transition zone. Based on those equations, the P_{cav} can be quickly determined by ρ_v, ρ_l, C_v, C_l. Define:

$$r_p = \frac{\rho_v}{\rho l} \tag{32}$$

2.8. Calculation of the Dynamic Coefficients

Based on Equation (12), the bearing forces are derived by:

$$F_x = k_{xx}x + k_{xy}y + b_{xx}\dot{x} + b_{xy}\dot{y} \tag{33}$$

$$F_y = k_{yx}x + k_{yy}y + b_{yx}\dot{x} + b_{yy}\dot{y} \tag{34}$$

And the eight coefficients are defined as:

$$k_{xx} = \frac{\partial F_x}{\partial x}, k_{xy} = \frac{\partial F_x}{\partial y}, k_{yx} = \frac{\partial F_y}{\partial x}, k_{yy} = \frac{\partial F_y}{\partial y} \tag{35}$$

$$b_{xx} = \frac{\partial F_x}{\partial \dot{x}}, b_{xy} = \frac{\partial F_x}{\partial \dot{y}}, b_{yx} = \frac{\partial F_y}{\partial \dot{x}}, b_{yy} = \frac{\partial F_y}{\partial \dot{y}} \tag{36}$$

These eight coefficients are defined as dynamic coefficients of the journal bearing, as shown in Figure 7. According to those references reviewed previously, there are two theoretical methods (IFPM and FPM) to calculate the dynamic coefficients. FPM is used in this study.

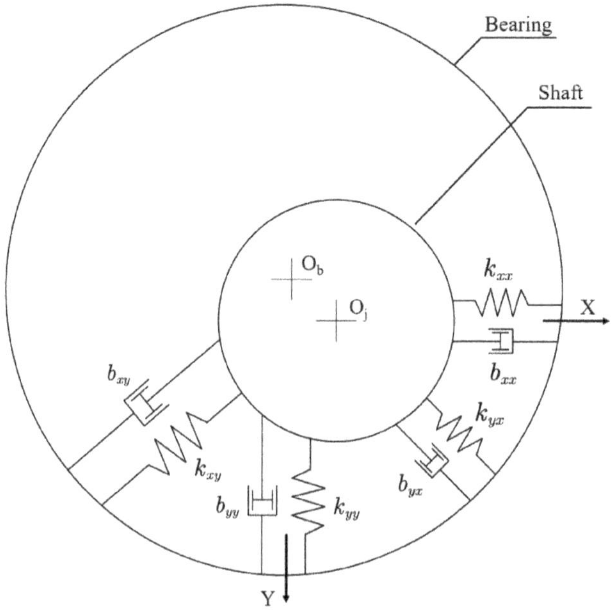

Figure 7. Pictorial representation of water–film force coefficients.

2.9. Finite Perturbation Method (FPM)

FPM perturbs the journal under the equilibrium position by the small Δx, Δy, $\Delta \dot{x}$ and $\Delta \dot{y}$, and the force finite difference is employed to describe the force coefficient. For the calculation accuracy, the perturbation is implemented in the positive and negative directions, and the partial differential in both directions is adopted to calculate the partial derivative. At the equilibrium position (x_o, y_o), the partial derivative $\frac{\partial F_x}{\partial x}$ (force coefficient k_{xx}) can be solved by [7]:

$$\begin{aligned} k_{xx} = \frac{\partial F_x}{\partial x} &\approx \frac{1}{2}\left[\frac{F_x(x_0+\Delta x, y_0, \dot{x}_0, \dot{y}_0) - F_x(x_0, y_0, \dot{x}_0, \dot{y}_0)}{\Delta x} + \frac{F_x(x_0, y_0, \dot{x}_0, \dot{y}_0) - F_x(x_0-\Delta x, y_0, \dot{x}_0, \dot{y}_0)}{\Delta x}\right] \\ &= \frac{F_x(x_0+\Delta x, y_0, \dot{x}_0, \dot{y}_0) - F_x(x_0-\Delta x, y_0, \dot{x}_0, \dot{y}_0)}{2\Delta x} \end{aligned} \quad (37)$$

At the steady state, $\dot{x}_o = \dot{y}_o = 0$; thus:

$$k_{xx} = \frac{F_x(x_0+\Delta x, y_0, 0, 0) - F_x(x_0-\Delta x, y_0, 0, 0)}{2\Delta x} \quad (38)$$

Similarly, the other stiffness coefficients can be calculated by:

$$k_{yx} = \frac{\partial F_y}{\partial x} \approx \frac{F_y(x_0+\Delta x, y_0, 0, 0) - F_y(x_0-\Delta x, y_0, 0, 0)}{2\Delta x} \quad (39)$$

$$k_{xy} = \frac{\partial F_x}{\partial y} \approx \frac{F_x(x_0, y_0+\Delta y, 0, 0) - F_x(x_0, y_0-\Delta y, 0, 0)}{2\Delta y} \quad (40)$$

$$k_{yy} = \frac{\partial F_y}{\partial y} \approx \frac{F_y(x_0, y_0+\Delta y, 0, 0) - F_y(x_0, y_0-\Delta y, 0, 0)}{2\Delta y} \quad (41)$$

The damping coefficients are acquired by:

$$b_{xx} = \frac{\partial F_x}{\partial \dot{x}} \approx \frac{F_x(x_0, y_0, \Delta \dot{x}, 0) - F_x(x_0, y_0, -\Delta \dot{x}, 0)}{2\Delta \dot{x}} \quad (42)$$

$$b_{yx} = \frac{\partial F_y}{\partial \dot{x}} \approx \frac{F_y(x_0, y_0, \Delta\dot{x}, 0) - F_y(x_0, y_0, -\Delta\dot{x}, 0)}{2\Delta\dot{x}} \quad (43)$$

$$b_{xy} = \frac{\partial F_x}{\partial \dot{y}} \approx \frac{F_x(x_0, y_0, 0, \Delta\dot{y}) - F_x(x_0, y_0, 0, \Delta\dot{y})}{2\Delta\dot{y}} \quad (44)$$

$$b_{yy} = \frac{\partial F_y}{\partial \dot{y}} \approx \frac{F_y(x_0, y_0, 0, \Delta\dot{y}) - F_y(x_0, y_0, 0, -\Delta\dot{y})}{2\Delta\dot{y}} \quad (45)$$

The FPM only needs to solve the bearing forces at eight perturbed positions: $(x_0 - \Delta x, y_0, 0, 0)$, $(x_0 + \Delta x, y_0, 0, 0)$, $(x_0, y_0 - \Delta y, 0, 0)$, $(x_0, y_0 + \Delta y, 0, 0)$, $(x_0, y_0, -\Delta\dot{x}, 0)$, $(x_0, y_0, \Delta\dot{x}, 0)$, $(x_0, y_0, 0, -\Delta\dot{y})$ and $(x_0, y_0, 0, \Delta\dot{y})$.

If the perturbation magnitude is large, there are some errors in the linear model. This is because the bearing force is a sophisticated nonlinear function. According to the theory, there is a maximum perturbation amplitude. When the amplitude is less than the maximum value, the calculation accuracy is high and acceptable. Applying FPM, the maximum perturbation amplitude can be determined by a given accuracy.

3. Numerical Procedure

Based on the numerical model developed in this study, the solution procedure is shown in Figure 8. In the dynamic part, a combination of FDM and FPM methods is applied. The equation of film thickness is displayed by the nodal global coordinate of the bearing surface. For the analysis of lubrication performances, the computer program is developed. The flowchart of the numerical procedure can be seen in Figure 8. The calculation procedure is designed as follows:

(1) Input the parameters of the stern bearing, initial film pressure, asperities contact forces and deformation of bearing. The assumption condition for the shaft center is given, and then the deformation of the shaft under external loads (the concentrated load of the propeller) is calculated;
(2) Calculating the water film shape according to the current journal position through Equation (20) and adding the value of bearing deformation into $h_{i\theta}$;
(3) Obtain the hydrodynamic pressure distribution by implementing the universal cavitation algorithm; Calculate the asperities contact forces;
(4) The trajectory position is acquired by the integral for the balance loads, hydrodynamic pressure and asperity contact force. In addition, the over-relaxation Newton–Raphson method is employed to accelerate convergence;
(5) For the deformation of the shaft under external loads, film pressure and asperity contact force and bearing deformation, the calculation needs to be resolved until they meet the convergence criterion:

$$\left| \frac{e^{k+1} - e^k}{e^k} \right| \leq 0.01 \text{ and } \left| \frac{\phi^{k+1} - \phi^k}{\phi^k} \right| \leq 0.01$$

(6) In order to determine the bearing force of each perturbation position, step (2)~(5) is recalculated by perturbing the position and velocity of the journal;
(7) According to Equations (38)–(45), the dynamic coefficients are acquired.

In the analysis, the prediction is mainly decided by the value of the volume modulus (κ). For the accurate solution, the algebraic equations are stiff because of the actual value of the bulk modulus of water. One reasonable explanation is that the slight variation of density has a significant effect on the hydrodynamic pressure at the full film area. Therefore, for the numerical model, the round-off errors appear [7]. The artificial low bulk modulus is adopted to avoid the problem of accuracy and convergence. As noted, its magnitude is less than the physical value. The excellent results are given by the stable numerical algorithm. The numerical range is 1/100 to 1/10 of the actual physical value.

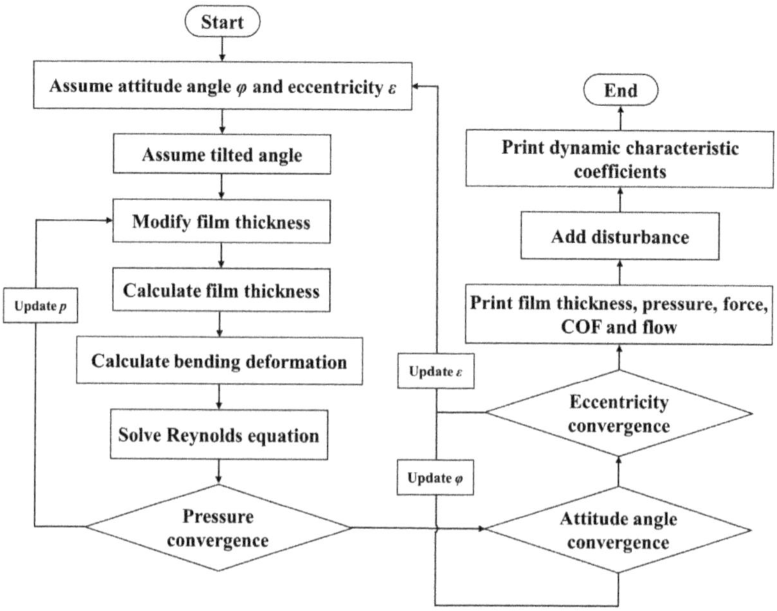

Figure 8. Flowchart of the calculation.

4. Results and Discussion

4.1. Validation

To validate the present numerical scheme, a result obtained from the present analysis is compared with the results obtained by the experimental data given by Su [34]. Table 2 lists the bearing parameters for the comparison case, and Figure 9 shows the film pressure predicted by the model developed in this study and the experimental results from [34]. As shown in this figure, compared with the results gotten from the Reynolds boundary, the film pressure predicted by the JFO boundary matches the experimental results much better. The pressure is recognized as vapor pressure in the Reynolds boundary condition. In addition, at the position of full film rupture, the circumferential gradient must be equal to zero. From the above analysis, compared with the experiment results, the error is large.

Table 2. Parameters of the water-lubricated bearing.

Item	Value
Radius of Bearing (m)	0.05
Radial Clearance (mm)	0.1455
Ratio of Bearing Length and Bearing Diameter	1.333
Eccentricity ratio	0.61
Angular Velocity (rad/s)	48.1
Lubrication Supply Pressure (Pa)	0
Cavitation Pressure (Pa)	−72,139.79

Numerical research on the water-lubricated bearing with cavitation is carried out to explore the influences of propeller deformation on the behaviors of the bearing. For the following situation, the water-lubricated stern bearing (shown in Figure 10) variables used in the computation are listed in Table 3. The weight of the screw propeller is 0.21498×10^6 N. According to the validation case, the JFO boundary condition is much closer to the real condition. Therefore, the following analysis is only based on the JFO boundary condition. Meanwhile, the deformations of the bearing and propeller are taken into consideration. The structure of the propeller is shown in Figure 10. The shaft is divided into two parts by the flange. According to the position of the bearings, the whole shaft can be divided into five segments, as shown in Figure 10. The length of each part is listed in Table 3, together with the bearings' parameters and material properties.

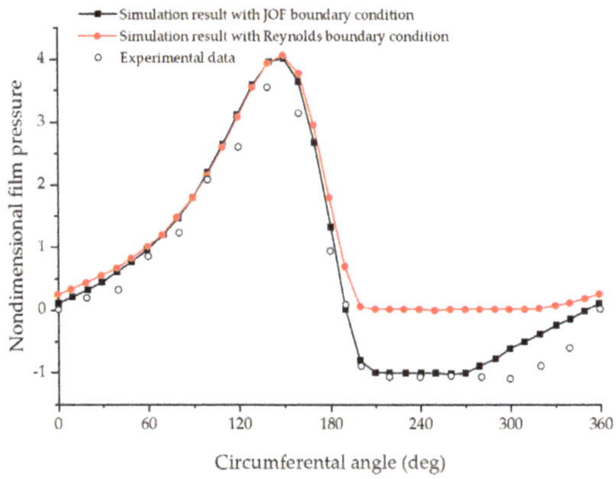

(**a**)

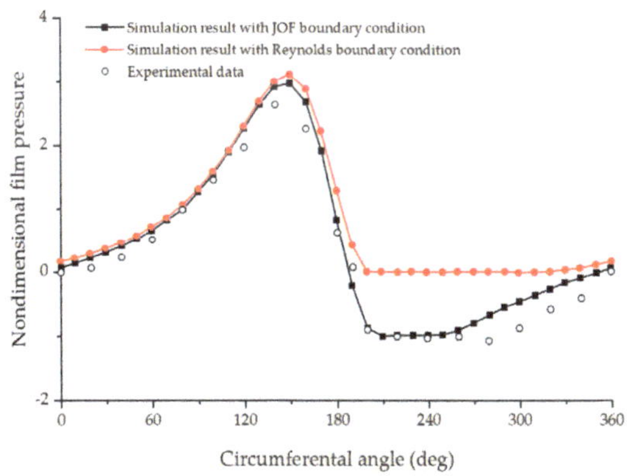

(**b**)

Figure 9. The film pressure distribution in different sections. (**a**) The section located at the place from middle section distance for 1/5 L; (**b**) The section located at the place from middle section distance for 3/5 L.

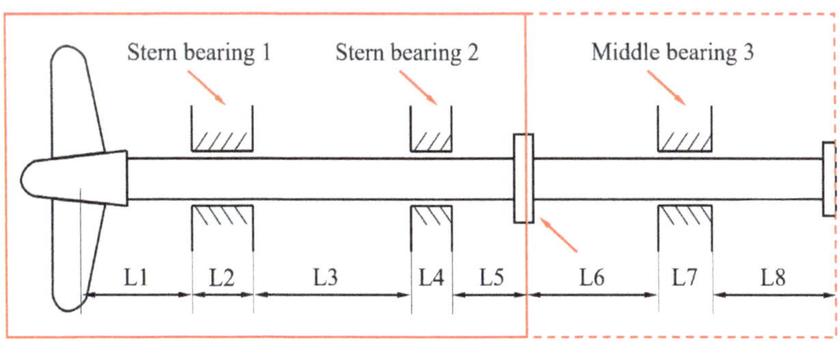

Figure 10. Model of the ship shafting.

Table 3. The parameters and material properties of the marine stern bearing.

Shaft Segment No.		Middle Shaft		Stern Shaft		
		1	2	3	4	5
Length (m)		2.9	3.6	1.55	3.45	2.05
Diameter of shaft (cm)		39	39	49.8	49.8	49.8
Young's Modulus of shaft(N/m^2)		2.1×10^{11}				
Density of shaft (kg/m^3)		7850				
Poisson Ratio of shaft		0.3				
Bearing material (bearing 1, bearing 2 and bearing 3)		Aluminium alloy PTFE				
Radial clearance(m)	Bearing 1	0.6×10^{-3}				
	Bearing 2	0.6×10^{-3}				
	Bearing 3	0.4×10^{-3}				
Roughness of shaft (m)		4.3×10^{-6}				
Roughness of bearing (m)		4.3×10^{-6}				
Shaft to Bearing contact friction coefficient		0.1				
Kinematic viscosity of lubricant ($N{\cdot}s/cm^2$)		0.15×10^{-6}				
Ratio of density r_p		2.31×10^{-5}				
Density of the liquid phase of lubricant (kg/m^3)		890				
Sound velocity of the sound in the pure vapor (m/s)		343				
Sound velocity of the sound in the pure liquid (m/s)		1450				

4.2. The Equivalent Stiffness

The dynamic stability and maximum vibration amplitude of the bearing-rotor system are affected by the cavitation to a large extent, especially for the dynamically loaded bearing. The stability of hydrodynamic bearings is generally addressed since the early design stages of a rotor-bearing system. The water film is recognized as the spring with the equivalent stiffness:

$$K_{ss} = m\omega^2 \tag{46}$$

Then from [35],

$$K_{ss} = \frac{k_{xx}b_{yy} + k_{yy}b_{xx} - k_{xy}b_{yx} - k_{yx}b_{xy}}{b_{yy} + b_{xx}} \tag{47}$$

And this equivalent stiffness is important in the stability analysis of hydrodynamic bearings. In stability calculation, if $K_{SS} < 0$, the system is absolutely unstable.

The absolute smooth surface does not exist. Generally, the surface roughness is identical to, or larger than the film thickness estimated by the smooth surface hydrodynamic lubrication theory. Based on the model developed by Patir and Cheng [36], the empirical pressure flow, shear flow factors and asperity contact factor are used in this study. The correlation between relative velocity and coefficient of friction of bearing-rotor system is constructed by Stribeck under constant load to describe the range of different operating conditions. The reference [37] has indicated the relationship between the asperity contact force and lubrication regime, and the bottom of the Stribeck curve is regarded as the transition point of mixed lubrication and hydrodynamic lubrication. Due to the concentrated load of the propeller, the bending deformation is taking place in the whole operating condition. Simultaneously, the journal deformation is affected by the hydrodynamic pressure and asperity contact force [38,39]. Furthermore, for a large width-bearing like the water-lubricated stern bearing in ships (as shown in Figure 10), in the analysis of the bearing-rotor system, the bending deformation of the shaft cannot be ignored. Figure 11 shows the Stribeck curve and the equivalent stiffness K_{SS} in different speeds of bearing 1. From Figure 11, we can see that under hydrodynamic lubrication, the equivalent stiffness K_{SS} decreases with speeds. As noted, the K_{SS} descends sharply under low speeds. However, with the increase in speeds, the trend is inclined to be stable. One reasonable explanation for it is that in hydrodynamic lubrication and the hydrodynamic film forms, it is enough to support the journal away from the bearing. Finally, the solid contact between the journal and bearing disappears, and the equivalent stiffness reduces rapidly and tends to be stable.

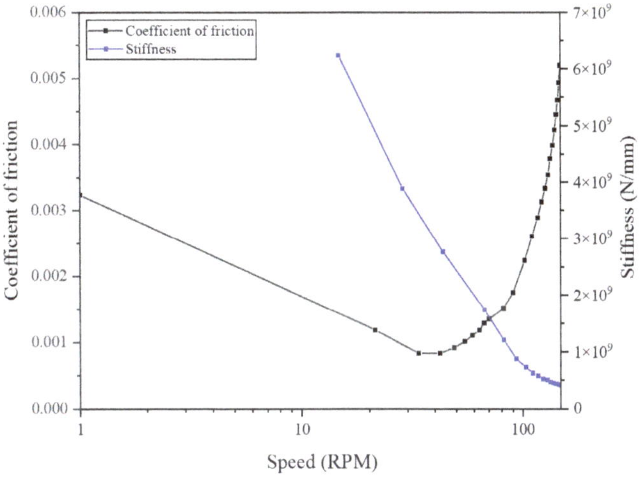

Figure 11. Friction coefficient and equivalent stiffness K_{SS} curves.

4.3. The Natural Frequency Analysis

Based on the method developed by Zhou [40], the natural frequency of the shaft is calculated in this study. Due to the dynamic parameters and water film, supporting positions are various with speed. The natural frequency of the shaft should be different at different speeds. The natural frequencies at 90 rpm, 110 rpm, 130 rpm and 150 rpm in different frequency ratios are listed in Table 4. It is clear that the speed has an influence on the natural frequency when lubrication is taken into consideration. For natural frequency in the present, it increases first with speed and then decreases, which means there is a critical speed for the natural frequency. From the working conditions of rotational speeds, the critical speed exists between 130 rpm and 150 rpm. It can be explained by the fact that in addition to external loads and tangential forces, the shaft is also subject to large centrifugal force and gyroscopic moment. Therefore, under so many influence factors, the situation of natural frequency is caused. Compared with the results by reference [40] (the predicted model used in reference [40] only considers the bending deformation of the shaft caused by a concentrated load of screw), the results predicted in this study are smaller. That means the lubrication cannot be neglected in the natural frequency calculation [41,42].

Table 4. Calculating natural frequencies with different operating conditions and frequency ratios.

Frequency Ratios	Natural Frequency (r/min)				
	Results by Ref. [34]	Different Operating Speed Considering Lubrication			
		90 rpm	110 rpm	130 rpm	150 rpm
h = 1	1022.59	950.33	963.22	969.23	969.15
h = 1/4	890.87	845.36	853.69	857.46	857.15
h = 0	849.89	811.17	818.31	821.55	821.29
h = −1/4	811.84	778.66	784.84	787.64	787.46
h = −1	709.71	693.80	698.07	700.05	700.12

5. Conclusions

The dynamic parameters of the stern bearing considering cavitation, as well as the bending deformation of the propeller, are investigated. Based on the displacement superposition method, the interaction between lubrication properties and bending de-formation has been taken into consideration. By implementing the validated lubrication model designed in this study, the dynamic parameters of bearing systems are presented by the finite perturbation method. As an important parameter in stability analysis, the property of equivalent stiffness in the hydrodynamic lubrication regime is given. Finally, some significant conclusions are obtained, as follows:

(1) For equivalent stiffness, due to the increase in hydrodynamic effect, it is more affected by the speed, especially at low speeds;
(2) For natural frequency, there is a critical speed between 130 rpm and 150 rpm, which makes the natural frequency strike the maximum value because of the comprehensive influencing factors (external loads, tangential forces, large centrifugal forces and gyroscopic moment).

Author Contributions: T.H.: conceptualization, methodology, software, validation, formal analysis, investigation, resources, data curation, writing—original draft, writing—review & editing. Z.X.: methodology, software, validation, formal analysis, investigation, resources, data curation, visualization, supervision, project administration, funding acquisition. Z.K.: software, validation. L.D.: software, validation. Y.L.: software, validation. C.M.: software, validation. J.J.: modification, check. All authors have read and agreed to the published version of the manuscript.

Funding: This research was funded by [National Natural Science Foundation of China] grant number [No. 52006156], Natural Science Basic Research Program of Shaanxi (Program No. 2022JM-003), the 2021 Joint Projects between Chinese and CEECs' Universities (No. 2021101), 2021101 (No. 2022A1515010864), the Fundamental Research Funds for the Central Universities (No. D5000220095). The authors would like to sincerely express their appreciation.

Data Availability Statement: Not applicable.

Conflicts of Interest: The authors declare no conflict of interest.

Abbreviations

IFPM	infinitesimal perturbation method
FPM	finite perturbation method

Nomenclature

k	liquid bulk-modulus	ΔH	deformation of bearing
ρ_{cav}	liquid density	E_1	Young's modulus of journal
α	density ratio	E_2	Young's modulus of bearing
P	film pressure	E_j	the elastic modulus for journal material
W	external load	P_{ml}	the bubble point pressure
ϕ_0	attitude angle	ρ_v	density of vapor phase
l	the length of the shaft	ρ_l	density of liquid phase
γ	the angle of journal misalignment	C_v	velocity of the sound in the pure vapor
F	the external force	C_l	velocity of the sound in the pure liquid
e	eccentricity	W	asperity contact forces
E'	composite Young's modulus of bearing and journal	Ω	cavitation region range
I	the inertial moment of cross-section of shaft	θ	starting angle of cavitation area
h	film thickness of water	M_x	product of density and velocity in x direction

References

1. Holmes, R. The vibration of a rigid shaft on short sleeve bearings. *J. Mech. Eng. Sci.* **1960**, *2*, 337–341. [CrossRef]
2. Sternlicht, B. Elastic and damping properties of cylindrical journal bearings. *J. Basic Eng.* **1959**, *91*, 101–109. [CrossRef]
3. Lund, J.W.; Sternlicht, B. Rotor-bearing dynamics with emphasis on attenuation. *J. Basic Eng.* **1962**, *84*, 491–502. [CrossRef]
4. Lund, J.W. Calculation of stiffness and damping properties of gas bearings. *J. Lubr. Technol.* **1968**, *90*, 793–803. [CrossRef]
5. Reinhardt, E.; Lund, J.W. The influence of fluid inertia on the dynamic properties of journal bearings. *J. Lubr. Technol.* **1975**, *97*, 159–165. [CrossRef]
6. Someya, T. *Journal-Bearing Databook*; Springer: Berlin, Germany, 1989.
7. Wang, L.H.; Tieu, A.K. A study on the static and dynamic characteristics of journal bearing. In *Research Report Tribology*; University of Wollongong: Wollongong, Australia, 1989.
8. Choy, F.K.; Braun, M.J.; Hu, Y. Nonlinear effects in a plain journal bearing, Part I: Analytical study. *J. Tribol.* **1991**, *113*, 555–562. [CrossRef]
9. Choy, F.K.; Braun, M.J.; Hu, Y. Nonlinear transient and frequency response analysis of a hydrodynamic journal bearing. *J. Tribol.-Trans. ASME* **1992**, *114*, 448–454. [CrossRef]
10. Rho, B.H.; Kim, K.W. A study of nonlinear frequency response analysis of hydrodynamic journal bearings with external disturbances. *Tribol. Trans.* **2002**, *45*, 117–121. [CrossRef]
11. Lin, Q.Y.; Wei, Z.Y.; Wang, N.; Chen, W. Analysis on the lubrication performances of journal bearing system using computational fluid dynamics and fluid-structure interaction considering thermal influence and cavitation. *Tribol. Int.* **2013**, *64*, 8–15. [CrossRef]
12. Vijayaraghavan, D.; Keith, T.G. Analysis of a finite grooved misaligned journal bearing considering cavitation and starvation effects. *J. Tribol.* **1990**, *112*, 60–67. [CrossRef]
13. Bouyer, J.; Fillon, M. An experimental analysis of misalignment effects on hydrodynamic plain journal bearing performances. *J. Tribol.* **2002**, *124*, 313–319. [CrossRef]
14. Sun, J.; Gui, C.L. Hydrodynamic lubrication analysis of journal bearing considering misalignment caused by shaft deformation. *Tribol. Int.* **2004**, *37*, 841–848. [CrossRef]
15. Hirani, H.; Verma, M. Tribological study of elastomeric bearings for marine propeller shaft system. *Tribol. Int.* **2009**, *42*, 378–390. [CrossRef]
16. Bou-Said, J.F.; Huebner, K.H. Application of finite element methods to lubrication: An engineering approach. *J. Lubr. Tech.* **1972**, *94*, 313–321.
17. San Andres, L.S. Effect of shaft misalignment on the dynamic force response of annular pressure seals. *Tribol. Trans.* **1993**, *36*, 173–182. [CrossRef]
18. Qiu, Z.L. A Theoretical and Experimental Study on Dynamic Characteristics of Journal Bearings. Ph.D. Thesis, University of Wollongong, Wollongong, Australia, 1995.
19. Feng, H.; Jiang, S.; Ji, A. Investigations of the static and dynamic characteristics of water-lubricated hydrodynamic journal bearing considering turbulent, thermohydrodynamic and misaligned effects. *Tribol. Int.* **2019**, *130*, 245–260. [CrossRef]
20. Hanawa, N.; Kuniyoshi, M.; Miyatake, M.; Yoshimoto, S. Static characteristics of a water-lubricated hydrostatic thrust bearing with a porous land region and a capillary restrictor. *Precis. Eng.* **2017**, *50*, 293–307. [CrossRef]
21. Lin, X.; Wang, R.; Zhang, S.; Jiang, S. Study on dynamic characteristics for high speed water-lubricated spiral groove thrust bearing considering cavitating effect. *Tribol. Int.* **2020**, *143*, 106022. [CrossRef]
22. Liang, X.; Yan, X.; Ouyang, W.; Wood, R.J.K.; Liu, Z. Thermo-Elasto-Hydrodynamic analysis and optimization of rubber-supported water-lubricated thrust bearings with polymer coated pads. *Tribol. Int.* **2019**, *138*, 365–379. [CrossRef]
23. Dowson, D.; Taylor, C.M. Fundamental Aspects of Cavitation in Bearings. *IMECHE* **1974**, *7*, 15–26.
24. Brewe, D.E.; Ball, J.H.; Khonsari, M.M. *Current Research in Cavitating Fluid Films*; STLE Special Publication SP-28; NASA: Washington, DC, USA, 1990; Volume 21, pp. 56–61.
25. Brewe, D.E. Theoretical modeling of the vapor cavitation in dynamincally loaded journal bearings. *J. Tribol.-Trans. ASME* **1986**, *108*, 628–637. [CrossRef]
26. Floberg, L. Cavitation Boundary Conditions with Regard to the Number of Streamers and Tensile Strength of the Liquid. *IMECHE* **1974**, *17*, 31–36.
27. Elrod, H.G.; Adams, M. A Computer Program for Cavitation and Starvation Problems. *IMECHE* **1974**, *1*, 37–42.
28. Elrod, H.G. A Cavitation Algorithm. *J. Lubr. Technol.* **1981**, *103*, 350–354. [CrossRef]
29. Vijayaraghavan, D.; Keith, T.G. Development and evaluation of a cavitation algorithm. *Tribol. Trans.* **1989**, *103*, 225–233. [CrossRef]
30. Priestner, C.; Allmaier, H.; Priebsch, H.H.; Forstner, C. Refined simulation of friction power loss in crank shaft slider bearings considering wear in the mixed lubrication regime. *Tribol. Int.* **2012**, *46*, 200–207. [CrossRef]
31. Hearn, E.J. *Mechanics of Materials 1*; Butterworth-Heinemann: Oxford, UK, 1997; Volume 14, pp. 26–31.
32. Choi, J.; Kim, S.S.; Rhim, S.S.; Choi, J.H. Numerical modeling of journal bearing considering both elastohydrodynamic lubrication and multi-flexible-body dynamics. *Int. J. Auto. Tech-KOR* **2012**, *13*, 255–261. [CrossRef]
33. Batada, G.; Chupin, L. Compressiable fluid model for hydrodynamic lubrication cavitation. *J. Tribol.* **2013**, *135*, 041702-1–041702-13. [CrossRef]
34. Su, H. Determining the value of pressure in the cavitation zone for steady-state journal bearing. *Chin. Mech. Eng.* **2007**, *5*, 88–89.

35. Chouchane, M.; Naimi, S.; Ligier, J.L. Stability analysis of hydrodynamic bearings with a central circumferential feeding groove. In Proceedings of the 13th World Congress in Mechanism and Machine Science, Guanajuato, Mexico, 19–23 June 2011.
36. Patir, N.; Cheng, H.S. Application of average flow model to lubrication between rough sliding surfaces. *J. Lubr. Technol.* **1979**, *101*, 220–230. [CrossRef]
37. He, T.; Zou, D.; Lu, X.; Guo, Y.; Wang, Z.; Li, W. Mixed-lubrication analysis of marine stern tube bearing considering bending deformation of stern shaft and cavitation. *Tribol. Int.* **2014**, *73*, 108–116. [CrossRef]
38. Xie, Z.; Wang, X.; Zhu, W. Theoretical and experimental exploration into the fluid structure coupling dynamic behaviors towards water-lubricated bearing with axial asymmetric grooves. *Mech. Syst. Signal Process.* **2022**, *168*, 108624. [CrossRef]
39. Xie, Z.; Zhu, W. Theoretical and experimental exploration on the micro asperity contact load ratios and lubrication regimes transition for water-lubricated stern tube bearing. *Tribol. Int.* **2021**, *164*, 107105. [CrossRef]
40. Zhou, H.J.; Lv, B.L.; Wang, D.H. Research and analysis of gyroscopic vibration dynamic response of shafting based on an Improved Fourier Series Method. *J. Ship Mech.* **2012**, *16*, 962–970.
41. Kuang, F.; Zhou, X.; Liu, Z.; Huang, J.; Liu, X.; Qian, K.; Gryllias, K. Computer-vision-based research on friction vibration and coupling of frictional and torsional vibrations in water-lubricated bearing-shaft system. *Tribol. Int.* **2020**, *150*, 106336. [CrossRef]
42. Xu, J.; Jiao, C.; Zou, D.; Ta, N.; Rao, Z. Study on the dynamic behavior of herringbone gear structure of marine propulsion system powered by double-cylinder turbines. *China Technol. Sci.* **2022**, *65*, 611–630. [CrossRef]

lubricants

MDPI

Article

Analysis of Nonlinear Time-Domain Lubrication Characteristics of the Hydrodynamic Journal Bearing System

Changgang Lin [1,2,3], Fan Jian [4], Shili Sun [4], Can Sima [1,2,3], Libo Qi [1,2,3] and Mingsong Zou [1,2,3,*]

1 China Ship Scientific Research Center, Wuxi 214082, China
2 Taihu Laboratory of Deep-Sea Technological Science, Wuxi 214082, China
3 State Key Laboratory of Deep-Sea Manned Vehicles, Wuxi 214082, China
4 College of Shipbuilding Engineering, Harbin Engineering University, Harbin 150001, China
* Correspondence: zoumings@126.com

Abstract: The nonlinear time-domain lubrication characteristics of the hydrodynamic journal bearing system are studied in this paper. The motion equation of the hydrodynamic journal bearing system is established based on the balance of the relationship among the water film force, journal inertia force, and external load. The water film pressure distribution of the sliding bearing is calculated by the finite difference method. Firstly, the variation law of the water film pressure distribution with time under the external periodic load is calculated considering the inertial force of the journal. The influence of the initial eccentricity on the orbit of the journal center is studied. Secondly, the maximum water film pressure, the orbit of the journal center, eccentricity, water film pressure, and the minimum water film thickness of the bearing under the action of circumferential and unidirectional periodic external loads are calculated, and the effects of inertial force and rotational speed on the dynamic characteristics of the bearing are analyzed. Finally, the water film dynamic characteristics under low speed and heavy load are studied. The result shows that the pressure of the dimensionless water film caused by inertial force is reduced by 7 to 10 percent at the rotational speed between 200 r/min and 800 r/min, which means that the influence of inertia force cannot be ignored.

Keywords: time-domain; lubrication; hydrodynamic; journal bearing; inertia force

Citation: Lin, C.; Jian, F.; Sun, S.; Sima, C.; Qi, L.; Zou, M. Analysis of Nonlinear Time-Domain Lubrication Characteristics of the Hydrodynamic Journal Bearing System. *Lubricants* **2023**, *11*, 145. https://doi.org/10.3390/lubricants11030145

Received: 1 February 2023
Revised: 14 March 2023
Accepted: 15 March 2023
Published: 17 March 2023

1. Introduction

Radial sliding bearing is a widely used bearing in the industry. It has the advantages of strong bearing capacity, stable operation, good vibration resistance, and long life, and it is widely used in various fields [1]. Traditional sliding bearings use greases as lubricants. High-viscosity lubricants generate more heat during operation, and the problem of lubricant leakage during operation is even more serious, especially for oil-lubricated bearings used in water machinery. Oil pollution seriously damages the water environment [2,3]. Using water instead of oil as a lubricating medium can not only save a lot of oil but also avoid environmental pollution caused by using oil as a lubricating medium [4,5]. At the same time, water-lubricated bearings have low cost, good flame retardance, easy maintenance, and can also reduce the friction, wear, vibration, noise, non-functional consumption, and other key problems of friction pairs. Therefore, water-lubricated bearings are widely used in ship stern shaft, water turbine, water pump, water turbine, steam turbine, and other equipment. For example, Canada's Thomson-Golden Co., Ltd. (Canada) adopted a water lubrication system in the support of the ship's stern shaft and compounded a layer of polymer material on the stainless-steel bearing as the bearing bush. Water-lubricated bearings are widely used in centrifugal pumps and marine centrifugal pumps in Japan. The Hayward Taylor Company of the United Kingdom has adopted water-lubricated sliding bearings in the structure of the packless pump. Vickers and Michell of Germany use water-lubricated rubber bearings in deep well pumps and submersible pumps [6].

The hydrodynamic journal bearing system could be considered to be a typical rotor-bearing system. Accurately grasping the lubrication characteristics of the rotor-bearing system has important guiding significance for the optimization of and vibration and noise design of the engineering machinery system [7,8]. When the mechanical system is running, the clearance of the rotor-bearing system is in a constantly changing state. Therefore, the lubrication state of the rotor-bearing system has strong time-varying characteristics, which are very complex to solve [9]. The commonly simplified method is to use the infinitely short bearing or infinitely long bearing theory to study nonlinear dynamic characteristics and the instability of the rotor-bearing system [10–12]. Brancati et al. studied the nonlinear dynamic characteristics of rigid unbalanced rotors supported by oil film bearings. A short bearing theory was adopted to solve the oil film force. The results showed that the rotor exhibited periodic and one-half periodic vibration under the nonlinear oil film force [13,14]. Li et al. adopted the infinite length bearing model and presented the approximate analytic expressions of the periodic solution of the non-linear rotor-bearing system [15]. Lin et al. established a comprehensive finite element model considering the nonlinear force of the water film and the flexibility of the propeller blade for a propeller-shaft system. The long-bearing approximation was adopted to calculate the nonlinear force of the water film. The research shows that the rotor will exhibit complex dynamic behavior under nonlinear oil film, such as bifurcation, stability problems, chaos, etc. [16,17]. Cai et al. studied the effects of wear and shaft-shape error defects on the tribo-dynamic responses of water-lubricated bearings under non-linear propeller disturbances using a numerical model. Scholars have adopted various methods to solve the water film force [18].

It is important to improve mathematical models for more accurate prediction of bearing characteristics. With the continuous development of numerical calculation technology, the lubrication characteristics can also be analyzed in detail. Some scholars have conducted research by establishing numerical models of water-lubricated bearings. Singh et al. used the two-dimensional Reynolds equation to study the changes in bearing film thickness, bearing capacity, and temperature field under different journal speeds [19]. Kornaeva et al. developed a simulation model to calculate the flow of viscous incompressible fluid in the annular channel under the influence of vibration. The effect of boundary flow of low-viscosity fluids was determined [20]. Xie et al. used finite difference method to study the lubrication mechanism and performance parameters of water-lubricated bearing considering wall sliding and inertia force and studied the influence of wall sliding and inertia force on performance parameters [21]. Gao et al. used computational fluid dynamics (CFD) to analyze the influence of the eccentricity ratio on the water film pressure distribution, and numerical analysis was carried out for different sizes of sliding bearings at different speeds. Based on the analysis, a reference has been provided for the selection of the initial diameter for the design of high-efficiency water-lubricated plain bearings under a given load and rotational speed [22]. Wang et al. combined computational fluid dynamics and fluid-structure coupling to study the effect of elastic deformation on load-carrying capacity [23]. Through experimental studies, some academics have shown that obvious bushing wear would occur at the edge of the water-lubricated bearing working under the fluid lubrication state [24,25]. Sun et al. established a transient model of a ship propeller rotor-bearing system to study the system response and lubrication performance [26].

Due to the action of load and the rotational speed, there form an eccentricity and the corresponding convergent and divergent water films between the bearing and the journal. In a convergent area, the pressure is very large, and the resultant force resists the external load, and correspondingly the external load will directly affect the dynamic pressure of the water film. It should be noted that the journal itself has an acceleration under the action of external loads, not only in the direction of velocity but also in the magnitude that both are changed when the external force varies with time. In this case, the presence of inertial forces is bound to have an impact on the operation of the bearing. At the same time, the influence of inertia force on the lubrication characteristics of water-lubricated bearings is studied, which is of great significance to improve the lubrication performance of bearings.

Compared with oil-lubricated bearings, water-lubricated bearings are more difficult to converge during time-domain simulation, especially when the inertial force is considered, which poses great challenges to the present paper.

The nonlinear time-domain lubrication characteristics of the hydrodynamic journal bearing system are studied in this paper. Firstly, on the basis of the Reynolds equation, when solving the force equilibrium equation, the relevant theoretical formulas were derived, taking into account the inertial force caused by acceleration and the centripetal force caused by circular motion. The variation law of the water film pressure distribution with time under the external periodic load was calculated considering the inertial force of the journal. The influence of the initial eccentricity on the orbit of the journal center was studied. Secondly, the maximum water film pressure, the orbit of the journal center, eccentricity, and the minimum thickness of the water film of the bearing under the action of circumferential and unidirectional periodic external loads were calculated, and the effects of inertial force and rotational speed on the dynamic characteristics of the bearing were analyzed. Finally, the water film dynamic characteristics under low speed and heavy load are studied.

2. Numerical Procedures

The sketch map of hydrodynamic journal bearing geometry is shown in Figure 1. The origin of the coordinate system is at the center of the bearing. The x-axis points to the left, and the y-axis points down. $h = c + e\cos\theta$ denotes the thickness of the water film, and the water film expands along θ, direction, and $s = \theta \cdot R$; the attitude angle, ϕ, is defined as the angle between the maximum thickness of the water film and the $-x$ axis; the angular velocity is ω; the distance between the journal center and the bearing center is the eccentricity, e; ξ and η point to the radial and tangential direction; the absolute clearance is $c = R - r$; and the journal center displacement in x and y directions is defined as $u = e\cos\phi, v = e\sin\phi$.

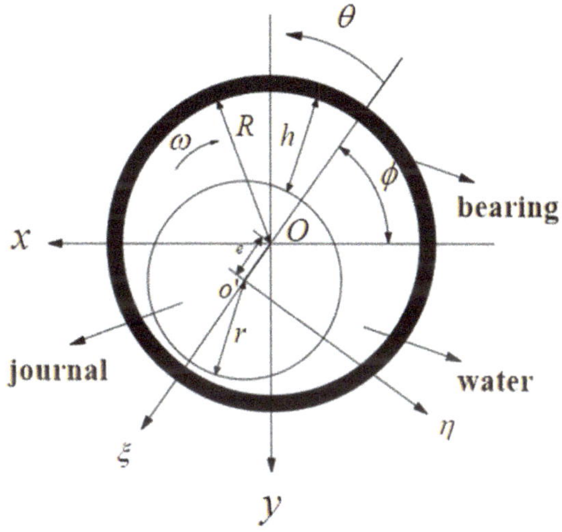

Figure 1. Sketch map of hydrodynamic journal bearing geometry.

2.1. Governing Equation and Boundary Conditions

According to the actual situation of water-lubricated bearings, we made the following assumptions:

(1) The film thickness is far smaller than the radius of curvature of the adjacent solid surface;
(2) The water film pressure in the thickness direction is constant;
(3) Only the velocity gradient in the film thickness direction is considered;

(4) The lubricant is in an isothermal and incompressible state.

The water film between the bearing and the journal satisfies the Reynolds equation and this equation can be written as

$$\frac{\partial}{\partial s}\left(\frac{h^3}{\eta_0}\frac{\partial p}{\partial s}\right)+\frac{\partial}{\partial z}\left(\frac{h^3}{\eta_0}\frac{\partial p}{\partial z}\right)=6\left[U\frac{\partial h}{\partial s}+2\frac{\partial h}{\partial t}\right] \quad (1)$$

Substituting the water film thickness $h = c + e\cos\theta$ into Equation (1) gives

$$\frac{\partial}{\partial s}\left(\frac{h^3}{\eta_0}\frac{\partial p}{\partial s}\right)+\frac{\partial}{\partial z}\left(\frac{h^3}{\eta_0}\frac{\partial p}{\partial z}\right)=6\left[\omega\frac{\partial h}{\partial \theta}+2\frac{\partial e}{\partial t}\cos\theta+2e\sin\theta\frac{\partial \phi}{\partial t}\right] \quad (2)$$

By using $\frac{\partial h}{\partial \theta} = -e\sin\theta$, Equation (2) can be further written as

$$\frac{\partial}{\partial s}\left(\frac{h^3}{\eta_0}\frac{\partial p}{\partial s}\right)+\frac{\partial}{\partial z}\left(\frac{h^3}{\eta_0}\frac{\partial p}{\partial z}\right)=6\left[\left(\omega-2\frac{\partial \phi}{\partial t}\right)\frac{\partial h}{\partial \theta}+2\frac{\partial e}{\partial t}\cos\theta\right] \quad (3)$$

In which, $\omega, \frac{\partial \phi}{\partial t}, \frac{\partial e}{\partial t}$, and η_0, respectively, denote angular velocity of the axial, variation rate of offset angel, variation rate of offset displacement, and dynamic viscosity.

2.2. Discretisation of the Governing Equation

In order to make the complex formula form compact and highlight the role of related factors, the Reynolds equation will be solved in a dimensionless form. The following dimensionless transformation is adopted:

$$\begin{aligned}
&s = \theta \cdot R \Rightarrow \theta = s/R\\
&z = \bar{z}(L/2) \Rightarrow \bar{z} = z/(L/2), -1 \le \bar{z} \le 1\\
&\bar{h} = h/c = 1 + \varepsilon\cos\theta;\\
&\bar{p} = p/p_0; p_0 = 6\eta_0(\omega - 2\partial\phi/\partial t)\cdot(R/c)^2\\
&q = 2\tfrac{\partial \varepsilon}{\partial t}/(\omega - 2\partial\phi/\partial t)
\end{aligned} \quad (4)$$

By substituting the equation $p = p_1 + q \cdot p_2$ into Equation (3), the following equation can be derived

$$\begin{aligned}
\frac{\partial}{\partial s}\left(\frac{h^3}{\eta_0}\frac{\partial p_1}{\partial s}\right)+\frac{\partial}{\partial z}\left(\frac{h^3}{\eta_0}\frac{\partial p_1}{\partial z}\right)&=6(\omega-2\partial\phi/\partial t)\frac{\partial h}{\partial \theta}\\
\frac{\partial}{\partial s}\left(\frac{h^3}{\eta_0}\frac{\partial p_2}{\partial s}\right)+\frac{\partial}{\partial z}\left(\frac{h^3}{\eta_0}\frac{\partial p_2}{\partial z}\right)&=6c(\omega-2\partial\phi/\partial t)\frac{h-c}{e}
\end{aligned} \quad (5)$$

Substituting the dimensionless form in Equation (4) into Equation (5) gives

$$\begin{aligned}
p_0\frac{c^3}{\eta_0 R^2}\frac{\partial}{\partial \theta}\left(\bar{h}^3\frac{\partial \bar{p}_1}{\partial \theta}\right)+p_0\frac{c^3}{\eta_0(L/2)^2}\frac{\partial}{\partial \bar{z}}\left(\bar{h}^3\frac{\partial \bar{p}_1}{\partial \bar{z}}\right)&=6c(\omega-2\partial\phi/\partial t)\frac{\partial \bar{h}}{\partial \theta}\\
p_0\frac{c^3}{\eta_0 R^2}\frac{\partial}{\partial \theta}\left(\bar{h}^3\frac{\partial \bar{p}_2}{\partial \theta}\right)+p_0\frac{c^3}{\eta_0(L/2)^2}\frac{\partial}{\partial \bar{z}}\left(\bar{h}^3\frac{\partial \bar{p}_2}{\partial \bar{z}}\right)&=6c(\omega-2\partial\phi/\partial t)\frac{\bar{h}-1}{\varepsilon}
\end{aligned} \quad (6)$$

The dimensionless Reynolds equation is obtained by simplifying Equation (6)

$$\begin{aligned}
\frac{\partial}{\partial \theta}\left(\bar{h}^3\frac{\partial \bar{p}_1}{\partial \theta}\right)+\frac{R^2}{(L/2)^2}\frac{\partial}{\partial \bar{z}}\left(\bar{h}^3\frac{\partial \bar{p}_1}{\partial \bar{z}}\right)&=\frac{\partial \bar{h}}{\partial \theta}\\
\frac{\partial}{\partial \theta}\left(\bar{h}^3\frac{\partial \bar{p}_2}{\partial \theta}\right)+\frac{R^2}{(L/2)^2}\frac{\partial}{\partial \bar{z}}\left(\bar{h}^3\frac{\partial \bar{p}_2}{\partial \bar{z}}\right)&=\frac{\bar{h}-1}{\varepsilon}
\end{aligned} \quad (7)$$

Letting $q = \overline{A}/\overline{B}$, then $\overline{A} = 2d\varepsilon/dt, \overline{B} = \omega - 2d\phi/dt$. Substituting it into Equation (7) gives

$$p(\theta) = \left[\overline{Bp1}(\theta) + \overline{Ap2}(\theta)\right]\eta_0\left(\frac{R}{c}\right)^2 \quad (8)$$

The finite difference method is used to solve the $\overline{p1}(\theta)$ and $\overline{p2}(\theta)$ in Equation (8). Expand the water film of the bearing bush into a rectangle along θ, direction, as shown in Figure 2. The θ direction is the circumferential direction of the water film, $\theta = 0$ is the starting edge of the water film, and $\theta = 2\pi$ is the end edge of the water film. We then divide the θ direction into m grids and divide the $\overline{y}$ direction into n grids. The width (step length) of each grid in the θ direction is $\Delta\theta = 2\pi/m$, and the width of each grid in the $\overline{y}$ direction is $\Delta\overline{y} = 2/n$. The dimensionless pressure value at node (i,j) is represented by $\overline{p}(i,j)$, and the pressure value at each node is used to form each order difference quotient, which approximately replaces the derivative in the Reynolds equation and then can be based on the four nodes $(i-1,j)$, $(i,j+1)$, $(i+1,j)$, and $(i,j-1)$ around the (i,j) node to calculate the pressure value of the intermediate node.

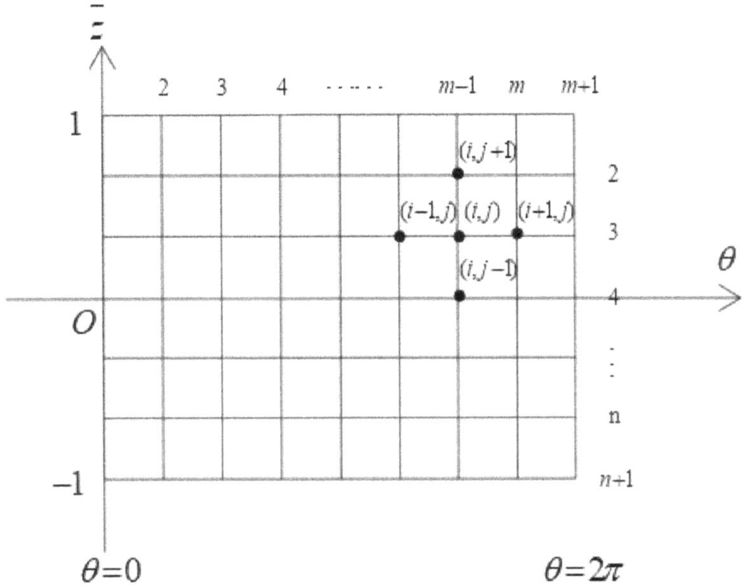

Figure 2. Dividing the water film.

At present, the mass conservation boundary conditions (also known as JFO boundary conditions) double Reynolds boundary conditions, and Reynolds boundary conditions proposed by Jakobsson, Floberg and Olsson based on hydraulic cavitation theory are the three boundary conditions commonly used in bearing numerical solutions [27]. Although the Reynolds boundary condition does not meet the conditions of conservation of lubricating oil quality at the boundary of oil film rupture and reformation, it meets the requirements of this paper because it is an easy-to-solve numerical method and the calculation accuracy is in line with the requirements of this paper. The boundary conditions used are Reynolds boundary conditions. When calculating the water film pressure distribution, each line is calculated point by point from the starting edge to the ending edge. If the calculated pressure value at a certain point is negative, the pressure value is zero. This point is taken as the approximate position of the natural rupture edge of the water film on the line, and the pressure at all points after this point is zero, making the approximate position of the rupture edge approach the natural boundary conditions.

The difference schemes for solving $\overline{p1}, \overline{p2}$ are as follows

$$\begin{aligned} \overline{p}_{1,i,j} &= \frac{A_{i,j}*\overline{p}_{1,i+1,j}+B_{i,j}*\overline{p}_{1,i-1,j}+C_{i,j}*\overline{p}_{1,i,j+1}+Di,j*\overline{p}_{1,i,j-1}-F_{1,i,j}}{E_{i,j}}; \\ \overline{p}_{2,i,j} &= \frac{A_{i,j}*\overline{p}_{2,i+1,j}+B_{i,j}*\overline{p}_{2,i-1,j}+C_{i,j}*\overline{p}_{2,i,j+1}+D_{i,j}*\overline{p}_{2,i,j-1}-F_{2,i,j}}{E_{i,j}} \end{aligned} \tag{9}$$

where

$$
\begin{aligned}
A_{i,j} &= 2\overline{h}^3_{i+\frac{1}{2},j};\\
B_{i,j} &= 2\overline{h}^3_{i-\frac{1}{2},j};\\
C_{i,j} &= \left(\frac{D}{L}\times\frac{\Delta\phi}{\Delta\lambda}\right)^2 * \overline{h}^3_{i,j+\frac{1}{2}};\\
D_{i,j} &= \left(\frac{D}{L}\times\frac{\Delta\phi}{\Delta\lambda}\right)^2 * \overline{h}^3_{i,j-\frac{1}{2}};\\
E_{i,j} &= A_{i,j}+B_{i,j}+C_{i,j}+D_{i,j};\\
F_{1,i,j} &= \Delta\phi * \frac{h_{i+1,j}-h_{i-1,j}}{2};\\
F_{2,i,j} &= \Delta\phi^2 * \cos\phi;
\end{aligned}
\tag{10}
$$

The carrying capacity of water film in x and y directions is

$$
\begin{aligned}
Wx &= -RL\eta_0\left(\frac{R}{c}\right)^2\left[\overline{B}\int\int\overline{p_1}\cos(\theta+\phi)d\theta d\overline{z}+\overline{A}\int\int\overline{p_2}\cos(\theta+\phi)d\theta d\overline{z}\right]\\
Wy &= RL\eta_0\left(\frac{R}{c}\right)^2\left[\overline{B}\int\int p_1\sin(\theta+\phi)d\theta d\overline{z}+\overline{A}\int\int p_2\sin(\theta+\phi)d\theta d\overline{z}\right]
\end{aligned}
\tag{11}
$$

2.3. Solution Procedure with Translational Inertial Force and Centripetal Force

When the inertia force of the journal is not considered, the external load is directly balanced with the water film force. The force balance equations in x and y directions are listed as follows

$$
\begin{aligned}
Wx + Gx &= 0\\
Wy + Gy &= 0
\end{aligned}
\tag{12}
$$

where Gx and Gy are external loads.

With $W_0 = 3RL\eta\left(\frac{R}{c}\right)^2$, there are

$$
\begin{aligned}
m_1 &= -W_0\cdot\int\int\overline{p_1}\cos(\theta+\phi)d\theta d\overline{z},\ m_2 = -W_0\cdot\int\int\overline{p_2}\cos(\theta+\phi)d\theta d\overline{z}\\
m_3 &= -W_0\cdot\int\int\overline{p_1}\sin(\theta+\phi)d\theta d\overline{z},\ m_4 = -W_0\cdot\int\int\overline{p_2}\sin(\theta+\phi)d\theta d\overline{z}
\end{aligned}
\tag{13}
$$

The force balance equation is simplified as

$$
\begin{bmatrix} m_1 & m_2\\ m_3 & m_4\end{bmatrix}\begin{bmatrix}\overline{B}\\ \overline{A}\end{bmatrix} = \begin{bmatrix}G_x\\ G_y\end{bmatrix}
\tag{14}
$$

When considering the journal inertial force and centripetal acceleration along the radial and tangential direction of the line connecting the axis and the maximum water film thickness, the inertial force balance equation is listed as follows

$$
\begin{aligned}
I_\xi &= Mc\frac{d^2\varepsilon}{dt^2} = W_\xi + G_\xi + Mc\varepsilon\left(\frac{d\phi}{dt}\right)^2\\
I_\eta &= Mc\varepsilon\frac{d^2\phi}{dt^2} = W_\eta + G_\eta
\end{aligned}
\tag{15}
$$

Using the force transformation relationship, the radial and tangential water film forces are decomposed into x and y directions, and the following equations can be obtained:

$$
\begin{aligned}
W_x &= W_\xi\cdot\cos\phi - W_\eta\cdot\sin\phi\\
W_y &= W_\xi\cdot\sin\phi + W_\eta\cdot\cos\phi
\end{aligned}
\tag{16}
$$

The relationships between eccentricity ratio and deflection angle and the acceleration of axial neck center are as follows:

$$
\begin{aligned}
\partial^2x/\partial t^2 &= c\cdot\partial^2\varepsilon/\partial t^2\cos\phi - e\partial^2\phi/\partial t^2\sin\phi\\
\partial^2y/\partial t^2 &= c\cdot\partial^2\varepsilon/\partial t^2\sin\phi + e\partial^2\phi/\partial t^2\cos\phi
\end{aligned}
\tag{17}
$$

The force balance equation considering journal inertia force is listed as

$$-\begin{bmatrix} M & 0 \\ 0 & M \end{bmatrix}\begin{bmatrix} \partial^2 x/\partial t^2 \\ \partial^2 y/\partial t^2 \end{bmatrix}+\begin{bmatrix} c_{xx} & c_{xy} \\ c_{yx} & c_{yy} \end{bmatrix}\begin{bmatrix} \partial x/\partial t \\ \partial y/\partial t \end{bmatrix}+\begin{bmatrix} k_{xx} & k_{xy} \\ k_{yx} & k_{yy} \end{bmatrix}\begin{bmatrix} x \\ y \end{bmatrix}=-\begin{bmatrix} G_x \\ G_y \end{bmatrix}-F_c\begin{bmatrix} \cos\phi \\ \sin\phi \end{bmatrix} \quad (18)$$
$$F_c = mc\varepsilon\left(\frac{d\phi}{dt}\right)^2$$

where F_c is centripetal force.

When the bearing is not subjected to external loads, the water film force in the x, y direction is decomposed into a radial force W_ξ and tangential force W_η

$$\begin{aligned} W_\xi &= \iint p\cos\theta ds \\ W_\eta &= \iint p\sin\theta ds \end{aligned} \quad (19)$$

At this point, the relationship between the water film force and the stiffness coefficient and journal center displacement is as follows

$$\begin{aligned} W_x &= W_\xi\cdot\cos\phi - W_\eta\cdot\sin\phi = k_{xx}\cdot e\cos\phi + k_{xy}\cdot e\sin\phi \\ W_y &= W_\xi\cdot\sin\phi + W_\eta\cdot\cos\phi = k_{yx}\cdot e\cos\phi + k_{yy}\cdot e\sin\phi \end{aligned} \quad (20)$$

After simplification, the dimensioned stiffness coefficient is expressed by radial and tangential water film forces:

$$\begin{aligned} k_{xx} &= W_\xi/e \\ k_{xy} &= -W_\eta/e \\ k_{yx} &= W_\xi/e \\ k_{yy} &= W_\eta/e \end{aligned} \quad (21)$$

In the process of solving the force balance equation considering the inertial force, the stiffness coefficient is solved according to the above equation, and the damping coefficient matrix is replaced by the following equation:

$$\begin{aligned} \begin{bmatrix} W_x \\ W_y \end{bmatrix} &= \begin{bmatrix} m_1 & m_2 \\ m_3 & m_4 \end{bmatrix}\begin{bmatrix} \omega \\ 0 \end{bmatrix}\Bigg|_{t=t_0} + 2\begin{bmatrix} -m_1 & m_2 \\ -m_3 & m_4 \end{bmatrix}\begin{bmatrix} \partial\phi/\partial t \\ \frac{\partial\varepsilon}{\partial t} \end{bmatrix} \\ &= \begin{bmatrix} m_1 & m_2 \\ m_3 & m_4 \end{bmatrix}\begin{bmatrix} \omega \\ 0 \end{bmatrix} + 2\begin{bmatrix} -m_1 & m_2 \\ -m_3 & m_4 \end{bmatrix}\begin{bmatrix} -\sin\phi/e & \cos\phi/e \\ \cos\phi/c & \sin\phi/c \end{bmatrix}\begin{bmatrix} \partial x/\partial t \\ \partial y/\partial t \end{bmatrix} \\ &= \begin{bmatrix} k_{xx} & k_{xy} \\ k_{yx} & k_{yy} \end{bmatrix}\begin{bmatrix} x \\ y \end{bmatrix} + \begin{bmatrix} c_{xx} & c_{xy} \\ c_{yx} & c_{yy} \end{bmatrix}\begin{bmatrix} \partial x/\partial t \\ \partial y/\partial t \end{bmatrix} \end{aligned} \quad (22)$$

In the time-domain calculation in this paper, the stiffness coefficient and damping coefficient is replaced by other variables, and then the time-domain iterative solution is performed.

Considering $\overline{A} = 2\frac{\partial\varepsilon}{\partial t}, \overline{B} = \omega - 2\partial\phi/\partial t$, replacing Equation (17) with Equation (18) gives

$$-\begin{bmatrix} M & 0 \\ 0 & M \end{bmatrix}\begin{bmatrix} c\cdot\partial^2\varepsilon/\partial t^2\cos\phi - e\cdot\partial^2\phi/\partial t^2\sin\phi \\ c\cdot\partial^2\varepsilon/\partial t^2\sin\phi + e\cdot\partial^2\phi/\partial t^2\cos\phi \end{bmatrix} + \begin{bmatrix} m_1 & m_2 \\ m_3 & m_4 \end{bmatrix}\begin{bmatrix} \overline{B} \\ \overline{A} \end{bmatrix} = -\begin{bmatrix} G_x \\ G_y \end{bmatrix} - F_c\begin{bmatrix} \cos\phi \\ \sin\phi \end{bmatrix} \quad (23)$$

Simplifying Equation (23) gives

$$\begin{aligned} &-c\cdot\partial^2\varepsilon/\partial t^2\begin{bmatrix} M & 0 \\ 0 & M \end{bmatrix}\begin{bmatrix} \cos\phi \\ \sin\phi \end{bmatrix} - e\cdot\partial^2\phi/\partial t^2\begin{bmatrix} M & 0 \\ 0 & M \end{bmatrix}\begin{bmatrix} -\sin\phi \\ \cos\phi \end{bmatrix} + \begin{bmatrix} m_1 & m_2 \\ m_3 & m_4 \end{bmatrix}\begin{bmatrix} \omega - 2\partial\phi/\partial t \\ 2\partial\varepsilon/\partial t \end{bmatrix} \\ &= -\begin{bmatrix} G_x \\ G_y \end{bmatrix} - F_c\begin{bmatrix} \cos\phi \\ \sin\phi \end{bmatrix} \end{aligned} \quad (24)$$

A more concise format is written as

$$\partial^2\varepsilon/\partial t^2\begin{bmatrix} -cM\cos\phi \\ -cM\sin\phi \end{bmatrix} + \partial^2\phi/\partial t^2\begin{bmatrix} eM\sin\phi \\ -eM\cos\phi \end{bmatrix} - \partial\phi/\partial t\begin{bmatrix} 2m_1 \\ 2m_3 \end{bmatrix} + \partial\varepsilon/\partial t\begin{bmatrix} 2m_2 \\ 2m_4 \end{bmatrix} + \begin{bmatrix} m_1 & m_2 \\ m_3 & m_4 \end{bmatrix}\begin{bmatrix} \omega \\ 0 \end{bmatrix} = -\begin{bmatrix} G_x \\ G_y \end{bmatrix} - F_c\begin{bmatrix} \cos\phi \\ \sin\phi \end{bmatrix} \tag{25}$$

Backward difference scheme is adopted to solve $\partial\varepsilon/\partial t$, $\partial\phi/\partial t$ in this paper. The difference scheme is as follows

$$\begin{aligned} \partial^2\varepsilon/\partial^2 t_{(s+1)} &= (\partial\varepsilon/\partial t_{(s+1)} - \partial\varepsilon/\partial t_{(s)})/dt \\ \partial^2\phi/\partial^2 t_{(s+1)} &= (\partial\phi/\partial t_{(s+1)} - \partial\phi/\partial t_{(s)})/dt \end{aligned} \tag{26}$$

Substituting the difference Equation (26) into the simplified force balance equation gives

$$\begin{aligned} &((\partial\varepsilon/\partial t_{(s+1)} - \partial\varepsilon/\partial t_{(s)})/dt)\begin{bmatrix} -cM\cos\phi \\ -cM\sin\phi \end{bmatrix} + ((\partial\phi/\partial t_{(s+1)} - \partial\phi/\partial t_{(s)})/dt)\begin{bmatrix} eM\sin\phi \\ -eM\cos\phi \end{bmatrix} \\ &-\partial\phi/\partial t_{(s+1)}\begin{bmatrix} 2m_1 \\ 2m_3 \end{bmatrix} + \partial\varepsilon/\partial t_{(s+1)}\begin{bmatrix} 2m_2 \\ 2m_4 \end{bmatrix} + \begin{bmatrix} m_1 & m_2 \\ m_3 & m_4 \end{bmatrix}\begin{bmatrix} \omega \\ 0 \end{bmatrix} = -\begin{bmatrix} G_x \\ G_y \end{bmatrix} - F_c\begin{bmatrix} \cos\phi \\ \sin\phi \end{bmatrix} \end{aligned} \tag{27}$$

The final difference forms for solving $\partial\varepsilon/\partial t$ and $\partial\phi/\partial t$ are written as

$$\begin{aligned} &\begin{bmatrix} -cM\cos\phi/dt + 2m_2 & eM\sin\phi/dt - 2m_1 \\ -cM\sin\phi/dt + 2m_4 & -eM\cos\phi/dt - 2m_3 \end{bmatrix}\begin{bmatrix} \partial\varepsilon/\partial t_{(s+1)} \\ \partial\phi/\partial t_{(s+1)} \end{bmatrix} \\ &= \begin{bmatrix} G_x \\ G_y \end{bmatrix} - \begin{bmatrix} m_1 & m_2 \\ m_3 & m_4 \end{bmatrix}\begin{bmatrix} \omega \\ 0 \end{bmatrix} + \partial\varepsilon/\partial t_{(s)}\begin{bmatrix} -cM\cos\phi/dt \\ -cM\sin\phi/dt \end{bmatrix} + \partial\phi/\partial t_{(s)}\begin{bmatrix} eM\sin\phi/dt \\ -eM\cos\phi/dt \end{bmatrix} \end{aligned} \tag{28}$$

Through Equation (28), the force balance equation considering inertial force is solved.

2.4. Iterative Method

In this paper, the iterative method is used for numerical calculation. Firstly, the bearing bush is expanded along the axis, with n nodes (n = 40) in the axis and m nodes in the circumferential direction. When calculating water film pressure, iterates from starting point to ending point. Assuming that the internal pressure of each point is zero, the pressure of each point is calculated by finite difference method. If a point pressure is negative, the water film rupture at this point is determined to be zero, and the first approximation of the water film pressure is obtained. The approximate value was used to recalculate the pressure value from the first point again. The overrelaxation iteration method was used to calculate the pressure value on the middle node by using the pressure value of four nodes around node (i,j). The node pressure was corrected until the iteration error range was met to obtain the water film pressure distribution of each node.

To determine whether the results of each iteration satisfy sufficient accuracy to determine whether the iteration terminated, the following relative convergence criteria were adopted:

$$\frac{\sum_{i=2}^{m}\sum_{j=2}^{n}\left|\overset{s+1}{P_{i,j}} - \overset{s}{P_{i,j}}\right|}{\sum_{i=2}^{m}\sum_{j=2}^{n}\left|\overset{s+1}{P_{i,j}}\right|} \le \delta \tag{29}$$

This allows the relative error value, δ, to generally take 10^{-3}~10^{-5}. In this paper, the relative error value takes 10^{-3}.

3. Analysis of Water Film Dynamic Characteristics Considering Journal Inertial Force

3.1. Validation

For the model in this paper, the convergence analysis of the number of meshes and time steps was carried out. The results of the grid number convergence verification are shown in Figure 3, and the time step convergence verification is shown in Figures 4–6. Figure 3 shows the comparison of dimensionless water film pressure under the working conditions of five groups of different grid numbers, and the second and third mesh quantities can better meet the convergence, and considering the problem of calculation efficiency, m = 60 and n = 40 are used to divide the grids in subsequent calculations. Selecting 10 μs for the time step can better meet the convergence requirements.

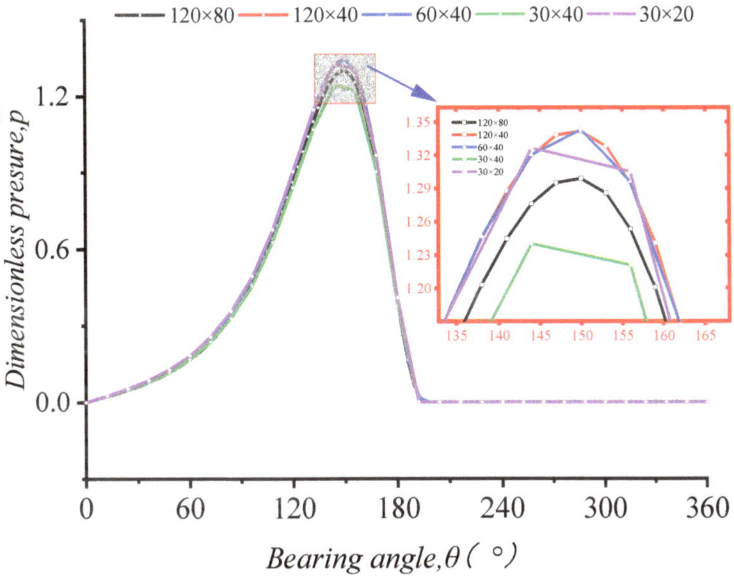

Figure 3. Effect of radial and circumferential mesh numbers on pressure.

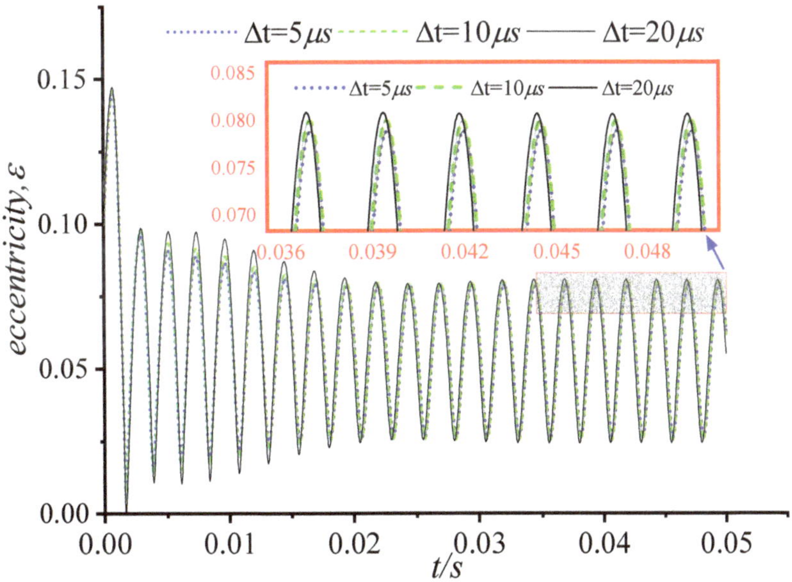

Figure 4. Comparison of eccentricity at different time steps.

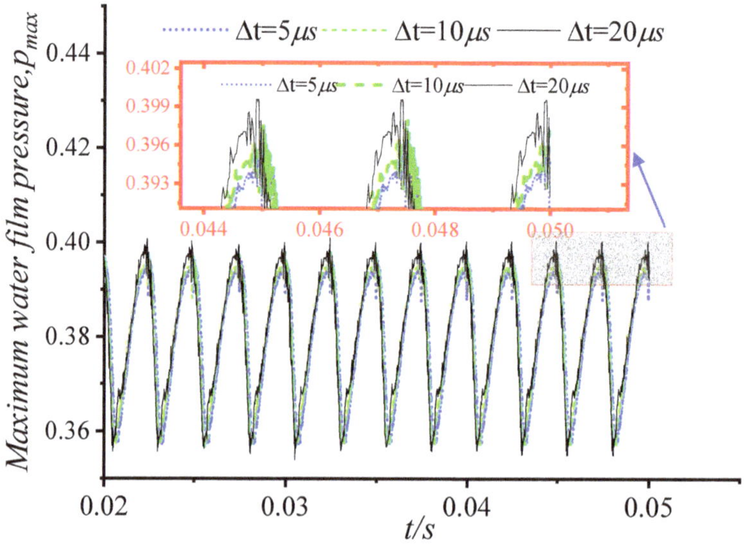

Figure 5. Comparison of maximum pressure at different time steps.

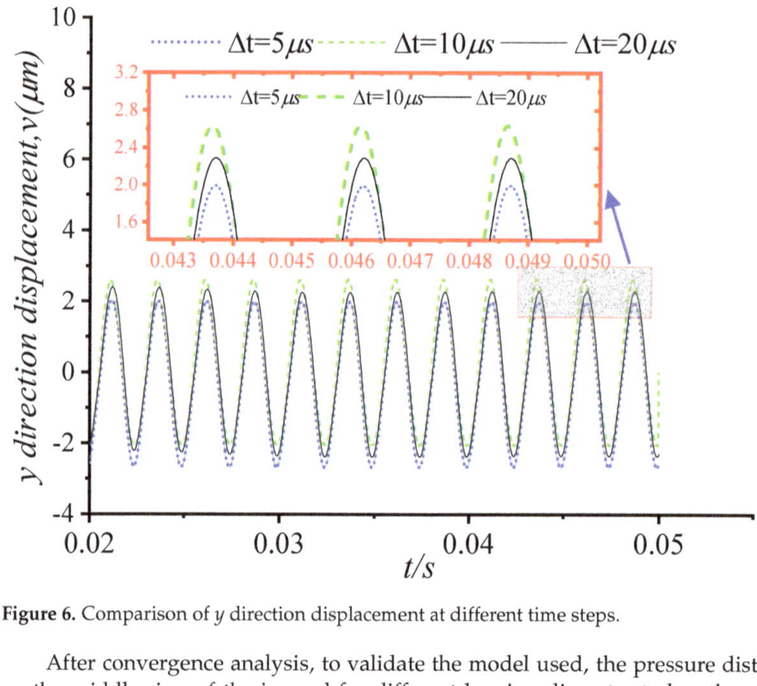

Figure 6. Comparison of *y* direction displacement at different time steps.

After convergence analysis, to validate the model used, the pressure distributions on the middle ring of the journal for different bearing diameter-to-length ratios were compared with the results calculated using a finite difference method [28], and the results were calculated using the CFD method [29] in dimensionless form. Table 1 shows the simulation parameters used in the research. The result and the comparison are shown in Figures 7 and 8. From Figure 7, it can be seen that for D/L = 0.5, 1, and 2, the pressures from $\theta = 0°$ to $180°$ computed in this work are in very good agreement with those computed using the method in [28,29]. However, the Reynolds boundary conditions do not predict the reformation of the film, and the pressures from the position where the film ruptures to θ = 360° are all set to zero.

Table 1. Parameters of the simulation by Zhang et al. [28] and Zhang et al. [29].

Parameters	Value
Bearing radius, R	40 mm
Bearing width	80 mm
Bearing clearance, c	0.04 mm
Rotate speed, r/min,	3000

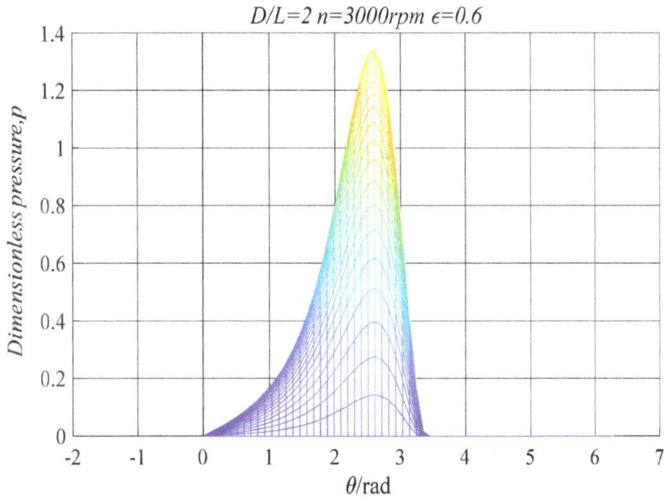

Figure 7. Dimensionless pressure (p/p_0) distribution at z = L/2 for D/L= 2, ε = 0.6.

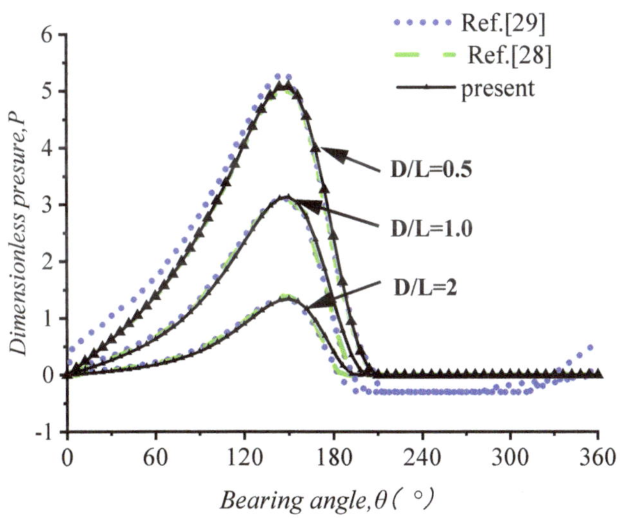

Figure 8. Comparison of dimensionless pressure (p/p_0) distribution at z = L/2 for D/L = 0.5, 1, and 2, ε = 0.6.

In addition, the finite bearing orbit was calculated under the condition that the external load of the journal is its own gravity and the inertia force of the journal is not considered, as shown in Figure 9. The calculation parameters refer to the parameters in [30]. Figure 9 shows a single trajectory starting from rest concentrically and spiraling gradually towards the equilibrium point under a downward gravity force. The orbit of Figure 9 is remarkably similar to one published by Booker [30] using the mobility method. Thus, the correctness of the calculation method in this paper is further verified.

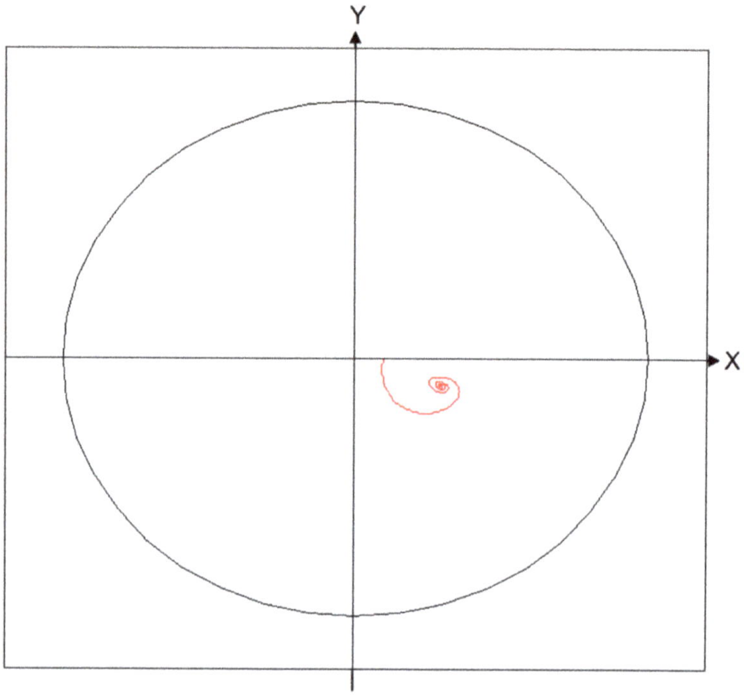

Figure 9. Finite bearing orbit plot.

3.2. *Effect of Inertial Force on Lubrication Characteristics*

The effects of different loads, rotational speeds, and initial eccentricity on the time-domain lubrication characteristics of water-lubricated bearings with and without inertial force are studied in this paper.

The model parameters used in this paper are shown in Table 2. Considering the effect of sinusoidal periodic load, the load expression is $G_x = 1.0E4\cos(800\pi t)$, $G_y = 1.0E4\sin(800\pi t)$. The initial eccentricity $\varepsilon = 0.1$. The magnitude of this load remains the same, and the direction of the force changes at a frequency of 400 Hz.

Table 2. Parameters of the numerical simulation.

Parameters	Value
Bearing radius, R	35.05 mm
Bearing width, L	70.09 mm
Bearing clearance, c	0.0845 mm
Rotate speed, n	545.6 r/min
Viscosity of water, η_0	0.001103 Pa·s
The density of water, ρ	1000 kg/m^3

Figures 10 and 11 show the water film pressure space distribution and water film thickness distribution at a certain time when subjected to periodic external loads.

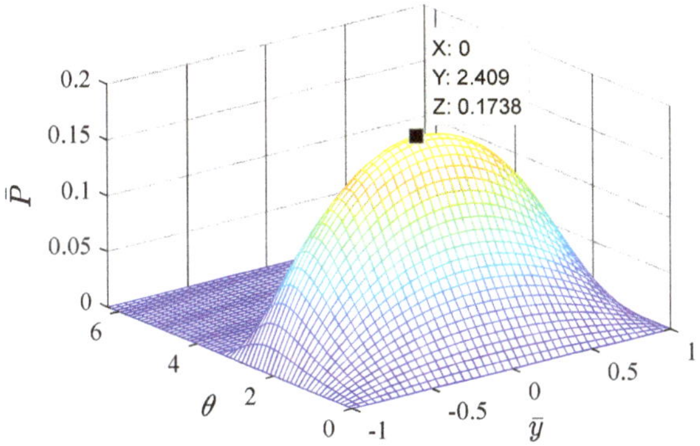

Figure 10. Dimensionless water film pressure distribution (t = 0.00625 s).

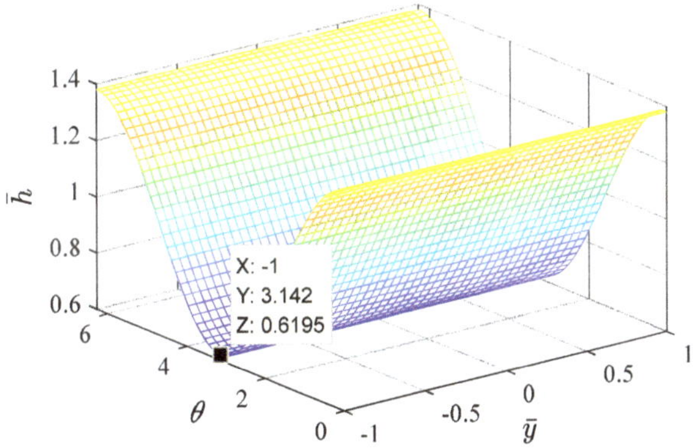

Figure 11. Dimensionless water film thickness distribution (t = 0.00625 s).

The three-dimensional distribution of the dimensionless water film pressure of the radial bearing is approximately a continuous parabolic distribution. In the interval where $0 \leq \theta \leq 2\pi$, the dimensionless water film pressure increases to the maximum pressure value and then decreases. The peak value of the water film pressure at the peak and valley points of the selected load appears at $\theta = 2.409rad = 138^o$. The water film pressure peak at the zero point of the load appears at $\theta = 2.304rad = 132^o$. The water film pressure is concentrated in $0 \leq \theta \leq \pi$. There is a zero pressure zone in the region where $\pi \leq \theta \leq 2\pi$ because the journal has a rotating effect and a squeezing effect on the water film. In the circumferential direction, due to the rotation of the journal, the water is drawn into the convergent wedge of $0 \leq \theta \leq \pi$, and a positive pressure is generated due to the extrusion. However, within the divergence wedge of $\pi \leq \theta \leq 2\pi$, the water is brought out, and a negative pressure area should be generated in theory. The liquid cannot bear the tensile stress, so the water film ruptures in the negative pressure area. Through numerical calculation, the boundary line of the non-zero pressure region can be accurately obtained, and the boundary line is near $\theta = \pi$. As can be seen from Figure 11, the thickness of the water film decreases continuously at first and reaches a minimum value when $\theta = \pi$.

The dynamic pressure lubrication state is the most ideal working state of the water-lubricated bearing. The lubrication characteristics of water-lubricated bearings depend not only on the selection of bearing materials and the design of structural parameters but also on the working conditions of the bearing during actual operation. When the load or rotational speed of the water-lubricated bearing changes, in order to meet the load-bearing requirements of the working condition, the eccentricity of the bearing will change accordingly. Therefore, the dynamic pressure lubrication characteristics of water-lubricated bearings are affected by the load, rotational speed, and eccentricity during operation.

The axial position of the bearing under a stable load is fixed in motion, and the thickness of the water film decreases with the increase in the load. The greater the load, the less reliable the bearing's working stability and the more prone to failure. However, in the unsteady load bearing with changes in the magnitude and direction of the load, the position of the axis changes rapidly with time, forming a complex orbit of the journal center curve. For the study of journal orbits, using mobility and impedance methods is computationally less expensive than solving the Reynolds equation (Booker, 2014).

Keeping the journal rotation speed constant, we took the initial eccentricities as 0.1, 0.5, and 0.9, respectively, and investigated the effect of the initial eccentricity on the orbit of journal center. We drew the orbit of the journal center and the displacement time–history curves in the x and y directions, as shown in Figures 12–17. It can be seen that the initial eccentricity set is different, which changes the initial position of the axis rotation. However, under the action of the circumferential sinusoidal periodic load, after the bearing rotates smoothly, the orbit of the journal center gradually converges to the center for a uniform circular motion. At the initial moment, due to the existence of the initial eccentricity, the axial displacement is disturbed. But after the external load and the water film force are balanced, the displacement in the x and y directions changes regularly and periodically. The displacement amplitude is the same, and the trajectory amplitude is about 3×10^{-5} m.

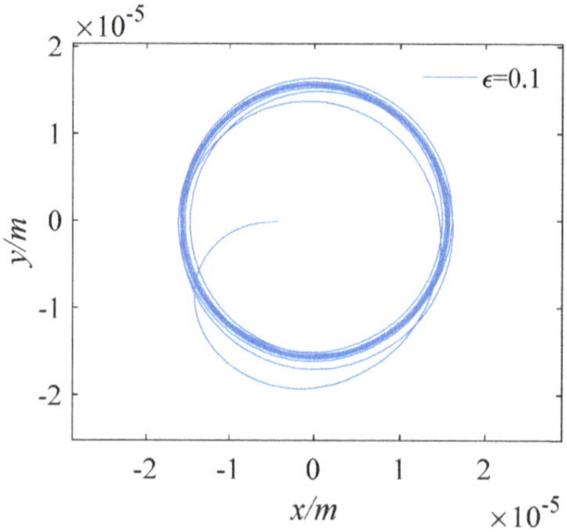

Figure 12. Orbit of journal center (ε = 0.1).

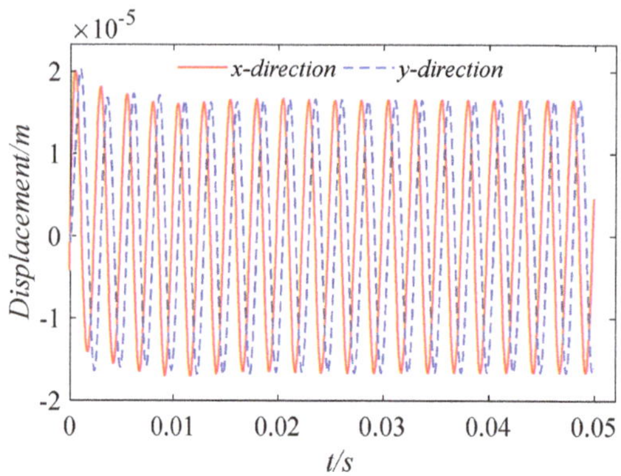

Figure 13. Displacement in x, y direction (ε = 0.1).

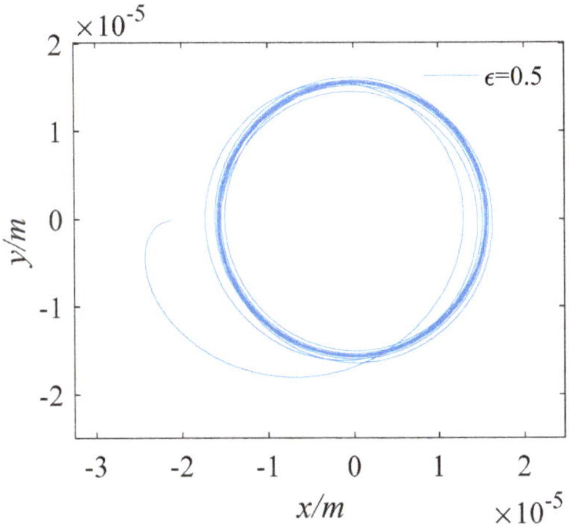

Figure 14. Orbit of journal center (ε = 0.5).

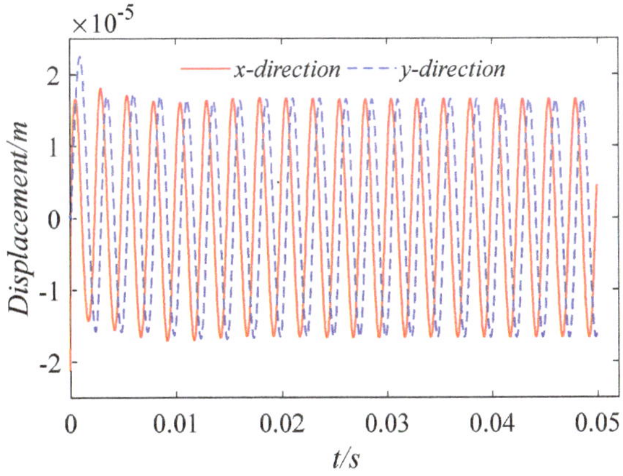

Figure 15. Displacement in x, y direction (ε = 0.5).

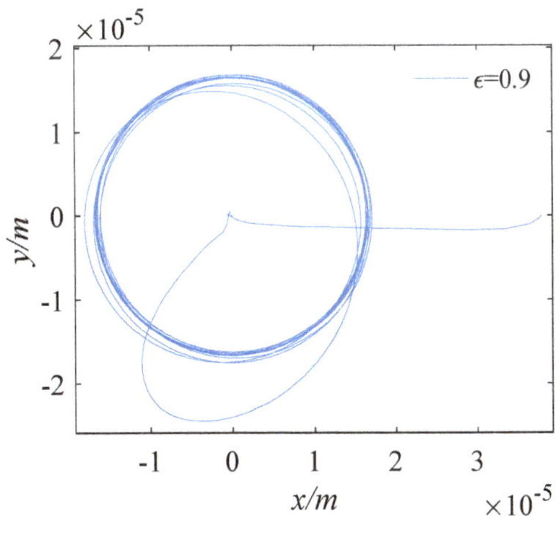

Figure 16. Orbit of journal center (ε = 0.9).

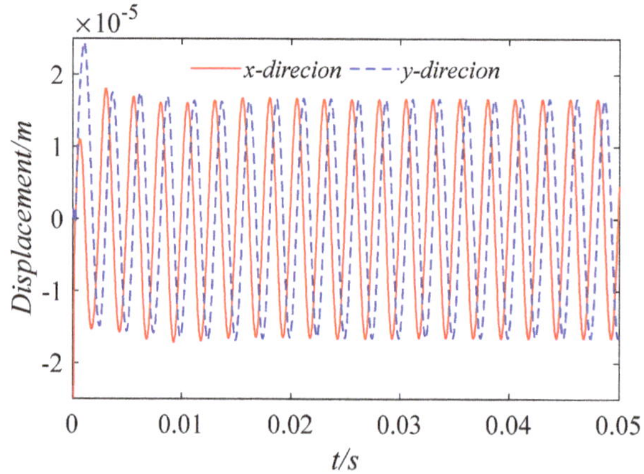

Figure 17. Displacement in x, y direction ($\varepsilon = 0.9$).

When the bearing rotates around the journal, there are inertial forces, I_x and I_y, caused by the acceleration of the journal; centripetal force, F_c, generated by uniform circular motion; and water film forces, W_x and W_y, in the x and y directions. These three forces are in balance with the external load under the superposition of the three forces. The three forces are decomposed along the x and y directions, and the time–history curve is shown in Figures 18 and 19. It can be seen from the figure that the water film force, inertial force, and centripetal force all change periodically during the rotation of the bearing under the action of the circumferential external load. The main component in balance with the external load is the water film force, and the centripetal force and inertial force account for about 2%. When the rotation tends to be stable, the tangential and normal inertial forces almost disappear.

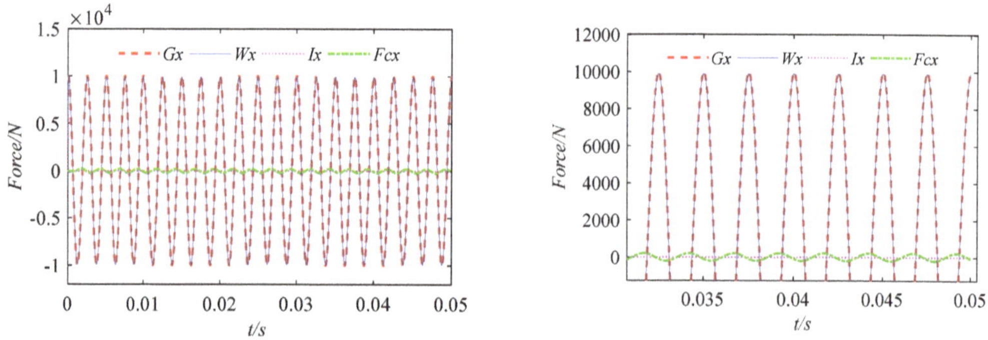

Figure 18. Time–history curves of four forces in the x direction.

Under the premise of not changing the bearing size and speed, we set the initial eccentricity as $\varepsilon = 0.1$. In order to analyze the influence of inertial force on the dynamic characteristics of the bearing, the orbit of the journal center image, the dimensionless maximum water film pressure, the time–history curves of eccentricity, and the deflection angle are shown in Figures 20–23. Under the circumferential external load, the inertial force had little influence on the axial trajectory. At the same speed, when the bearing rotation tends to stabilize, the shaft trajectory is slightly smaller than when the inertia force is not taken into account. Before the bearing motion is balanced, the convergence speed is faster when the inertial force of the journal is considered. The overall trend of changes in the time–history curves when the inertial force is considered are roughly the same. There is a large fluctuation at the initial unbalanced moment, and the value is smaller than that when the inertial force is not considered. When the rotation tends to be stable, the dimensionless water film force is small when the inertial force is considered. Comparing the time–history curves of the eccentricity and deflection angle, the difference is consistent with the water film pressure. However, the minimum water film thickness is relatively large.

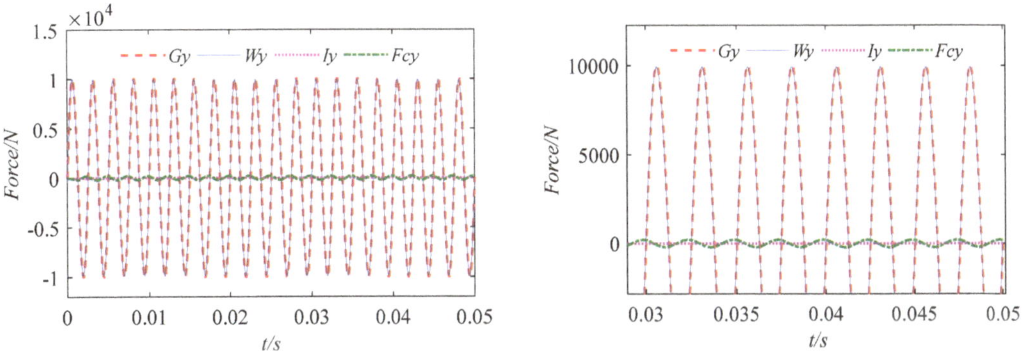

Figure 19. Time–history curves of four forces in the y direction.

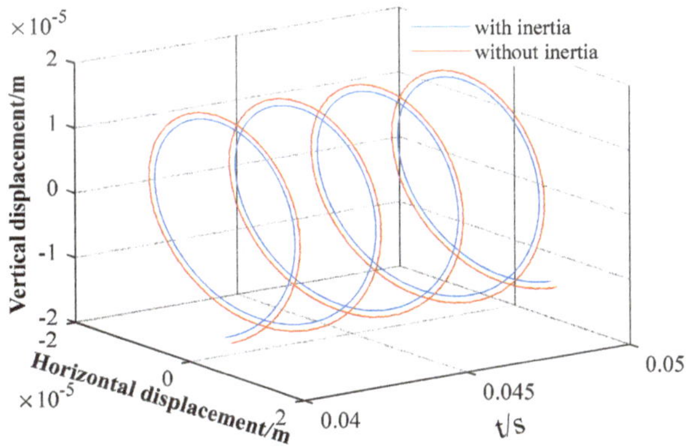

Figure 20. Comparison of orbits of journal center with and without inertia force.

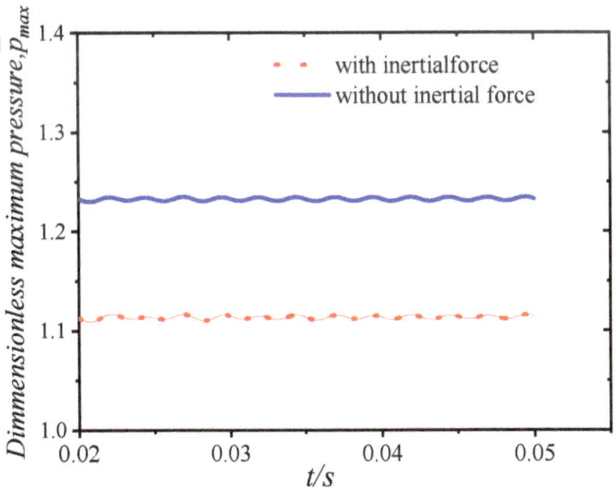

Figure 21. Comparison of dimensionless maximum water film pressure with and without inertia force.

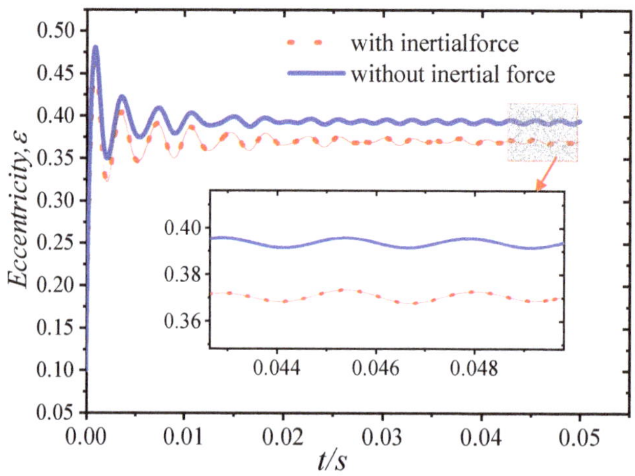

Figure 22. Comparison of eccentricity time–history curve with and without inertia force.

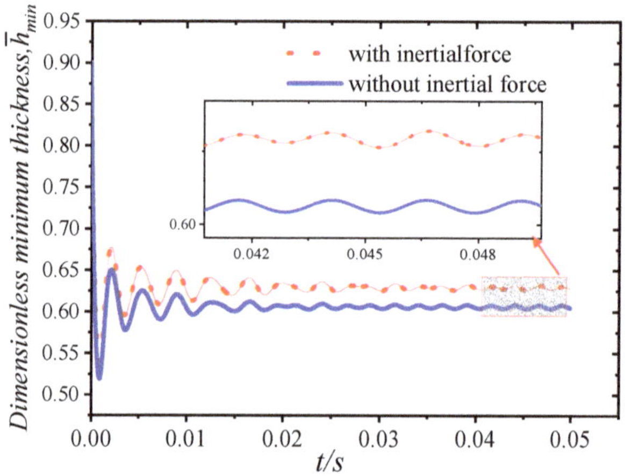

Figure 23. Comparison of deflection angle time–history curves with and without inertia force.

Since the external load applied in the above example is a circumferential load, after the rotation approached a uniform circular motion, the tangential and normal inertial force components caused by the acceleration of the journal were very small. To continue analyzing the influence of inertial forces, the bearing size was not changed, and only a unidirectional load was applied. The load expression is $G_y = 1.0E4\sin(80\pi t)$, and the initial eccentricity is $\varepsilon = 0.1$. The rotational speeds were set to be 545.6 r/min, 1091.2 r/min, and 1636.8 r/min (the linear speeds are 2 m/s, 4 m/s and 6 m/s, respectively) in order to study the influence of inertial forces at different speeds on the axis trajectory when unidirectional loads are applied. The results in 0.45 s–0.5 s are shown in Figures 24–29. It can be seen that under the action of a one-way external load, with the increase in rotational speed, regardless of whether the inertial force is considered, the axis trajectory gradually diverges. At the same speed, when considering the journal inertial force, the axis trajectory is more convergent, mainly because the inertial force and the water film pressure participate in the balance of external load. When the inertial force is not considered, the water film force is directly balanced with the external load, resulting in divergence of the axis trajectory.

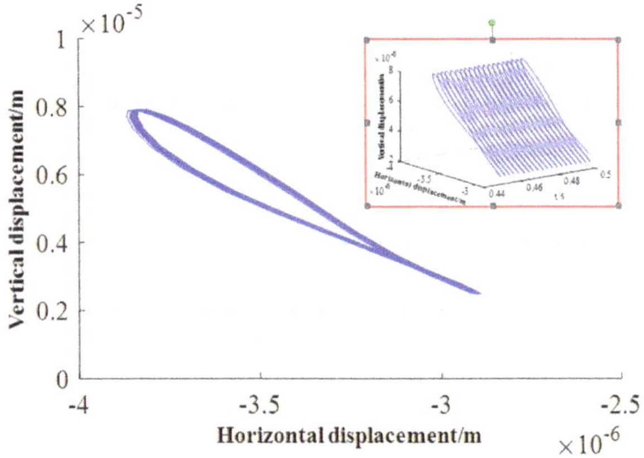

Figure 24. Orbit of journal center (with inertial force, n = 545.6 r/min).

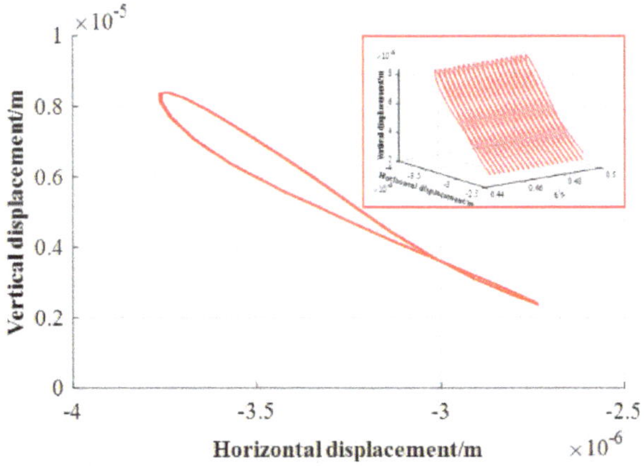

Figure 25. Orbit of journal center (without inertial force, n = 545.6 r/min).

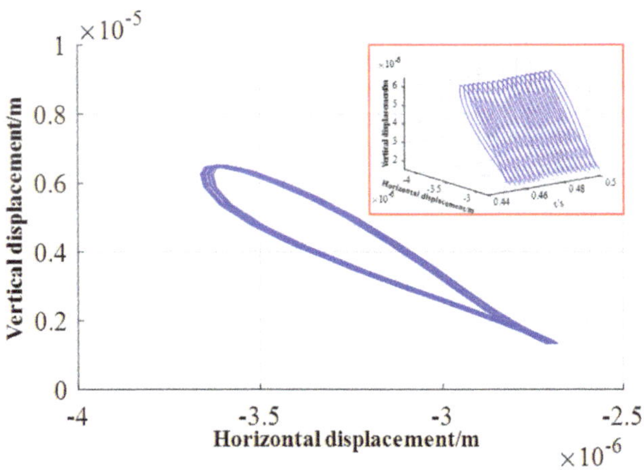

Figure 26. Orbit of journal center (with inertial force, n = 1091.2 r/min).

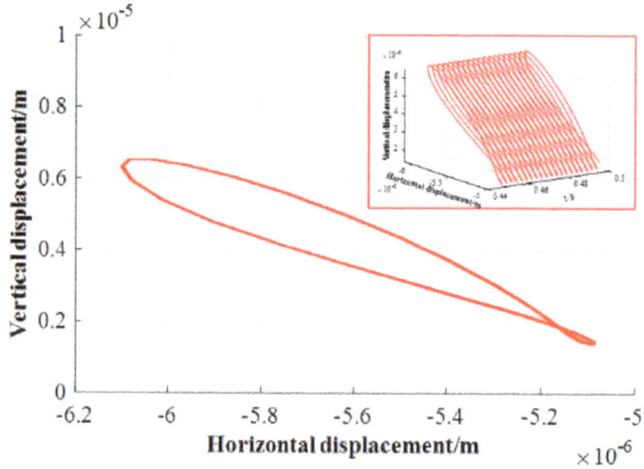

Figure 27. Orbit of journal center (without inertial force, n = 1091.2 r/min).

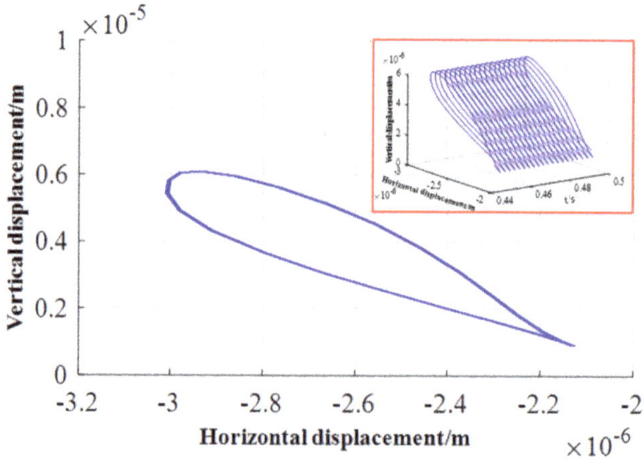

Figure 28. Orbit of journal center (with inertial force, n = 1636.8 r/min).

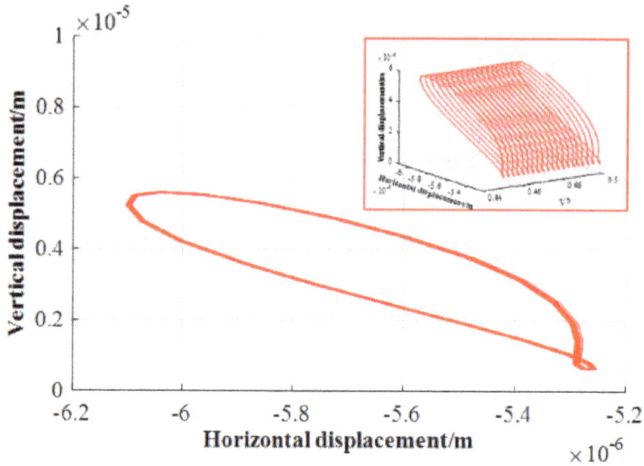

Figure 29. Orbit of journal center (without inertial force, n = 1636.8 r/min).

In fluid mechanics, the state of motion of a fluid is divided into a state of laminar motion and a state of turbulent motion. The Reynolds number $\text{Re} = \frac{\rho V h}{\eta_0}$ (which represents the ratio of inertial force to viscous force) is often used to characterize the state of motion of a fluid. In order to distinguish between these two states of motion, it is only necessary to find the critical Reynolds number that transitions between the two. The comparison of the Reynolds number and minimum thickness with different rotational speeds is shown in Figure 30. We can clearly see that with the increase in rotation speed, the trend of decreasing the minimum water film thickness shows clear linearity, and the growth trend of the Reynolds number gradually slows down. At the same speed, the difference between the thickness of the dimensionless minimum water film when considering and not considering the inertial force is basically unchanged, while the Reynolds number interpolation gradually increases.

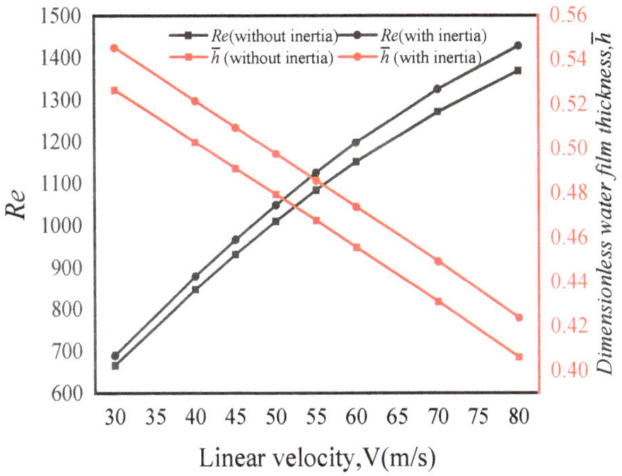

Figure 30. Comparison of Reynolds number and minimum thickness.

4. Analysis of Water Film Dynamic Characteristics under Low Speed and Heavy Load

In all kinds of mechanical equipment, many equipment failures are related to the running state of the bearing. According to statistics, in rotating machinery using bearings, about 30% of the failures are caused by the bearings. It can be seen that the quality of the bearings has a great influence on the working conditions of the machine. Bearings operate under long-term heavy load conditions, and various internal structures produce certain stress concentrations at stress points under long-term stress. If it is operated in such a state of stress concentration for a long time, the bearing stress point will be damaged or deformed to a certain extent, which will lead to the failure of the entire bearing component. Low-speed heavy-duty bearings bear larger loads, and it is more necessary to study the dynamic characteristics of water-lubricated low-speed heavy-duty bearings.

We applied a constant load, $F1$, in the y direction while applying the load. The load was applied in a linear manner at first, and was fully applied after 0.1 s.

The constant load was applied as:

$$\begin{array}{l} G_x = 1.0E4\cos(800\pi t) \\ G_y = 1.0E4\sin(800\pi t) + F1 \\ m = \begin{cases} 100t(t \leq 0.01s) \\ 1 \quad (t > 0.01s) \end{cases} \end{array} \tag{30}$$

We set the rotational speed to 200 r/min, 400 r/min, and 800 r/min, while the constant load was $F1$ = 1000 N, 2000 N, 4000 N, respectively. Taking the calculation result of the speed of 200 as an example, Figures 31 and 32 show the time–calendar curves of the dimensionless maximum pressure and minimum film thickness, respectively. When a constant load is applied, the time calendar of pressure and film thickness also shows strong periodicity. With the increase in constant load, the water film pressure is the opposite trend to the film thickness. In order to analyze the proportion of inertial force, the average water film pressure of nine groups of working conditions after the combination of three speeds and three loads was counted, as shown in Figures 33–35. From the figure, it can be observed that the increase in speed and external load leads to an increase in pressure, and the pressure of the water film is relatively small when considering the inertial force. Figure 36 shows the proportion of inertial force obtained after statistics. From the figure, it can be observed that the increase in speed and external load leads to an increase in pressure, and the pressure of the water film is relatively small when considering the inertial force. Figure 36 shows the proportion of inertial force obtained after statistics. When the constant load remains unchanged, the increase in speed leads to an increase in the inertial force, and at the same speed, the increase in the external load will also lead to an increase in the inertial force, but the proportion of inertial force in the total pressure will decrease.

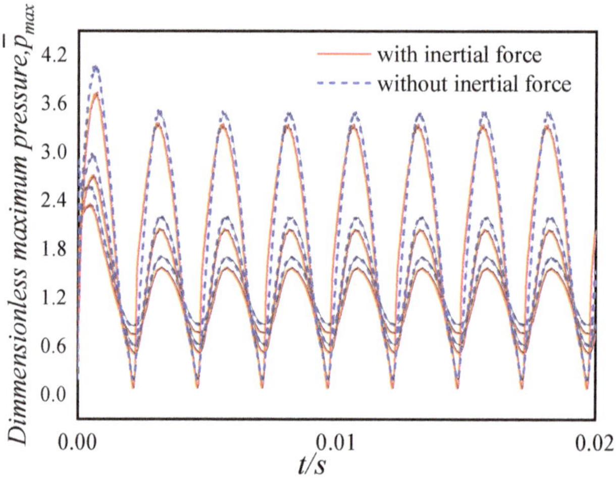

Figure 31. Dimensionless maximum water film pressure (n = 200 r/min, F_1 = 1000, 2000, 4000 N).

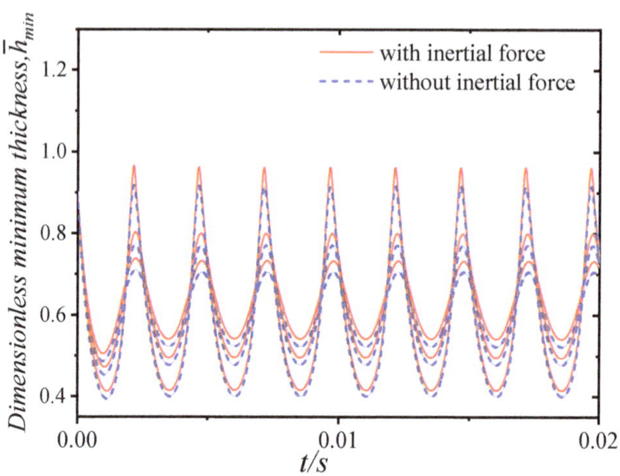

Figure 32. Dimensionless minimum water film thickness (n = 200 r/min, F_1 = 1000, 2000, 4000 N).

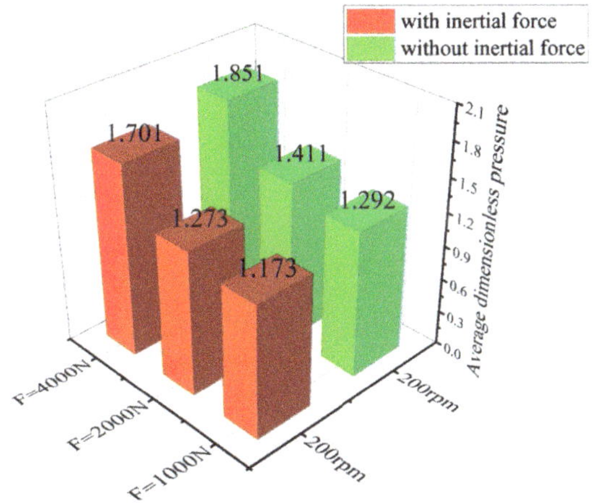

Figure 33. Dimensionless average water film pressure (n = 200 r/min, F_1 = 1000, 2000, 4000 N).

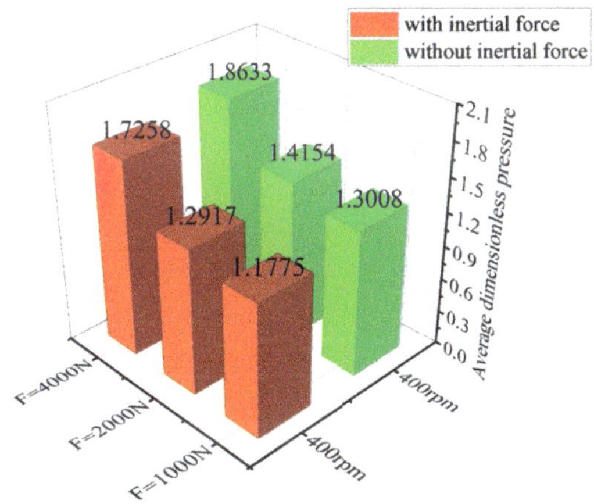

Figure 34. Dimensionless average water film pressure (n = 400 r/min, F_1 = 1000, 2000, 4000 N).

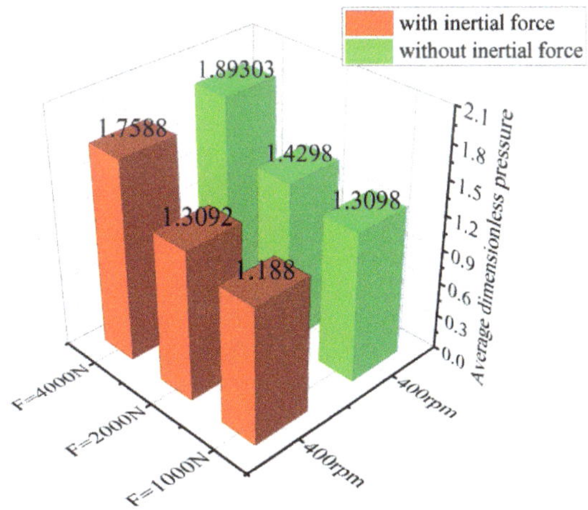

Figure 35. Dimensionless average water film pressure (n = 800 r/min, F_1 = 1000, 2000, 4000 N).

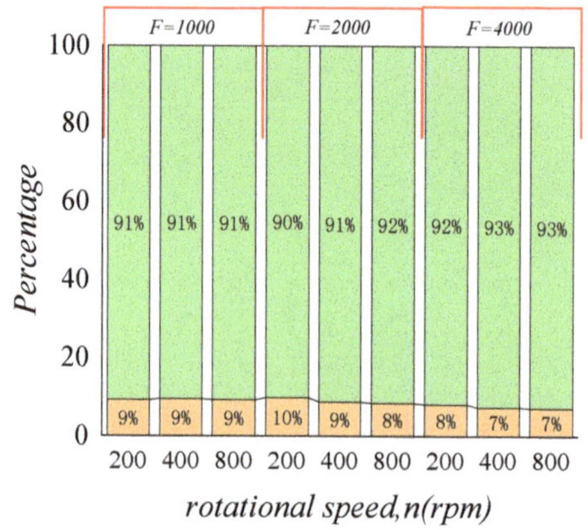

Figure 36. The proportion of inertial force.

We set the static load amplitude to 20,000 N and compared the dynamic characteristics at different speeds when inertial force is considered, as shown in Figures 37–40.

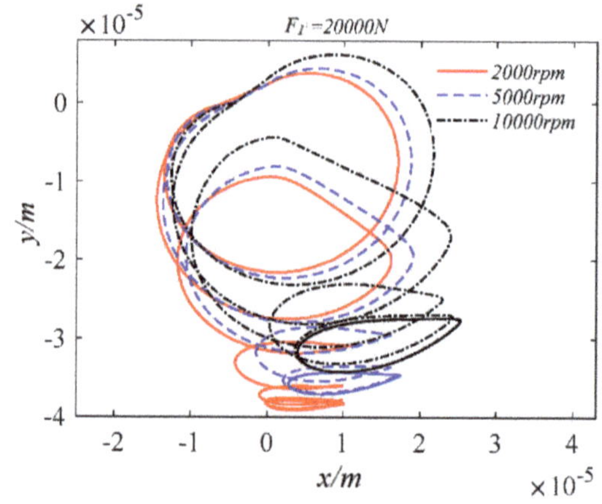

Figure 37. Orbits of journal center at different speeds.

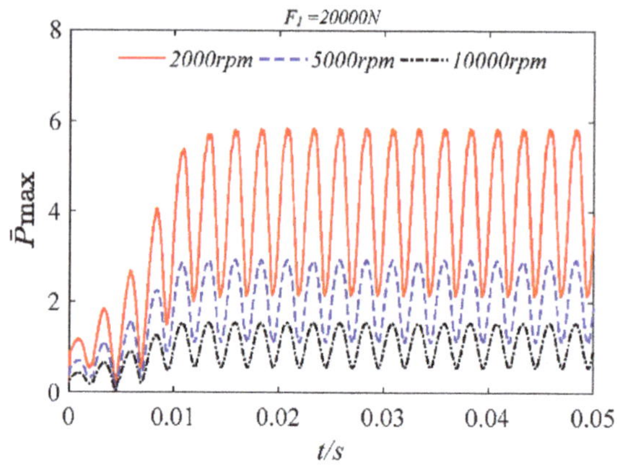

Figure 38. Dimensionless maximum water film pressures at different speeds.

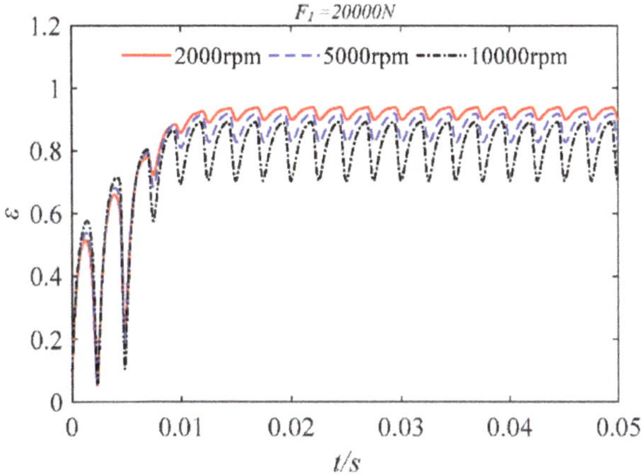

Figure 39. Eccentricity time–history curves at different speeds.

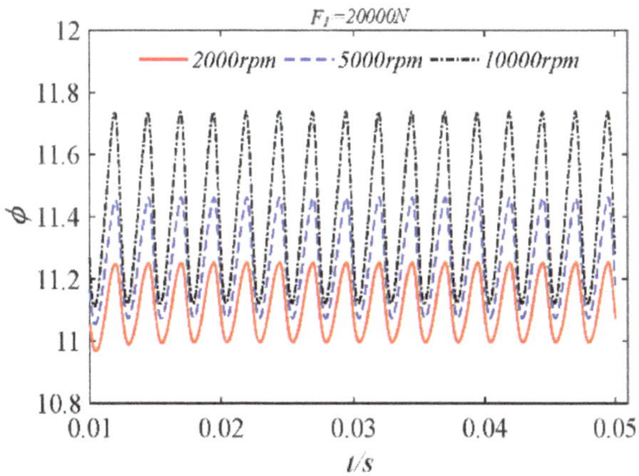

Figure 40. Deflection angle time–history curves at different speeds.

As shown in the figure, when the bearing is subjected to a large static load, the orbit of the journal center is elliptical. The increase in rotational speed causes the orbit of the journal center to diverge. The decrease in rotational speed leads to an increase in the peak pressure of the dimensionless maximum water film. After increasing the static load, the eccentricity fluctuates significantly during its application. When the rotation of the bearing tends to be stable, the higher the bearing speed, the greater the range of eccentricity changes. The reason for this phenomenon is that when the static load is large, the higher speed will cause the bearing to produce a larger centripetal force, which will produce large fluctuations during rotation, resulting in the magnitude of the eccentricity change.

5. Conclusions

In this paper, the Reynolds equation for the incompressible water-lubricated motion of the bearing and the force balance equation considering the inertial force of the shaft diameter have been established. Based on the finite difference method, the water film force distribution, eccentricity time–history curve, and orbit of the journal center of the bearing under static load and sinusoidal periodic load have been calculated. At the same time, the influence of rotational speed and external load on inertial force has been studied. Through the analysis of the obtained calculated data, the following conclusions are drawn:

(1) By analyzing the water film pressure distribution curve, it was obtained that the circumferential water film pressure first increases and then decreases, and reaches the maximum value at a certain point, showing nonlinearity. In practical situations, if water or oil is selected as the lubricant, a negative pressure area will be generated after positive pressure, and the liquid film cannot withstand too much negative pressure and rupture. The axial water film pressure is parabolically distributed, possibly due to the leakage of water from both ends of the bearing as the bearing rotates.

(2) Although the bearing capacity of the water-lubricated bearing is low, it can run at a lower speed when subjected to a constant amplitude circumferential load and when the boundary line of the non-zero pressure region of the water film is near $\theta = \pi$. For the application of periodic loads, when considering the inertial force, the axial trajectory is more convergent, the water film pressure is smaller than the eccentricity, and the thickness of the water film is larger, which means that the load-carrying capacity of the water-lubricated bearing is greater when the inertial force is considered.

(3) At the same time, the increase in speed and external load will lead to an increase in inertial force, but the increase in external load will lead to a decrease in the proportion of inertial force. Increasing the bearing speed is conducive to improving the bearing performance of water-lubricated bearings.

(4) In addition to the periodic external load in the circumferential direction, applying a constant load will cause the trajectory of the water-lubricated bearing to deviate. Considering the inertial force of the shaft diameter, the variation range of the eccentricity becomes larger, and the maximum pressure of the water film also becomes larger. In the calculation of low-speed heavy load in this paper, the speed range is between 200 r/min and 800 r/min, and the pressure of the dimensionless water film caused by inertial force is reduced by 7 to 10 percent, which means that the influence of inertia force cannot be ignored, and the lubrication performance of the bearing can be evaluated more accurately considering the inertia force.

Author Contributions: Conceptualization, S.S.; methodology, C.L. and S.S.; software, F.J.; validation, C.L., F.J. and S.S.; formal analysis, C.L. and F.J.; data curation, F.J.; writing—original draft preparation, C.L. and F.J.; writing—review and editing, C.L.; visualization, F.J. and L.Q.; supervision, M.Z.; project administration, C.S.; funding acquisition, S.S., M.Z. and L.Q. All authors have read and agreed to the published version of the manuscript.

Funding: This work is supported by the Excellent Youth Project of Heilongjiang Natural Science Foundation (YQ2019E010), the National Natural Science Foundation of China (Grant Nos. 12172338 and 51909246), and the Natural Science Foundation of Jiangsu Province of China (Grant No. BK20220044).

Data Availability Statement: The data used to support the findings of this study are available from the corresponding author upon request.

Conflicts of Interest: The authors declare no conflict of interest.

Nomenclature

ε	Eccentricity ratio
D	Bearing diameter, m $D = 2R$
d	Journal diameter, m $d = 2r$
L	Bearing width, m
ϕ	Attitude angle
ω	Angular velocity, rad
c	Radial clearance, m
h	Water film thickness, m
ρ	Density of liquid phase.
e	Eccentricity, m
η_0	Viscosity of water, Pa·s
u, v	x, y direction displacement, m
n	Rotate speed, rpm

Gx, Gy	x, y direction external load, N
M	Mass of the journal, kg
p	Pressure, Pa
v	Poisson's ratio
$k_{xx}, k_{xy}, k_{yx}, k_{yy}$	Stiffness factor
$c_{xx}, c_{xy}, c_{yx}, c_{yy}$	Damping factor
I_x, I_y	Inertial forces, N
F_c	Centripetal force, N
$\mathrm{Re}_I, \mathrm{Re}_J$	Reynolds number
θ, z	Circumference, axial direction
ξ, η	Radial, tangential direction
i, j	Circumference, axial node

References

1. Xie, Z.L.; Jiao, J.; Yang, K. A state-of-art review on the water-lubricated bearing. *Tribol. Int.* **2023**, *180*, 108276. [CrossRef]
2. Mallya, R.; Shenoy, S.B.; Pai, R. Steady state characteristics of misaligned multiple axial groove water-lubricated journal bearing. *J. Eng. Tribol.* **2015**, *229*, 712–722. [CrossRef]
3. Lin, C.G.; Yang, Y.N.; Chu, J.L.; Sima, C.; Liu, P.; Qi, L.B.; Zou, M.S. Study on nonlinear dynamic characteristics of propulsion shafting under friction contact of stern bearings. *Tribol. Int.* **2023**, *183*, 108391. [CrossRef]
4. Cai, J.L.; Xiang, G.; Li, S. Mathematical modeling for nonlinear dynamic mixed friction behaviors of novel coupled bearing lubricated with low viscosity fluid. *Phys. Fluids* **2022**, *34*, 093612. [CrossRef]
5. Lin, C.G.; Zou, M.S.; Sima, C.; Liu, S.X.; Jiang, L.W. Friction-induced vibration and noise of marine stern tube bearings considering perturbations of the stochastic rough surface. *Tribol. Int.* **2019**, *131*, 661–671. [CrossRef]
6. Zhang, X.; Wang, X.R.; Niu, G.L.; Zhang, L.J. The Research Actuality and Development Trends of Water Lubricated Bearings. *Equip. Manuf. Technol.* **2008**, *1*, 101–102.
7. Xie, Z.L.; Jiao, J.; Yang, K. Theoretical and experimental study on the fluid-structure-acoustic coupling dynamics of a new water lubricated bearing. *Tribol. Int.* **2023**, *177*, 107982. [CrossRef]
8. Zou, M.S.; Tang, H.C.; Liu, S.X. Modeling and calculation of acoustic radiation of underwater stiffened cylindrical shells treated with local damping. *Mar. Struct.* **2023**, *88*, 103366. [CrossRef]
9. Xie, Z.L.; Jiao, J.; Wrona, S. The fluid-structure interaction lubrication performances of a novel bearing: Experimental and numerical study. *Tribol. Int.* **2023**, *179*, 108151. [CrossRef]
10. Wang, J.G.; Zhou, J.Z.; Dong, D.W.; Yan, B.; Huang, C.R. Nonlinear dynamic analysis of a rub-impact rotor supported by oil film bearings. *Arch. Appl. Mech.* **2012**, *83*, 413–430. [CrossRef]
11. Brown, R.D.; Addison, P.; Chan, A.H.C. Chaos in the unbalance response of journal bearings. *Nonlinear Dyn.* **1994**, *5*, 421–432. [CrossRef]
12. Jing, J.P.; Meng, G.; Sun, Y.; Xia, S.B. On the non-linear dynamic behavior of a rotor-bearing system. *J. Sound Vib.* **2004**, *274*, 1031–1044. [CrossRef]
13. Brancati, R.; Rocca, E.; Russo, M.; Russo, R. Journal orbits and their stability for rigid unbalanced rotors. *J. Tribol.* **1995**, *117*, 709–716. [CrossRef]
14. Brancati, R.; Russo, M.; Russo, R. On the stability of periodic motions of an unbalanced rigid rotor on lubricated journal bearings. *Nonlinear Dyn.* **1996**, *10*, 175–185. [CrossRef]
15. Li, D.X.; Xu, J.X. A method to determine the periodic solution of the non-linear dynamics system. *J. Sound Vib.* **2004**, *275*, 1–16. [CrossRef]
16. Lin, C.G.; Zou, M.S.; Sima, C.; Qi, L.B.; Yu, Y. Non-Linear coupled dynamics of a flexible propeller-shaft system supported by water film bearings. *J. Vib. Acoust.* **2020**, *142*, 031008. [CrossRef]
17. Sun, W.P.; Yan, Z.M.; Tan, T.; Zhao, D.L.; Luo, X.Q. Nonlinear characterization of the rotor-bearing system with the oil-film and unbalance forces considering the effect of the oil-film temperature. *Nonlinear Dyn.* **2018**, *92*, 1119–1145. [CrossRef]
18. Cai, J.L.; Han, Y.F.; Xiang, G.; Wang, J.X.; Wang, L.W. Effects of wear and shaft-shape error defects on the tribo-dynamic response of water-lubricated bearings under propeller disturbance. *Phys. Fluids* **2022**, *34*, 077118. [CrossRef]
19. Singh, U.; Roy, L.; Sahu, M. Steady-state thermo-hydrodynamic analysis of cylindrical fluid film journal bearing with an axial groove. *Tribol. Int.* **2008**, *13*, 1135–1144. [CrossRef]
20. Kornaeva, E.; Kornaev, A.; Savin, L.; Galichev, A.; Babin, A. Theoretical premises of a vibro-inertial method of viscosity measurement. *Vibroengineering Procedia* **2016**, *8*, 440–445.
21. Xie, Z.L.; Ta, N.; Rao, Z.S. The lubrication performance of water lubricated bearing with consideration of wall slip and inertial force. *J. Hydrodyn.* **2017**, *29*, 52–60. [CrossRef]
22. Gao, G.Y.; Yin, Z.W.; Jiang, D.X.; Zhang, X.L. Numerical analysis of plain journal bearing under hydrodynamic lubrication by water. *Tribol. Int.* **2014**, *75*, 31–38. [CrossRef]
23. Wang, Y.Z.; Yin, Z.W.; Jiang, D.; Gao, G.Y.; Zhang, X.L. Study of the lubrication performance of water-lubricated journal bearings with CFD and FSI method. *Ind. Lubr. Tribol.* **2016**, *68*, 341–348. [CrossRef]

24. Wojciech, L. Influence of local bush wear on water lubricated sliding bearing load carrying capacity. *Tribol. Int.* **2016**, *103*, 352–358.
25. Xiang, G.; Yang, T.Y.; Guo, J.; Wang, J.X. Optimization transient wear and contact performances of water-lubricated bearings under fluid-solid-thermal coupling condition using profile modification. *Wear* **2022**, *502–503*, 204379. [CrossRef]
26. Sun, F.X.; Zhang, X.B.; Wei, Y.S.; Wang, X.; Wang, D. Stability analysis of rubber-supported thrust bearing in a rotor-bearing system used in marine thrusters under disturbing moments. *Tribol. Int.* **2020**, *151*, 106356. [CrossRef]
27. Jakobsson, B. The finite journal bearing, considering vaporization. *Trans. Chalmers Univ. Technol.* **1957**, *190*, 1–116.
28. Zhang, Z.M. *Theory of Hydrodynamic Lubrication of Sliding Bearings*; Higher Education Press: Beijing, China, 1986.
29. Zhang, X.L.; Yin, Z.W.; Gao, G.Y.; Li, Z. Determination of stiffness coefficients of hydrodynamic water-lubricated plain journal bearings. *Tribol. Int.* **2015**, *85*, 37–47. [CrossRef]
30. Booker, J.F. Mobility/impedance methods: A guide for application. *J. Tribol.* **2014**, *136*, 024501. [CrossRef]

lubricants

Article

The Difference in Tribological Characteristics between CFRPEEK and Stainless Steel under Water Lubrication in Friction Testing Machine and Axial Piston Pump

Donglin Li [1,*], Xianshuai Ma [1], Shuai Wang [1], Junhua Wang [1], Fang Yang [1,2] and Yinshui Liu [3]

1 School of Mechatronics Engineering, Henan University of Science and Technology, Luoyang 471003, China
2 Science and Technology Innovation Center for Completed Set Equipment, Longmen Laboratory, Luoyang 471003, China
3 School of Mechanical Science and Engineering, Huazhong University of Science and Technology, Wuhan 430074, China
* Correspondence: lidonglin@haust.edu.cn

Abstract: A water lubricating axial piston pump (WLPP) is the core power component of a green and environmentally friendly water hydraulic system. The friction and wear of the friction pairs of a WLPP are the key factors that restrict its development. In order to explore the friction and wear mechanism of materials, the tribological properties of CFRPEEK against 316L and 1Cr17Ni2 under water lubrication were investigated in a friction testing machine and an axial piston pump, respectively. An environmental scanning electron microscope (ESEM), confocal laser scanning microscopy and a surface profiler were used to analyze the morphology of the samples. In a friction testing machine, two different metals are paired with CFRPEEK, and the friction coefficient and wear rate barely show any differences. The wear rate of CFRPEEK is two orders of magnitude higher than that of metal. In the WLPP, 316L can hardly be paired with CFRPEEK, while 1Cr17Ni2 works well. The wear of 1Cr17Ni2 in the WLPP is greater than that of CFRPEEK. The high-pressure water film lubrication friction pairs cause the wear of the metal and show the difference in these two test methods. The wear mechanism is mainly abrasive wear. Improving the wear resistance of metals is very important for the development of WLPP.

Keywords: CF/PEEK; axial piston pump; water hydraulics; water lubrication; wear

Citation: Li, D.; Ma, X.; Wang, S.; Wang, J.; Yang, F.; Liu, Y. The Difference in Tribological Characteristics between CFRPEEK and Stainless Steel under Water Lubrication in Friction Testing Machine and Axial Piston Pump. *Lubricants* **2023**, *11*, 158. https://doi.org/10.3390/lubricants11040158

Received: 20 February 2023
Revised: 19 March 2023
Accepted: 24 March 2023
Published: 26 March 2023

1. Introduction

Water hydraulics, which uses tap/sea water instead of traditional mineral oil as the working medium, has the advantages of environmental protection, non-combustion, compatibility with the environment and accessibility. This technology is widely used in seawater desalination [1,2], submersible buoyancy regulation [3,4], fire protection [5] and other fields. A water lubricating axial piston pump (WLPP), which increases the pressure of the working medium, is the core component of a water hydraulic system.

Figure 1 shows the structure of a WLPP. Its working principle is as follows. The rotating shaft drives the slipper to slide on the surface of the swashplate. The piston also rotates with the shaft and reciprocates in the cylinder sleeve. At the same time, the rotating shaft drives the floating plate to slide on the surface of the valve plate. Under the action of this motion and the hydraulic pressure, it forms three major friction pairs of the water pump, namely, the slipper/swashplate pair, the piston/cylinder sleeve pair and the valve plate/floating plate pair. These three friction pairs perform bearing and sealing functions [6]. However, compared with mineral oil, water has low viscosity, strong corrosion, high vaporization pressure and other physical and chemical properties [7], which make the friction pairs of a WLPP face serious friction and lubrication problems.

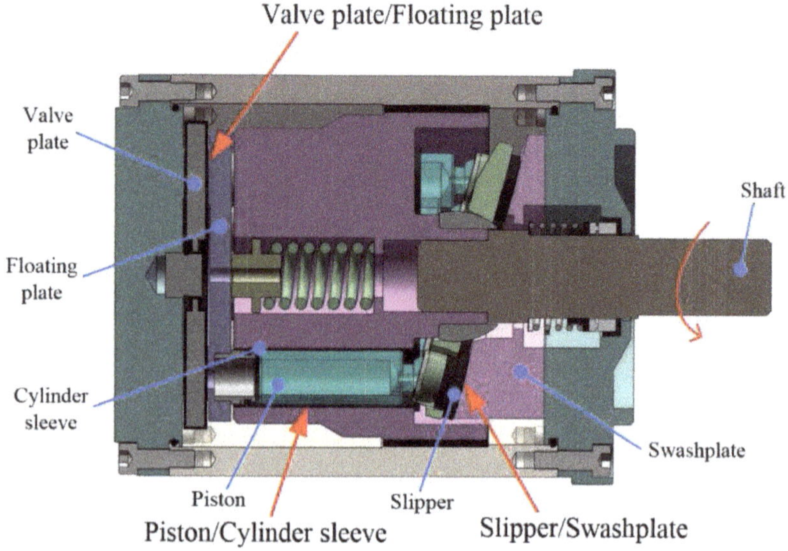

Figure 1. Structure of WLPP.

Therefore, the materials of water-lubricated friction pairs have been researched. Generally, friction pair materials for a WLPP require corrosion resistance, good mechanical properties and tribological properties [8,9]. At present, the materials of water-lubricated friction pairs are mainly stainless steel [10], engineering plastics [11] and ceramics [12,13]. Ceramics are characterized by high strength, large elastic modulus, wear resistance and corrosion resistance, and are suitable to use under high-speed and heavy-load conditions. Nevertheless, the toughness and fatigue resistance of ceramics are poor. The main failure modes of ceramic parts are brittle fracture and fatigue spalling [14]. Furthermore, ceramics are difficult to process and are not the best material for making the friction pairs of a WLPP. In comparison, engineering plastics and stainless steel are more suitable for the friction pairs.

Among many engineering plastics, Poly ether ether ketone (PEEK) is a promising polymer with high mechanical and tribological properties, such as high temperature resistance, corrosion resistance, self-lubrication and dimensional stability, which is widely used in the fields of automobiles, aerospace, nuclear power and medical equipment [15–18]. Many researchers have studied the friction and wear properties of PEEK/modified PEEK under water lubrication. Chen et al. [19] studied the tribological behaviors of CF/PEEK under seawater lubrication, with sliding speeds of 0.5 m/s–1 m/s and normal loads of 200 N–600 N. The results showed that the incorporation of CF can greatly improve the wear resistance of PEEK. Li et al. [20] investigated the effect of different temperatures (0–50 °C) on the friction and wear properties of CFRPEEK against AISI 431 steel under water lubrication, using a disc-on-disc tester under sliding speeds of 0.68–1.36 m/s and loads of 0.61–1.83 MPa. It was found that the increasing lubricant temperature will lead to the deterioration of tribological properties of the materials. Davim et al. [21–23] used a pin-on-disc machine to research the effects of the PV factors (0.5–3 MPa·m/s) and the sliding distance on the tribological behaviors of CF/PEEK and AISI 316L under dry friction and water lubrication. Li et al. [24] investigated the friction and wear behaviors of PEEK filled with short carbon fibers and SiO_2 against GCr15 under dry sliding conditions, with PV factors ranging from 1 to 12 MPa·m/s. Moreover, under high hydrostatic pressure, the friction and wear behavior of materials have been studied by scholars. Liu et al. [25] Studied the tribological behavior of different polymer materials under 0–40 MPa hydrostatic pressure. It was found that the wear behavior of thermoplastic polymers sliding in seawater is strongly dependent on the hydrostatic pressure. Wu et al. [26] have developed a disc-on-disc friction testing machine that can simulate hydrostatic pressure up to 80 MPa. In their follow-up work [27], the friction and wear characteristics of CF/PEEK against 431 stainless steel under high hydrostatic pressure were investigated. The results showed that hydrostatic pressure helps reduce the wear rates of CF/PEEK and 431 stainless steel.

The research on the tribological properties of PEEK and its poly composites focus on sliding speed, loads, lubricating temperature, material modification and hydrostatic pressure. The test method is usually conducted with a friction and wear testing machine with normal-pressure water lubrication. These studies focus more on the tribological characteristics of PEEK, and suggest that the strength and hardness of the metals are much greater than PEEK, while ignoring the wear issue of the metal materials. However, the WLPP works at 14 MPa and 1500 rpm, which makes the load of its friction pairs as high as 6.7 m/s and 4.1 MPa. Moreover, the friction pairs of the WLPP are lubricated by high-pressure water, which lead to the materials' deformation [28]. Hence, the wear mechanism of PEEK and steel will be different when the lubrication conditions change. It is necessary to study the tribological behaviors of PEEK and stainless steel under real conditions for the design of WLPP.

In this work, the difference in tribological characteristics between CFRPEEK and stainless steel, under water lubrication in a friction testing machine and an axial piston pump, was studied. The friction coefficients and wear rates of CFRPEEK and stainless steel were measured, and the worn surface morphology was obtained. In particular, the wear mechanism of metals was analyzed using a friction testing machine and high-pressure lubrication in the WLPP. This study has instructive significance for the design and material selection of the friction pair of the water hydraulic pump.

2. Experimental

2.1. Sample Preparation

2.1.1. Sample for Friction Testing Machine

The materials of the upper samples are 316L (ASTM A240M-15a) and 1Cr17Ni2 (GB/T 1221-2007). The 316L sample is austenitic stainless steel, which is corrosion-resistant and is generally used in a corrosive environment. It has low strength and its hardness cannot be improved via heat treatment. Its hardness is usually lower than 187 HB. The 1Cr17Ni2 sample is martensitic stainless steel with high strength, and is widely used in ships, steam turbines and other harsh environments. After heat treatment, its hardness is HRC41–44 (~400 HB). These two materials have good water corrosion resistance and are commonly used to make parts of water hydraulic piston pumps [20,29]. The main mechanical properties of the two stainless steel materials are shown in Table 1.

Table 1. Mechanical properties of the two stainless steels.

Properties	Unit	316L	1Cr17Ni2
density	g/cm^3	7.98	7.75
tensile strength	MPa	480	1080
yield strength	MPa	177	-
tensile modulus	GPa	210	210
hardness	HB	187	400

CFRPEEK was selected to make the lower sample, which is made of 70% PEEK matrix and 30% carbon fibers. This material was commercially obtained from Ensinger Co., Ltd. (Nufringen, Germany). The main performance parameters of CFRPEEK are shown in Table 2.

Table 2. Properties of CFRPEEK [30].

Properties	Unit	Value
density	g/cm^3	1.40
water absorption	%	0.06
tensile strength	MPa	265
compressive strength	MPa	300
tensile modulus	GPa	28
hardness	Shore D	87.5

2.1.2. Samples of WLPP

As shown in Figure 2, the friction surface of slippers is coated with CFRPEEK by injection molding process, and its roughness and flatness are not greater than 0.8 μm and 0.003 mm, respectively. The swashplate parts are made of 316L and 1Cr17Ni2, and the heat treatment process is the same as that of the upper samples. The roughness and flatness of swashplate parts are less than 0.8 μm and 0.01 mm, respectively.

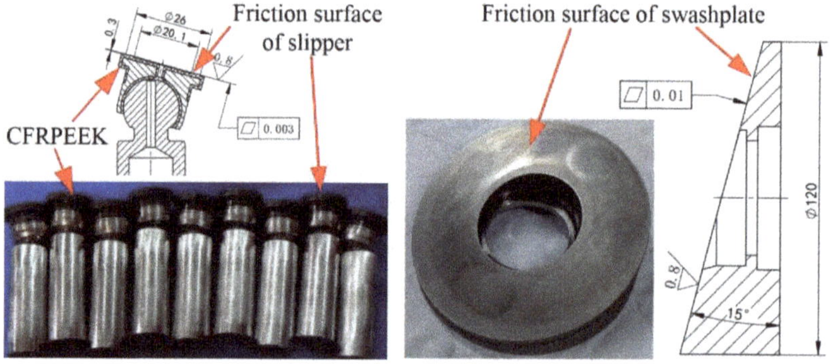

Figure 2. Slipper and swashplate parts of WLPP.

2.2. Tribology Tests

2.2.1. Tribology Tests in Friction Testing Machine

A pin-on-disc friction testing machine (MMU–10, SHIJIN, Jinan, China) was utilized for tribological tests. As shown in Figure 3, the upper sample (stainless steel pin) and the lower sample (CFRPEEK disc) were installed in the sample box. Circulating water flowed into the sample box from the inlet and flowed out from the outlet for lubrication. The temperature of water was kept between 22–25 °C.

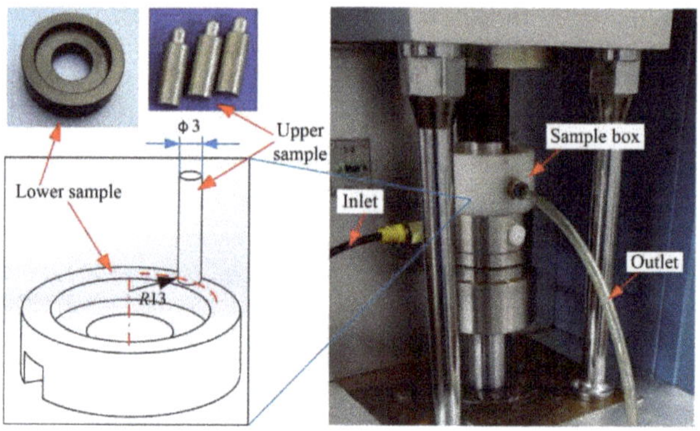

Figure 3. Pin-on-disc friction testing machine.

The surfaces of pin-on-disc samples were polished with abrasive papers. The surface roughnesses of the pin-on-disc samples were less than 0.1 μm and 0.2 μm. Before and after the test, the samples were ultrasonically cleaned in industrial alcohol for 5 min and dried with a blower for 3 min. Then, the mass of samples was measured by an analytical balance (MS105, 0.01 mg, Mettler-toledo, Zurich, Switzerland). The weight difference of the samples before and after the test was the wear amount. The wear rate (w) and wear amount of the sample could be calculated by the following formula:

$$w = \frac{\Delta m}{\rho N L} \quad (1)$$

where Δm is the mass loss of the sample (g), ρ is the density (g/mm^3), N is the load (N), and L is the sliding distance (m). Repeated tests were carried out for each group.

2.2.2. Tribology Tests in WLPP

Figure 4 shows the principle of the experimental system for testing the slipper/swashplate pairs of the WLPP. Slipper/swashplate pairs of different materials were installed in the tested pump to evaluate their tribological performance under real working conditions. A safety valve (4) was used to prevent system overpressure. The outlet pressure of the tested pump was regulated by the throttle valve (9), and measured by the pressure sensor (8, range: 0–25 MPa, accuracy: ±0.25% FS). The outlet flow was measured by flowmeter (10, range: 1–10 m^3/h, accuracy: ±1% FS). The input speed and torque of the tested pump were measured by the tacho-torquemeter (5, range: 0–200 Nm, 0–5000 rpm, accuracy: ±0.1% FS). The change in pump volumetric efficiency could indirectly represent the wear of internal parts [31,32]. Its volumetric efficiency (η_v) could be expressed as follows:

$$\eta_v = \frac{q}{q_t} \quad (2)$$

where q is the outlet flow of the WLPP (L/min), q_t is the theoretical flow of the WLPP (L/min), $q_t = 85n/1000$, and n is the speed of the WLPP (r/min). Moreover, the wear of parts, including the macro size and macro morphology, was measured directly by measuring instruments.

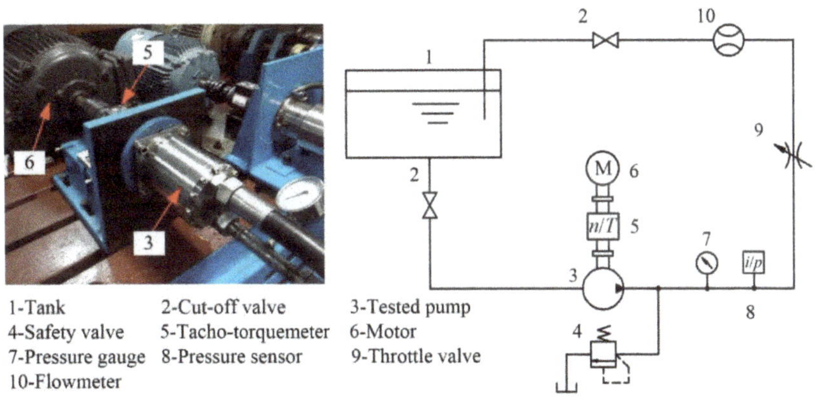

Figure 4. Principle of the experimental system for testing the slipper/swashplate pairs of WLPP.

3. Results and Discussions

3.1. *Tribological Behaviors and Surface Topography within Friction Testing Machine*

3.1.1. Tribological Behaviors of the Samples

In order to observe the friction and wear of materials in a short timeframe, the testing conditions in the friction testing machine were set to a contact pressure of 10 MPa and a sliding speed of 2 m/s. Each group of tests was carried out continuously for 180 min. The change in friction coefficients of the 316L/CFRPEEK and 1Cr17Ni2/CFRPEEK pairs, with time elapsed, is shown in Figure 5a. Both pairs have a running-in stage, after which the friction coefficient is stable. In order to compare the friction coefficient difference between the two materials, the average value of multiple sets of data is calculated, as shown in Figure 5b. Although the materials matched with CFRPEEK are different, the friction coefficients of the two pairs are slightly different. This means that different metal materials have little influence on the friction coefficient within the friction testing machine.

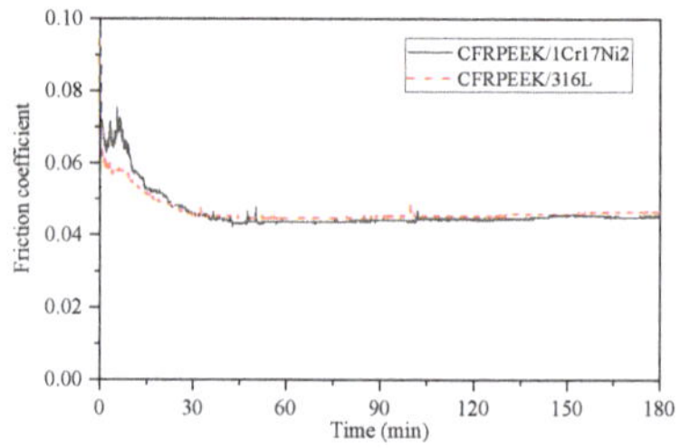

(**a**)

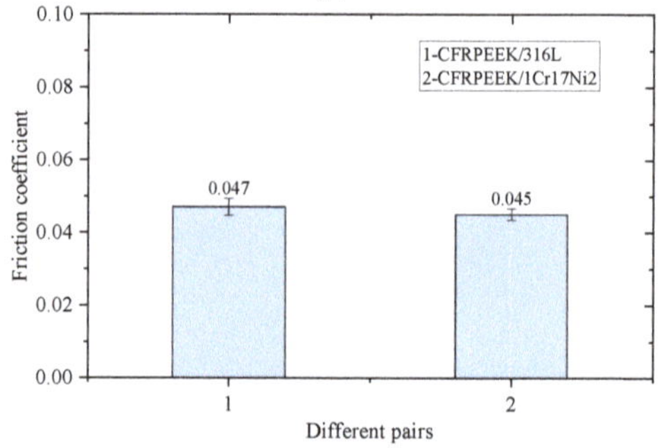

(**b**)

Figure 5. Friction coefficient of different pairs. (**a**) Friction coefficients with sliding time; (**b**) average friction coefficients.

Furthermore, the wear rates of CFRPEEK and the two metals are also tested, as shown in Figure 6. In these two pairs, the wear rates of CFRPEEK have little difference, with the order of 10^{-7}, which is similar to the results from previous studies [20,33]. Similarly, the wear rates of the two metals are similar, but are far less than the difference in CFRPEEK, with the order of 10^{-9}. This means that the two metals are more wear-resistant than CFRPEEK.

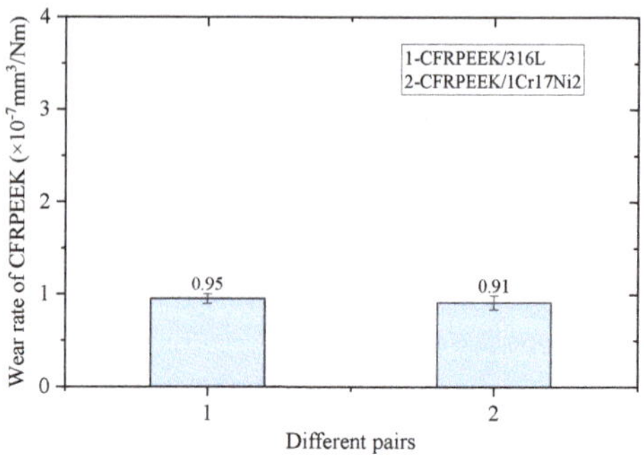

(**a**)

Figure 6. *Cont.*

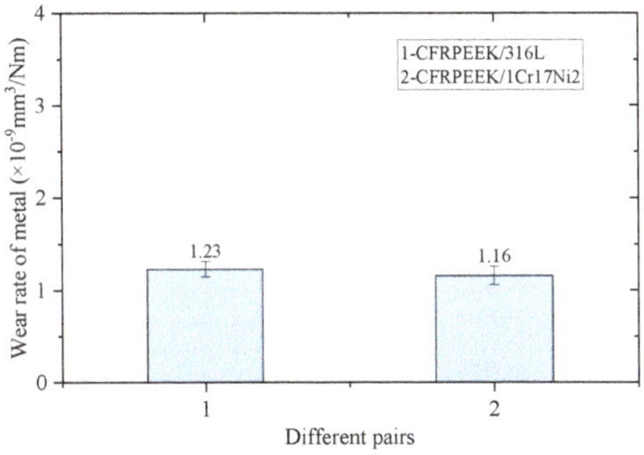

(**b**)

Figure 6. Wear rates of different pairs. (**a**) Wear rates of CFRPEEK; (**b**) wear rates of metals.

3.1.2. Surface Topography of the Samples

The surface morphology of the 316L/CFRPEEK pair is similar to that of the 1Cr17Ni2/CFRPEEK pair after testing. An environmental scanning electron microscope (ESEM, FEI Quanta 200) was used to analyze the morphology of the samples. Due to the non-conductivity of CFRPEEK, the CFRPEEK samples were subjected to a gold-plating treatment before the morphology was calculated. Only the ESEM image of the 1Cr17Ni2/CFRPEEK pair is analyzed below. As shown in Figure 7a, there are some cavities on the surface of the unused CFRPEEK, which may be formed by the fracture and extraction of carbon fibers caused by mechanical cutting and the sanding with the abrasive paper. Compared with the worn CFRPEEK surface (see Figure 7b), the cavities on unused CFRPEEK surface disappear due to wear. Furthermore, the carbon fibers thin and break due to wear. More significantly, the carbon fibers are separated and pulled out from the PEEK matrix [20]. The wear mechanisms of CFRPEEK are mainly surface fatigue wear and adhesive wear [34].

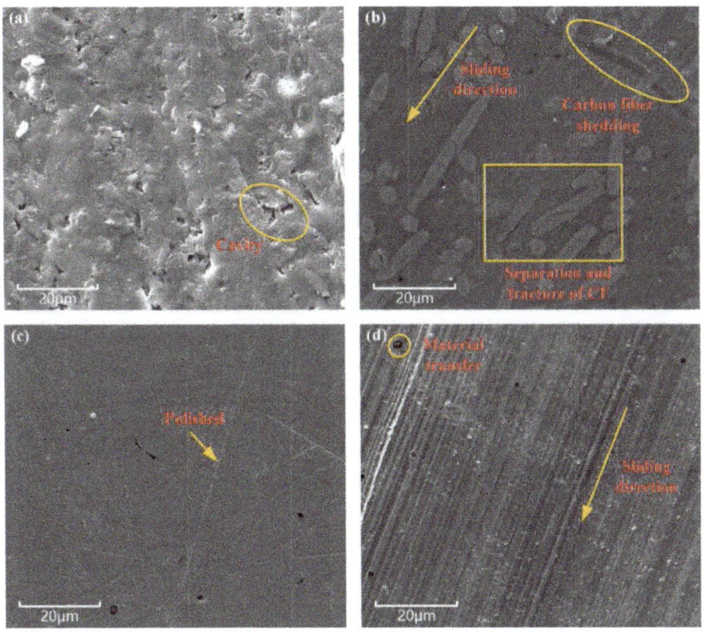

Figure 7. Surface topography of the samples. (**a**) Unused CFRPEEK; (**b**) worn CFRPEEK; (**c**) unused 1Cr17Ni2; (**d**) worn 1Cr17Ni2.

The wear morphology of 1Cr17Ni2 is different from that of CFRPEEK. As shown in Figure 7c, the surface of the unused 1Cr17Ni2 sample is flat, and there are some crisscross, fine textures present after polishing. As shown in Figure 7d, there are clear abrasion marks along the sliding direction on the worn 1Cr17Ni2 surface. At the same time, some PEEK materials have adhered to the 1Cr17Ni2 metal surface. The wear mechanism of CFRPEEK and 1Cr17Ni2 metal is as follows. The mechanical properties of the PEEK matrix are increased by filling carbon fibers; that is, carbon fibers improve the bearing capacity of the PEEK matrix. However, under high contact pressure and sliding speed, the carbon fiber is worn and peeled off. Due to the high hardness and strength of carbon fiber, it causes metal surface wear. The PEEK matrix is worn off and transferred to the 1Cr17Ni2 metal surface in a small amount, which reduces wear to a certain extent. These wear mechanisms also verify previous research [35,36].

3.2. Wear Analysis of Slipper/Swashplate Pair of WLPP

3.2.1. Volumetric Efficiency of WLPP

In the test system shown in Figure 4, the 316L and 1Cr17Ni2 swashplates are installed in the tested pump to test volumetric efficiency. At the beginning of the test, the tested pump speed was adjusted to 1500 rpm, and the outlet pressure was set to 14 MPa. Then, the tested pump was operated under this working condition. The tested pump with the 316L swashplate was operated continuously for 20 min at 1500 rpm and with no load (~0.5 MPa). The temperature of the pump shell was high and the vibration and noise were abnormal. Therefore, the volumetric efficiency of the pump could not be tested at high outlet pressure.

For the tested pump with the 1Cr17Ni2 swashplate, the vibration, noise and shell temperature were normal during the test. Under the working conditions of 14 MPa and 1500 rpm, the tested pump was worked for 6–8 h every day, with a total final testing time of 500 h. Its volumetric efficiency was tested. The volumetric efficiency of the pump changed with time elapsed, as shown in Figure 8. The volumetric efficiency of the pump decreased gradually with the increase in operation time, from 92.5% at the beginning to 88.6% at 500 h. This indirectly reflected the wear of the moving parts, which increased the fit or seal clearance, resulting in increased leakage and reduced volumetric efficiency [37].

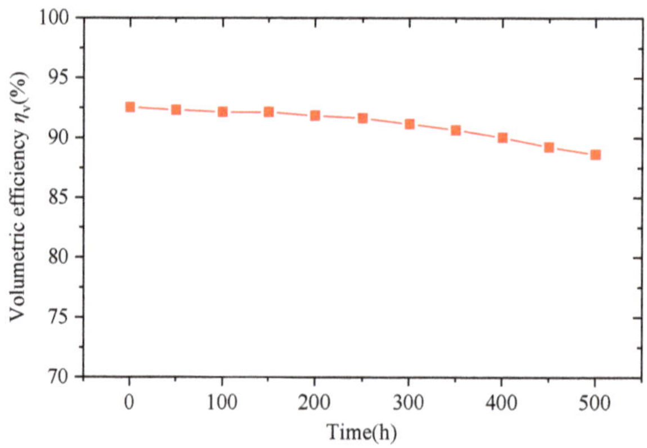

Figure 8. The volumetric efficiency of WLPP (installed with 1Cr17Ni2 swashplate) changes with time.

During the test of the WLPP, the 1Cr17Ni2/CFRPEEK and 316L/CFRPEEK friction pairs showed obvious differences. The results showed that the 316L/CFRPEEK friction pair cannot be used in the WLPP, while the 1Cr17Ni2/CFRPEEK friction pair can operate for a long time at a high speed and with a heavy load. This is significantly different from the tribological characteristics of the two in the friction testing machine.

3.2.2. Wear Analysis of Slipper and Swashplate Pairs

In order to study the wear mechanism of the swashplate and slipper friction pairs made of two materials, the worn appearances of the swashplate and the slipper were measured. A camera was used to take photos of the friction surface in order to observe the wear from a macro perspective. The wear profile of the swashplate was measured along the normal direction of the sliding direction with a profiler. This model is the MarSurf LD130 (Mahr, Gottingen, Germany), with a resolution of 0.8 nm. The three-dimensional morphology of the swashplate wear area was measured by confocal laser scanning microscopy (LSM800, Zeiss, Oberkochen, Germany). The wear of the friction surface of the swashplate parts is shown in Figure 9.

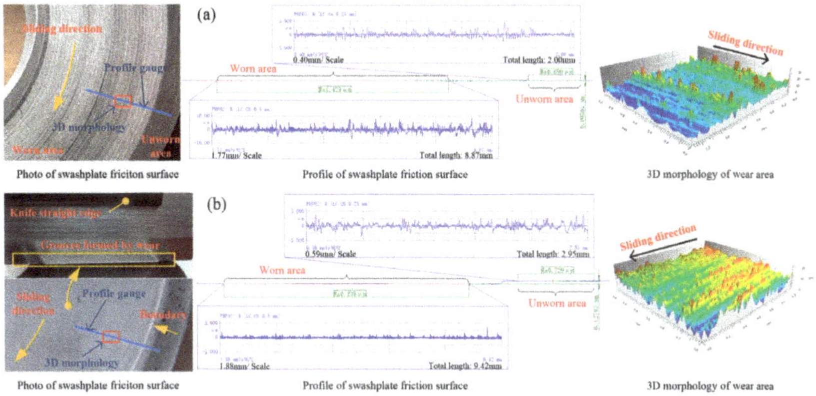

Figure 9. Wear of friction surface of swashplate parts. (**a**) The 316L swashplate; (**b**) the 1Cr17Ni2 swashplate.

The wear photo shows that the 316L swashplate friction surface (see Figure 9a) is severely worn after a short period of no-load operation. Along the sliding direction of the slipper, the wear forms a dense groove, which makes a clear boundary between the wear area and the non-wear area. From the profile measurement results showing the sliding direction of the slipper, it can be seen that the surface roughness increases significantly, from Ra0.69 μm before the test to Ra1.43 μm after the test. The height difference between the worn area and the non-worn area is 3 μm, which is the depth of the wear marks on the 316L swashplate surface. The 3D morphology of the wear area shows that the maximum height of the peaks and troughs formed by the wear mark is more than 30 μm, and the surface roughness is Sa2.99 μm.

By comparison, the friction surface of the 1Cr17Ni2 swashplate is much smoother, as shown in Figure 9b; however, due to the wear caused by long-time operation under high pressure, there are an grooves on the swashplate surface. Similarly, a clear boundary (step) is formed between the worn area and the non-worn area. The profile of the 1Cr17Ni2 swashplate was tested with the same method and position. The wear area is smoother than the non-wear area, and the roughness is reduced from Ra0.729 μm to Ra0.238 μm. This means that the 1Cr17Ni2 swashplate is polished by CFRPEEK. The 3D morphology observation results also show that the maximum contour height difference is less than 20 μm, and the surface roughness is Sa2.35 μm. Nevertheless, the depth of the wear marks on the swashplate surface reaches 0.128 mm, and the metal is considerably worn away, which is obviously different from the wear rate results obtained using the friction testing machine.

The wear of the CFRPEEK slippers matched with the different stainless steels was evaluated and analyzed. A micrometer was used to measure the height difference of the slipper friction surface before and after the test. Each slipper friction surface was measured at three points at equal intervals and averaged to reduce error. The CFRPEEK slipper matched with the 316L swashplate was worn 0.005 mm, and it was worn 0.103 mm when matched with the 1Cr17Ni2 swashplate. Although the wear of the former was much less than that of the latter, the wear rate of the former was also much greater than that of the latter. This is because the former had been operated for nearly 20 min under no load (~0.5 MPa), and the latter had been operated for 500 h under a rated load (14 MPa).

The wear morphology of the CFRPEEK slippers matched with different stainless steels was measured, as shown in Figure 10. When matched with the 316L swashplate, the slipper friction surface becomes very rough after a short time of operation (see Figure 10a). The friction surface is covered with dense and staggered grooves. As shown in Figure 10b, the PEEK matrix is embedded with a certain amount of metal debris. At the same time, the carbon fibers are broken or even pulled out, which causes some cavities on the surface of CFRPEEK. This is because the strength and hardness of carbon fiber are much higher than that of 316L stainless steel. When the two of them rub against each other, the carbon fibers cut 316L metal to form large abrasive debris. Due to the low strength and hardness of the PEEK matrix, it is embedded with metal debris under the action of the load and speed [33,34].

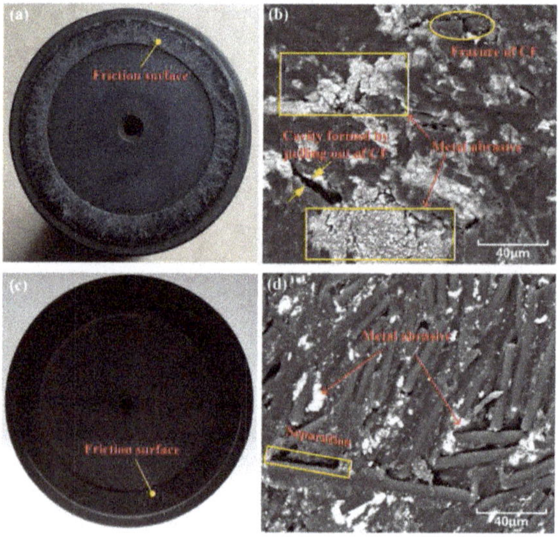

Figure 10. Wear of friction surface of slipper parts. (**a**) Photo of slipper after sliding with 316L swashplate; (**b**) ESEM morphology of slipper after sliding with 316L swashplate; (**c**) photo of slipper after sliding with 1Cr17Ni2 swashplate; (**d**) ESME morphology of slipper after sliding with 1Cr17Ni2 swashplate.

As shown in Figure 10c, when matched with the 1Cr17Ni2 swashplate, the slipper friction surface is relatively flat after high-pressure operation for a long time. The enlarged scanning electron microscope shows (Figure 10d) that the carbon fibers are densely distributed on the surface, and a few parts are separated from the PEEK matrix. Moreover, wear debris of a small size is embedded into the PEEK matrix around the carbon fiber.

However, when CFRPEEK slippers are paired with the 316L and 1Cr17Ni2 swashplate in the WLPP, the wear rates of the metals are very different. The wear rate of 316L is much greater than that of the 1Cr17Ni2. At the same time, the depth of the the 1Cr17Ni2 swashplate wear mark is greater than that of the CFRPEEK slipper. This means that CFRPEEK is more wear-resistant than the metals. These results are completely different from the results of the friction testing machine.

In summary, due to the low strength and hardness of 316L, the swash plate made of this material is microcut by carbon fibers with a high hardness. The wear mechanisms of 316L are mainly severe abrasive wear [34] and slight adhesive wear. However, with the higher strength and hardness of 1Cr17Ni2, the swash plate made of this material has a certain polishing effect. Its wear mechanism is slight abrasive wear. As for CFRPEEK, carbon fibers break and fall off on the surface, and the PEEK matrix is embedded with metal abrasive debris. The wear mechanisms of CFRPEEK are mainly abrasive wear [23], surface fatigue wear and adhesive wear [20,38].

3.3. Analysis of Tribological Differences under Two Test Methods

The cause of the difference in the friction tests between CFRPEEK and the two stainless steels in the friction testing machine and the axial piston pump is the pressure of the lubricating water. Atmospheric pressure water lubrication is used in the friction testing machine, while high-pressure (~14 MPa) water lubrication is used in the WLPP. The wear mechanisms of CFRPEEK and the metals under the two test methods are shown in Figure 11.

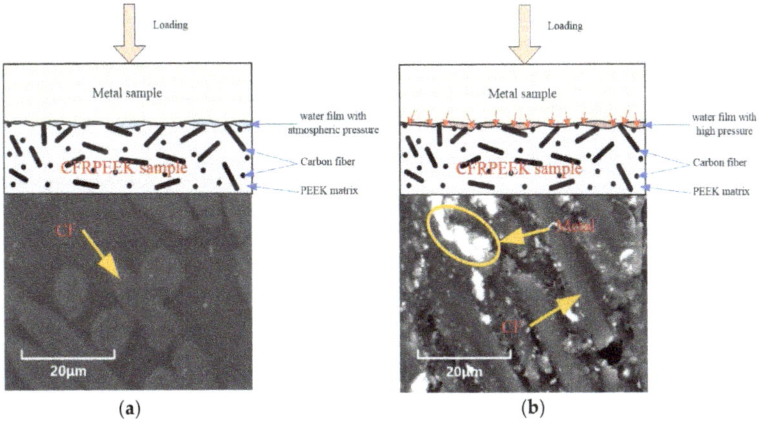

Figure 11. Wear mechanisms of CFRPEEK and metals. (**a**) Atmospheric pressure water lubrication in friction testing machine; (**b**) high-pressure water lubrication in WLPP.

Figure 11a shows the wear mechanism of CFRPEEK against two stainless steels under atmospheric pressure water lubrication, which is used to analyze the wear condition in the friction testing machine. Under the action of the test load, contact stress is generated between the metal sample and the CFRPEEK sample. The carbon fibers and the PEEK matrix in the contact area jointly bear the contact stress. Because the lubricating water is at atmospheric pressure, it cannot cause microdeformation on the surface of the carbon fibers or the PEEK matrix, but it still plays the role of lubrication and cooling. With the low hardness and strength of CFRPEEK, its wear rate is larger than that of metal. The worn surface of CFRPEEK is also relatively flat.

Figure 11b shows the wear mechanism of CFRPEEK against two stainless steels under high-pressure water lubrication, which is used to analyze the wear condition in the WLPP. The tensile strength and modulus of carbon fibers exceed 3.0 GPa and 230 GPa, respectively. Its hardness is second only to diamond, and more than 10 times that of the metals [39]. However, the hardness and elastic modulus of the PEEK matrix are 85 (shore D) and 3.5 GPa [30], respectively. Based on the diameter of the carbon fibers (~8 μm) [30], the deformations of the carbon fibers and the PEEK matrix are 0.5×10^{-3} μm and 32×10^{-3} μm, respectively, under the pressure of 14 MPa. Therefore, under the condition of high-pressure water lubrication, the deformation of the PEEK matrix is larger than that of the carbon fibers. The difference deformation increases the amount of the carbon fibers' contact with the metal and makes them bear a greater load. Due to the extremely high hardness of carbon fibers, the wear of the metal increases when rubbed against them. In this case, the strength and hardness of the metal materials affect their wear rates. Generally, the higher the strength and hardness of metal materials, the lower the wear rates. Due to the low strength and hardness of 316L, the wear of the 316L swashplate is serious. With the high strength and hardness of 1Cr17Ni2, the surface of the 1Cr17Ni2 swashplate is smooth after rubbing against carbon fibers. The micro surface of CFRPEEK matched with the metal is relatively rough. Metal wear debris is embedded into the PEEK matrix with low hardness between carbon fibers.

4. Conclusions

In this work, the tribological characteristics of CFRPEEK sliding against 316L and 1Cr17Ni2 stainless steel under water lubrication in a friction testing machine and an axial piston pump were investigated. Based on the study, the following conclusions can be drawn:

(1) Metal (316L or 1Cr17Ni2) is more wear-resistant than CFRPEEK within a friction testing machine. When tested with a friction testing machine, the friction coefficient and wear rates of the CFRPEEK/316L and CFRPEEK/1Cr17Ni2 friction pairs have no significant difference. The wear rate of CFRPEEK is on the order of 10^{-7}, while that of metal is on the order of 10^{-9}.

(2) The wear of 1Cr17Ni2 is greater than the wear of CFRPEEK in a WLPP. When tested in a water-lubricated axial piston pump, the CFRPEEK/316L and CFRPEEK/1Cr17Ni2 friction pairs show great differences. The wear rates of CFRPEEK against 316L are very high, and show that the pair can hardly be used in the pump. However, CFRPEEK against 1Cr17Ni2 can be used as friction pair materials in the pump for a long time. After 500 h in a high-pressure test, the CFRPEEK slipper and the 1Cr17Ni2 swashplate were worn by 0.103 mm and 0.128 mm, respectively. This is obviously different from the wear rate test results of metal and CFRPEEK in the friction testing machine.

(3) The amount of load borne by carbon fibers is the main reason for the difference in tribological properties between the different metals and CFRPEEK. The friction interface between metal (316L or 1Cr17Ni2) and CFRPEEK is lubricated by atmospheric pressure water in the friction testing machine, but it is lubricated by high-pressure water in the pump. This is the biggest difference between the two. The high-pressure water film causes the microdeformation of the PEEK matrix at the interface of CFRPEEK, which makes the carbon fibers bear more loads. High strength and high hardness carbon fibers cut metal to form grinding, resulting in the abrasive wear of the metal and CFRPEEK. The wear mechanism of CFRPEEK is mainly surface fatigue wear. The 316L with low strength and hardness rubs against carbon fibers, causing serious wear. The wear of 1Cr17Ni2 with higher strength and hardness is better.

Based on the above findings, it is necessary to study wear resistance methods for metals in the future to improve the performance of WLPPs.

Author Contributions: Methodology and writing—original draft preparation, D.L. and X.M.; test and data processing, S.W. and J.W.; project administration, F.Y.; checking and supervision, Y.L. All authors have read and agreed to the published version of the manuscript.

Funding: This research was funded by National Natural Science Foundation of China, grant number 52105054, Key scientific and technological projects in Henan Province, grant number 222102220012 and 222102220073, Key scientific research projects of colleges and universities in Henan Province, 22A460004 and 22A460017.

Data Availability Statement: The data-supported results are included within this article.

Conflicts of Interest: The authors declare no conflict of interest.

References

1. Lou, F.L.; Nie, S.L.; Yin, F.L.; Lu, W.; Ji, H.; Ma, Z.H.; Kong, X.L. Numerical and experimental research on the integrated energy recovery and pressure boost device for seawater reverse osmosis desalination system. *Desalination* **2022**, *523*, 115408. [CrossRef]
2. Yin, F.L.; Nie, S.L.; Ji, H.; Lou, F.L. Numerical study of structure parameters on energy transfer and flow characteristics of integrated energy recovery and pressure boost device. *Desalin. Water Treat.* **2018**, *131*, 141–154. [CrossRef]

3. Zhao, X.F.; Liu, Y.S.; Han, M.X.; Wu, D.F.; Li, D.L. Improving the performance of an AUV hovering system by introducing low-cost flow rate control into water hydraulic variable ballast system. *Ocean Eng.* **2016**, *125*, 155–169. [CrossRef]
4. Liu, Y.S.; Li, D.L.; Tang, Z.Y.; Deng, Y.P.; Wu, D.F. Thermodynamic modeling, simulation and experiments of a water hydraulic piston pump in water hydraulic variable ballast system. *Ocean Eng.* **2017**, *138*, 35–44. [CrossRef]
5. Liu, Y.S.; Jiang, Z.; Wang, D.; Li, X.H. Experimental research on the water mist fire suppression performance in an enclosed space by changing the characteristics of nozzles. *Exp. Therm. Fluid Sci.* **2014**, *52*, 174–181. [CrossRef]
6. Zhang, J.H.; Lyu, F.; Xu, B.; Huang, W.D.; Wu, W.; Guo, Z.M.; Xu, H.G.; Huang, X.C. Simulation and experimental investigation on low wear rate surface contour of piston/cylinder pair in an axial piston pump. *Tribol. Int.* **2021**, *162*, 107127. [CrossRef]
7. Chen, B.B.; Wang, J.Z.; Yan, F.Y. Friction and Wear Behaviors of Several Polymers Sliding Against GCr15 and 316 Steel Under the Lubrication of Sea Water. *Tribol. Lett.* **2011**, *42*, 17–25. [CrossRef]
8. Yin, F.L.; Kong, X.L.; Ji, H.; Nie, S.L.; Lu, W. Research on the pressure and flow characteristics of seawater axial piston pump considering cavitation for reverse osmosis desalination system. *Desalination* **2022**, *540*, 18. [CrossRef]
9. Liu, Y.S.; Cheng, Q.; Wang, Z.Y.; Pang, H.; Deng, Y.P.; Zhou, X.P.; Luo, X.H.; Cui, Y.; Wu, D.F. Seawater hydraulics: From the sea surface to depths of 11000 meters. *Sci. China-Technol. Sci.* **2022**, *65*, 2178–2189. [CrossRef]
10. Gao, C.P.; Guo, G.F.; Zhao, F.Y.; Wang, T.M.; Jim, B.; Wetzel, B.; Zhang, G.; Wang, Q.H. Tribological behaviors of epoxy composites under water lubrication conditions. *Tribol. Int.* **2016**, *95*, 333–341. [CrossRef]
11. Yin, F.L.; Ji, H.; Nie, S.L. Tribological behavior of various ceramic materials sliding against CF/PTFE/graphite-filled PEEK under seawater lubrication. *Proc. Inst. Mech. Eng. Part J-J. Eng. Tribol.* **2019**, *233*, 1729–1742. [CrossRef]
12. Brookes, C.A.; Fagan, M.J.; James, R.D.; Connachie, J.M. The Selection and Performance of Ceramic Components in a Sea-water Pump. In Proceedings of the JFPS International Symposium on Fluid Power, Yokohama, Japan, 12 November 1996.
13. Chen, M.; Kato, K.; Adachi, K. Friction and wear of self-mated SiC and Si3N4 sliding in water. *Wear* **2001**, *250*, 246–255. [CrossRef]
14. Yin, Z.B.; Huang, C.Z.; Yuan, J.T.; Zou, B.; Liu, H.L.; Zhu, H.T. Cutting performance and life prediction of an Al2O3/TiC micro nano-composite ceramic tool when machining austenitic stainless steel. *Ceram. Int.* **2015**, *41*, 7059–7065. [CrossRef]
15. Zhang, Y.H.; Tao, W.; Zhang, Y.; Tang, L.; Gu, J.W.; Jiang, Z.H. Continuous carbon fiber/crosslinkable poly(ether ether ketone) laminated composites with outstanding mechanical properties, robust solvent resistance and excellent thermal stability. *Compos. Sci. Technol.* **2018**, *165*, 148–153. [CrossRef]
16. Koh, Y.G.; Lee, J.A.; Kang, K.T. Prediction of Wear on Tibial Inserts Made of UHMWPE, PEEK, and CFR-PEEK in Total Knee Arthroplasty Using Finite-Element Analysis. *Lubricants* **2019**, *7*, 14. [CrossRef]
17. Yan, M.X.; Tian, X.Y.; Peng, G.; Li, D.C.; Zhang, X.Y. High temperature rheological behavior and sintering kinetics of CF/PEEK composites during selective laser sintering. *Compos. Sci. Technol.* **2018**, *165*, 140–147. [CrossRef]
18. Shukla, D.; Negi, Y.S.; Sen Uppadhyaya, J.; Kumar, V. Synthesis and Modification of Poly(ether ether ketone) and their Properties: A Review. *Polym. Rev.* **2012**, *52*, 189–228. [CrossRef]
19. Chen, B.B.; Wang, J.Z.; Yan, F.Y. Comparative investigation on the tribological behaviors of CF/PEEK composites under sea water lubrication. *Tribol. Int.* **2012**, *52*, 170–177. [CrossRef]
20. Li, D.; Liu, Y.; Deng, Y.; Fang, M.; Wu, D. The Effect of Different Temperature on Friction and Wear Properties of CFRPEEK against AISI 431 Steel under Water Lubrication. *Tribol. Trans.* **2018**, *61*, 357–366. [CrossRef]
21. Davim, J.P.; Cardoso, R. Thermo-mechanical model to predict the tribological behaviour of the composite PEEK-CF30/steel pair in dry sliding using multiple regression analysis. *Ind. Lubr. Tribol.* **2005**, *57*, 181–186. [CrossRef]
22. Davim, J.P.; Cardoso, R. Tribological behaviour of the composite PEEK-CF30 at dry sliding against steel using statistical techniques. *Mater. Des.* **2006**, *27*, 338–342. [CrossRef]
23. Davim, J.P.; Cardoso, R. Effect of the reinforcement (carbon or glass fibres) on friction and wear behaviour of the PEEK against steel surface at long dry sliding. *Wear* **2009**, *266*, 795–799. [CrossRef]
24. Li, G.; Qi, H.; Zhang, G.; Zhao, F.; Wang, T.; Wang, Q. Significant friction and wear reduction by assembling two individual PEEK composites with specific functionalities. *Mater. Des.* **2017**, *116*, 152–159. [CrossRef]
25. Liu, H.; Wang, J.Z.; Jiang, P.F.; Yan, F.Y. Hydrostatic pressure-dependent wear behavior of thermoplastic polymers in deep sea. *Polym. Adv. Technol.* **2018**, *29*, 2410–2415. [CrossRef]
26. Wu, D.; Guan, Z.; Cheng, Q.; Wei, G.; Tang, M.; Liu, Y. Development of a Friction Test Apparatus for Simulating the Ultra-High Pressure Environment of the Deep Ocean. *Wear* **2020**, *452–453*, 203294. [CrossRef]
27. Guan, Z.; Wu, D.; Cheng, Q.; Wang, Z.; Tang, M.; Liu, Y. Friction and Wear characteristics of CF/PEEK against 431 stainless steel under high hydrostatic pressure water lubrication. *Mater. Des.* **2020**, *196*, 109057. [CrossRef]
28. Pelosi, M.; Ivantysynova, M. A Geometric Multigrid Solver for the Piston-Cylinder Interface of Axial Piston Machines. *Tribol. Trans.* **2012**, *55*, 163–174. [CrossRef]
29. Liang, Y.N.; Gao, D.R.; Zhao, J.H. Tribological Properties of Friction Pair between 316L Stainless Steel and CF/PEEK with Nonsmooth Surface under Seawater Lubrication. *Tribol. Trans.* **2020**, *63*, 658–671. [CrossRef]
30. Ensinger. Available online: https://www.ensingerplastics.com/en/shapes/products/peek-tecapeek-cf30-black (accessed on 3 February 2023).
31. Chao, Q.; Zhang, J.H.; Xu, B.; Wang, Q.N.; Lyu, F.; Li, K. Integrated slipper retainer mechanism to eliminate slipper wear in high-speed axial piston pumps. *Front. Mech. Eng.* **2022**, *17*, 13. [CrossRef]

32. Ma, J.M.; Chen, J.; Li, J.; Li, Q.L.; Ren, C.Y. Wear analysis of swash plate/slipper pair of axis piston hydraulic pump. *Tribol. Int.* **2015**, *90*, 467–472. [CrossRef]
33. Zhang, Z.H.; Nie, S.L.; Yuan, S.H.; Liao, W.J. Comparative Evaluation of Tribological Characteristics of CF/PEEK and CF/PTFE/Graphite Filled PEEK Sliding against AISI630 Steel for Seawater Hydraulic Piston Pumps/Motors. *Tribol. Trans.* **2015**, *58*, 1096–1104. [CrossRef]
34. Wang, Z.Q.; Gao, D.R. Friction and wear properties of stainless steel sliding against polyetheretherketone and carbon-fiber-reinforced polyetheretherketone under natural seawater lubrication. *Mater. Des.* **2014**, *53*, 881–887. [CrossRef]
35. Mimaroglu, A.; Unal, H.; Ozel, A. Tribological Performance of Polyetheretherketone and its Composites under Water Environment. *Macromol. Symp.* **2013**, *327*, 108–113. [CrossRef]
36. Tang, Q.G.; Chen, J.S.; Andrea, V. Tribological Behaviors of Carbon Fiber-Reinforced PEEK Sliding on Ion-Nitrided 2Cr13 Steel Lubricated with Tap Water. *Tribol. Trans.* **2015**, *58*, 691–697. [CrossRef]
37. Nie, S.L.; He, H.; Ji, H.; Nie, S.; Yan, X.P.; Yin, F.L. Failure analysis of auxiliary support bearing/shaft tribopair in seawater hydraulic axial piston pump. *Eng. Fail. Anal.* **2023**, *146*, 16. [CrossRef]
38. Wang, Z.; Fu, Q.; Wood, R.J.; Wu, J.; Wang, S. Influence of bionic non-smooth surface texture on tribological characteristics of carbon-fiber-reinforced polyetheretherketone under seawater lubrication. *Tribol. Int.* **2020**, *144*, 106100. [CrossRef]
39. Newcomb, B.A. Processing, structure, and properties of carbon fibers. *Compos. Part A Appl. Sci. Manuf.* **2016**, *91*, 262–282. [CrossRef]

lubricants

MDPI

Article

Numerical Analysis of the Mixed-Lubrication Performance of Staved Stern Tube Bearings Lubricated with Water

Dongxing Tang [1,2], Yanfeng Han [1,2,*], Lei Yin [3,*] and Yi Chen [4]

1 College of Mechanical and Vehicle Engineering, Chongqing University, Chongqing 400044, China
2 State Key Laboratory of Mechanical Transmission, Chongqing University, Chongqing 400044, China
3 School of Mechanical and Power Engineering, Chongqing University of Science and Technology, Chongqing 401331, China
4 School of Education Technology, Northwest Normal University, Lanzhou 730070, China
* Correspondence: yfhan@cqu.edu.cn (Y.H.); 2022027@cqust.edu.cn (L.Y.)

Abstract: The present study aims to establish a mixed lubrication model for staved stern tube bearings lubricated with water, in which the average Reynolds equation and a KE elastic–plastic contact model are introduced to calculate the hydrodynamic pressure and contact pressure, respectively. The difference in the mixed lubrication behaviors between circular- and flat-staved bearings is compared; moreover, the effects of the number of staves on the mixed-lubrication performance of these two kinds of staved bearing are investigated. The mechanism of action of the number of staves in staved bearings on the mixed-lubrication performance is revealed. The numerical results show that the number of staves has a significant effect on the mixed-lubrication performance in circular- and flat-staved bearings. Furthermore, there is an optimal value for the number of staves, shown to be 30 in the current simulation, for improving the mixed-lubrication performance of flat-staved stern tube bearings lubricated with water.

Keywords: staved bearing; stave number; water-lubricated; mixed lubrication

Citation: Tang, D.; Han, Y.; Yin, L.; Chen, Y. Numerical Analysis of the Mixed-Lubrication Performance of Staved Stern Tube Bearings Lubricated with Water. *Lubricants* **2023**, *11*, 168. https://doi.org/10.3390/lubricants11040168

Received: 28 February 2023
Revised: 29 March 2023
Accepted: 1 April 2023
Published: 8 April 2023

1. Introduction

In recent years, with people's increasing awareness of environmental protection, researchers have paid water-lubricated bearings more attention, especially in the case of stern bearings used in marine ships. As is well known, the water-lubricated bearings typically operate under mixed-lubrication conditions, especially during the start and shut-down stage or under severe load, because of the low viscosity of water [1,2]. Moreover, in the case of mixed lubrication, the contact behavior can lead to increased friction and wear, which may exacerbate the vibration as well as the noise emitted by the propulsion system. Therefore, it is of great significance to investigate the mixed-lubrication performance of water-lubricated stern tube bearings.

He et al. [3] developed a mixed lubrication model to study the lubrication performance of marine stern tube bearings in consideration of the bending deformation of the stern shaft and cavitation. Lv et al. [4] presented a mixed lubrication model considering two-dimensional wall slip and journal misalignment to study the tribological properties of stern bearings. Both of the above studies employed the average Reynolds equation proposed by Patir and Cheng [5] to calculated the hydrodynamic pressure of the journal bearing. Deepak et al. [2] modified the average Reynolds equation to investigate the effect of non-Gaussian surface topography on the mixed-lubrication performance of water-lubricated journal bearings. Yang et al. [6] numerically investigated the performance of aft stern tube bearings during ship turning maneuvers using the Computation Fluid Dynamics (CFD) method. Huang et al. [7] proposed a numerical model to study the lubrication performance with misalignment of the shaft, and the effect of parameters including misalignment angle, eccentricity ratio, rotation speed, lubricant viscosity and center-line eccentricity angle on lubrication performance were investigated. Georgios et al. [8] compared two bearing modification designs, and determined the optimal geometric parameters of a double slope aft stern tube bearing necessary to maximize the effective pressure area while also minimizing the maximum local pressure. Xie et al. [9] theoretically and experimentally investigated the micro asperity contact load ratios and lubrication states of water-lubricated stern tube bearings. Liu et al. [10] experimentally demonstrated the phenomenon of pressure bias and bidirectional shaft inclination; furthermore, they numerically investigated the effects of eccentricity ratio, shaft inclination angle and bearing structure parameters on lubrication, as well as the dynamic characteristics of water-lubricated stern bearings. He et al. [11] established a mixed lubrication model for marine stern bearings in consideration of cavitation and the bending deformation of the shaft; moreover, they conducted a theoretical investigation on the effect of bending deformation on the dynamic characteristics. Zhu et al. [12] built a coupling model of shaft longitudinal vibration and bearing lubrication, in which the shaft axial shock and misalignment were considered, in order to investigate the effect of explosion shock on bearing lubrication performance. Wodtke et al. [13] conducted an investigation on the thermal effects of water-lubricated stern tube bearings with both experimental and theoretical methods. Zhang et al. [14] studied the effect of operating conditions on bearing temperature under hydrodynamic lubrication. In order to improve the lubrication performance as well as wear resistance of the water-lubricated stern bearings, Zhou et al. [15] studied the effects of different nano-fillers on friction, wear, mechanical and vulcanization properties of hydrogenated nitrile rubber used on ship stern shaft.

In addition to the research described above, many studies [16–28] on wear, dynamic characteristics, friction-induced vibration and noise of water-lubricated stern tube bearings have also been conducted by scholars. Han et al. [16] experimentally studied the mechanism of action of stick–slip nonlinear friction-induced vibration of water-lubricated stern rube bearing. To investigate the mechanism of friction-induced vibration and noise of marine stern tube bearings, a three-degree-of-freedom model considering perturbations of a stochastic rough surface was proposed by Lin et al. [17], whose results showed that the friction coefficient and the contact stiffness significantly influenced the stability of the system. Qin et al. [18] conducted an experimental investigation, in which a high-speed camera was used to track the vibration motion and a machine vision technique was employed to extract the friction-induced vibration of the bearing. Their analysis results showed that bearing vibration was mainly induced by the stick–slip phenomenon. A coupled vibration model, in which the lateral vibration response of the shaft journal under normal load was applied to the dynamic friction force, was proposed by Jin et al. [19] to investigate the interaction of the lateral vibration and friction vibration of a marine water-lubricated bearing. Subsequently, they updated the model to consider the movement of the stern shaft journal [20]. Lin et al. [21] pointed out that different parameters, including the rotation speed of the shaft, contact pressure, friction coefficient, system damping and stiffness, all have an important influence on the nonlinear friction-induced vibration characteristics of water-lubricated stern bearings. Ning et al. [22] proposed a wear monitoring method based on the amplitude spectrum of the ultrasonic reflection coefficient, and used this method to analyze the effects of bearing parameters, objective function and algorithm parameters on the identification results. A full-sized test rig of a water-lubricated stern bearing with an elevation adjustment device was built by Ouyang et al. [23], in which the eddy current sensor and high-precision wireless telemetry were used to collect the axis orbit signal and circumferential film pressure, respectively. In addition, the effect of bias load, rotational speed, and hull deformation on bearing lubrication characteristics were experimentally studied. Cai et al. [24] numerically explored the effects of wear and shaft shape error defects on the tribo-dynamic responses of water-lubricated bearings under non-linear propeller disturbance, and the results showed that the nonlinear disturbance had a significant influence on the dynamic behavior, and the use of reasonable wear parameters can improve the lubrication performance. Subsequently, they updated the tribo-dynamic model in order to consider mass conservation cavitation, and conducted an investigation on micro-groove water-lubricated bearings [25]. Recently, they proposed a new mathematical modeling technique to investigate the nonlinear dynamic mixed friction behaviors of journal-thrust-coupled bearings lubricated with water [26]. In order to improve the transient wear and asperity contact performance of the water-lubricated bearing, Xiang et al. [27] proposed a surface profile modification method under fluid–solid–thermal coupling conditions. Chen et al. [28] numerically investigated the effect of an imperfect journal, including journal error with respect to the amplitude as well as the waviness, on the tribo-dynamic behaviors of water-lubricated bearings during start-up. The studies on stern tube water-lubricated bearings described above all adopted a continuous circular bush, which is usually difficult to replace and maintain, while staved stern tube bearings lubricated with water can adapt to this situation better, with their convenient maintenance, because failed staves can be changed separately, instead of needing to replace the whole bearing. Recently, Anatoliy et al. [29] estimated the hydrodynamic pressure of water-lubricated staved bearings with circular and flat staves using an asymptotic technique, the results of which were compared with those obtained using Finite Element Method (FEM). However, to the best of authors' knowledge, studies on staved stern tube water-lubricated bearings are still few in number. Therefore, it is necessary to conduct research on this topic.

The purpose of the present study is to establish a mixed lubrication model for staved stern tube bearings lubricated with water in order to investigate their mixed lubrication behaviors. Additionally, the effect of the number of staves on the mixed-lubrication performance of circular-staved and flat-staved bearings are analyzed comparatively in order to reveal the mechanism of effect of the number of staves. Furthermore, an optimization of the number of staves with respect to the mixed-lubrication performance of a flat-staved bearing is conducted, the results of which can serve as a reference for designers.

2. Methods and Material

2.1. Governing Equation

2.1.1. Hydrodynamic Pressure Model

In the present study, to investigate the mixed-lubrication performance of staved stern tube water-lubricated bearings, the average Reynolds equation developed by Patir et al. [5], which has been widely employed by researchers, was used to calculate the hydrodynamic pressure using the Finite Difference Method (FDM), which can be described as follows:

$$\frac{\partial}{R_{J1}^2\partial\theta}\left(\phi_\theta\frac{\rho h_J^3}{12\eta}\frac{\partial p_{hJ}}{\partial\theta}\right)+\frac{\partial}{\partial z}\left(\phi_z\frac{\rho h_J^3}{12\eta}\frac{\partial p_{hJ}}{\partial z}\right)=\frac{\omega}{2}\left(\phi_c\frac{\partial\rho h_J}{R_{J1}\partial\theta}+\rho\sigma_s\frac{\partial\phi_s}{R_{J1}\partial\theta}\right) \tag{1}$$

where θ and z are the circumferential and axial directions, respectively; R_{J1} is the radius of the journal bearing, ϕ_θ and ϕ_z are the flow factors in the circumferential and axial directions, respectively; ϕ_c and ϕ_s are the contact factor and the shear factor, respectively; η and ρ are the viscosity and density of the water, respectively; h_J and p_{hJ} are the lubrication gap and the hydrodynamic pressure of the journal bearing, respectively; ω is the angular velocity of the journal; and σ_s is the composite roughness of journal and bearing.

2.1.2. Geometrical Model of a Staved Bearing

Staved bearings with circular and flat staves were investigated in the present simulation, the respective structures of which are shown in Figure 1. The staves are installed in dovetail grooves on the inner surface of the bearing; moreover, there are grooves between two adjacent staves to allow any silt in the water to pass through, thus reducing abrasive wear on the bearing surface resulting from hard particles in water.

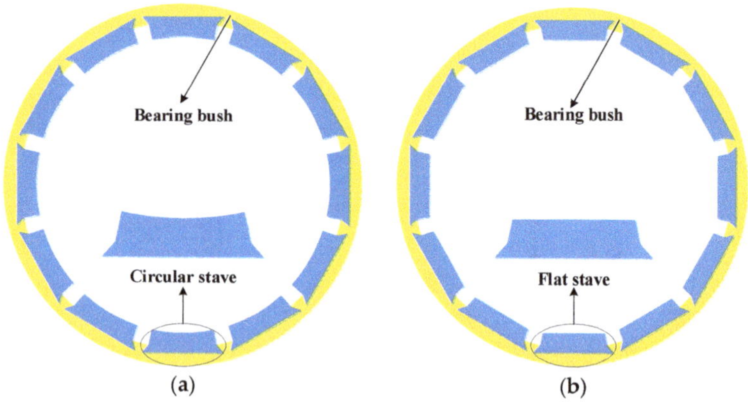

Figure 1. Bearings with (**a**) circular staves and (**b**) flat staves.

As shown in Figure 2, AC is the lubrication gap, denoted by $h_J(\theta, z)$; AB is the geometrical lubrication gap of a bearing with radius of R_{J1}, which can be calculated by $C(1 + \varepsilon\cos(\theta - \psi))$; and BC can be calculated by $OC-OB$, denoted by $G_{BC}(\theta, z)$; furthermore, the elastic deformation $\delta_J(\theta, z)$ is also used in the present study; therefore, the lubrication gap can be written as follows:

$$h_J(\theta, z) = C(1 + \varepsilon\cos(\theta - \psi)) + G_{BC}(\theta, z) + \delta_J(\theta, z) \tag{2}$$

where C and ε are the radius clearance and the eccentricity ratio, respectively, based on the center of the flat; and θ and ψ are the circumferential angle and the attitude angle, respectively.

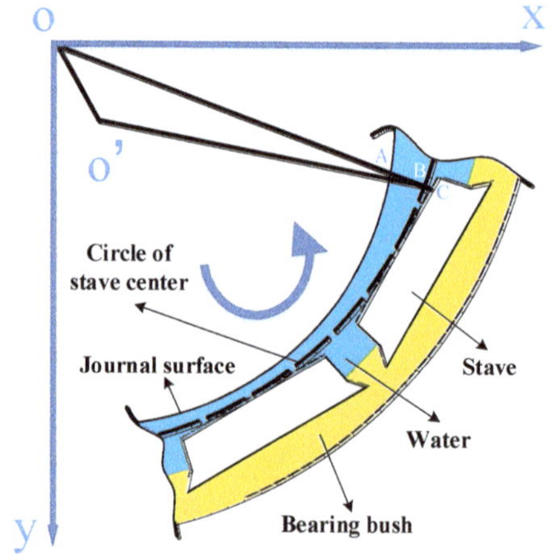

Figure 2. Geometry model for calculating the lubrication gap of the staved bearing.

2.1.3. Asperity Contact Model

The widely accepted elastic–plastic contact model for Gaussian surfaces (KE model) proposed by Kogut and Etsion [30,31] was used in the present study to calculate the contact pressure, which can be written as follows:

$$\begin{aligned} P_c = {} & \tfrac{2}{3}\pi\sigma\beta DKH_B\omega_c^*\Big(\int_{h^*}^{h^*+\omega_c^*} I_c^{1.5} + 1.03\int_{h^*+\omega_c^*}^{h^*+6\omega_c^*} I_c^{1.425} \\ & +1.4\int_{h^*+6\omega_c^*}^{h^*+110\omega_c^*} I_c^{1.263} + \tfrac{3}{K}\int_{h^*+110\omega_c^*}^{\infty} I_c^{1}\Big) \end{aligned} \tag{3}$$

where D and β are the density and composite curvature radius of asperity, respectively; K is a dimensionless parameter, which can be calculated by $K = 0.454 + 0.41v_B$; H_B is the hardness of the bearing; and * denotes a dimensionless parameter, where all dimensionless parameters are normalized to σ, which corresponds to the surface height. According to Beheshti and Khonsari [32], the value of $\sigma\beta D$ was reasonably set to 0.04 in the present study. h^* is the dimensionless lubrication gap, where the critical interference ω_c^* marks the transition from elastic to elastic–plastic deformation, which can be written as follows:

$$\omega_c^* = \left(\frac{\pi K H_B}{2E_c}\right)^2 \beta/\sigma \tag{4}$$

where E_c is the composite elastic modulus of the journal and bearing. Furthermore, in Equation (3), I_C^b is an integral operation, which can be written as follows:

$$I_c^b = \left(\frac{z^* - h^*}{\omega_c^*}\right)^b \phi^*(z^*)dz^* \tag{5}$$

where $\phi^*(z^*)$ is the dimensionless asperity probability density function, which can be described as follows:

$$\phi^*(z^*) = \frac{1}{\sqrt{2\pi}}\frac{\sigma_s}{\sigma}\exp\left[-0.5\left(\frac{\sigma_s}{\sigma}\right)^2 (z^*)^2\right] \tag{6}$$

where σ_s/σ is set to 1, in line with the recommendation of Kogut and Etsion [30].

2.2. Boundary Conditions

Considering the cavitation effect generated in the divergent area along the rotational direction, the Reynolds boundary condition is applied in the present study, which can be written as follows:

$$\begin{cases} p_{hJ}(\theta,0) = 0 \\ p_{hJ}(\theta_{J0},z) = 0, \partial p_{hJ}(\theta_{J0},z)/\partial\theta = 0 \end{cases} \tag{7}$$

where θ_{J0} indicates the positions at which the cavitation effect occurs along the circumferential direction of the bearing. Furthermore, the hydrodynamic pressures at the entrance and exit edges of the bearing are set to zero.

2.3. Numerical Procedure

As shown in Figure 3, it can be observed that the flow chart includes four calculation steps and three convergence judgments. The first calculation step is performed to calculate the discontinuous lubrication gap of the staved stern tube bearing. In the second calculation step, the hydrodynamic pressure is obtained by solving the average Reynolds equation using the Finite Difference Method (FDM). In the third calculation step, the contact pressure is obtained using the KE model. After obtaining the hydrodynamic and contact pressure, a fourth step is performed in which the deformation of the bearing is calculated. Then, the calculation steps are subjected to the process of convergence judgment, and the calculation ends only when the fluid pressure, contact pressure and attitude angle all converge; otherwise, the fluid pressure or lubrication gap are recalculated. The parameters used for analysis in the present study are listed in Table 1.

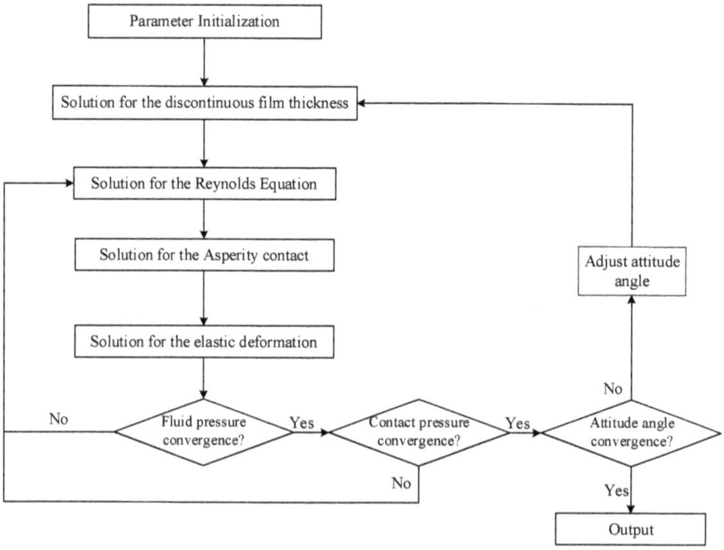

Figure 3. Flow chart for calculation.

Table 1. Parameters for calculation.

Parameters	Value
Journal bearing inner radius, R_{J1}	57.5 mm
Bearing width, L	230 mm
Radius clearance, C	0.5 mm
Journal bearing elastic modulus, E_J	2.32 GPa
Journal bearing Poisson ratio, ν_J	0.327
Composite roughness of journal and bearing, σ_s	1.2 μm
Water viscosity, η	8.49×10^{-4} Pa·s
Water density, ρ	1000 kg/m^3
Shaft elastic modulus, E_s	210 GPa
Shaft Poisson ratio, ν_s	0.3
Surface orientation, γ	1

3. Results and Discussion

3.1. Verification

To verify the appropriateness as well as the accuracy of the model used in the present study, a comparison between the fluid pressure obtained using the present model and that obtained using the model of Anatoliy et al. [29] was performed. The parameters of the staved bearing for the purposes of verification are as follows: journal diameter R_{J1} = 0.25 m, radius clearance C = 1 mm, rotation velocity ω = 100 rpm, bearing width L = 1 m, number of circular staves N_{st} = 12. It can be seen in Figure 4 that the results calculated using the present model are in good agreement with those calculated using the model of Anatoliy et al.

3.2. *Performance Comparison between Bearings with Circular and Flat Staves*

3.2.1. Load Capacity of the Staved Bearings

In Section 3.2, to compare the mixed-lubrication performance between circular- and flat-staved bearings, the numbers of staves are all set to be 12, and the rotation speed is set to 400rpm. As illustrated in Figure 5a, the load capacity of the circular-staved bearing is significantly stronger than that of the flat-staved bearing. In the hydrodynamic lubrication stage, the flat-staved bearing generates lower load capacity than the circular one. In fact, compared with the circular-staved bearing, the hydrodynamic effect generated by the flat-staved bearing is weak. Meanwhile, after moving into the mixed lubrication stage, the contact load of the flat-staved bearing is greater than that of the circular one, as shown in Figure 5b.

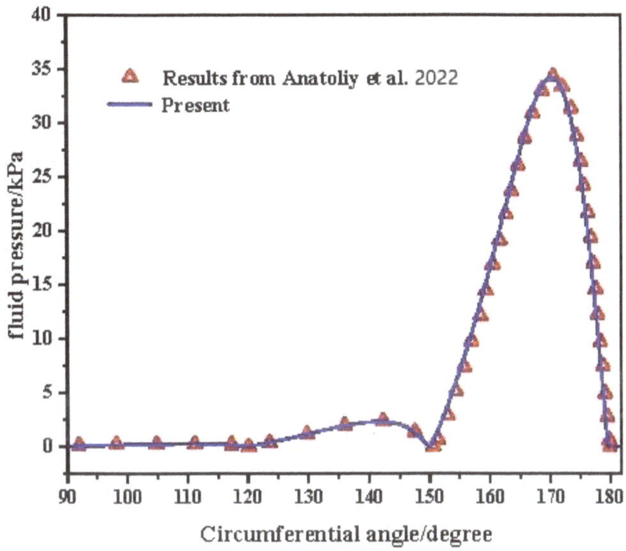

Figure 4. Verification of the fluid pressure [29].

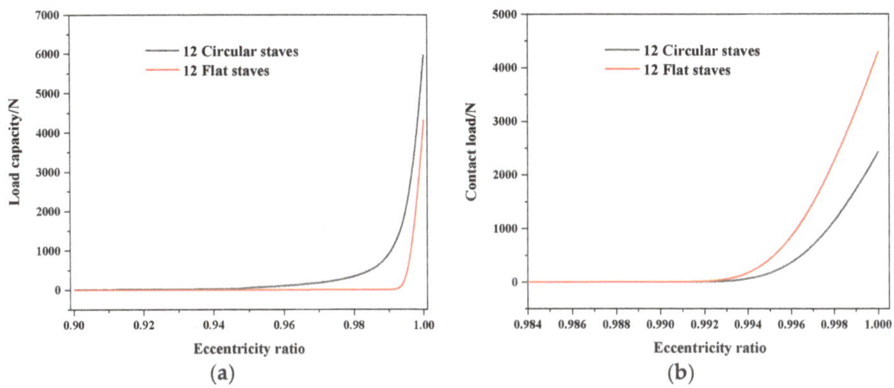

Figure 5. Comparison of the (**a**) load capacity and (**b**) contact load between circular- and flat-staved bearings with variation of eccentricity ratio at ω = 400 rpm.

3.2.2. Mixed Lubrication Distributions of Staved Bearings

In order to further investigate the difference in performance between circular- and flat-staved bearings under mixed-lubrication conditions, the fluid distributions and contact pressures under an eccentricity ratio of 0.995 and a rotation velocity of 400rpm were calculated, and the results are shown in Figure 6. It can be seen in Figure 6a,b that the circular-staved bearing is superior to the flat-staved one, regardless of whether in terms of the peak value or the distribution area of the fluid pressure. Furthermore, as shown in Figure 6c,d, although the peak values of contact pressure are equal, the contact pressure distribution area is much smaller for the circular-staved bearing than for the flat-staved bearing. Therefore, it can be summarized that, in the mixed-lubrication stage, the load capacity of the circular-staved bearing is mainly caused by fluid pressure, while in the flat-staved bearing, contact pressure plays a dominant role. In conclusion, under the current operating conditions, the mixed-lubrication performance of the flat-staved bearing is worse than that of the circular-staved one.

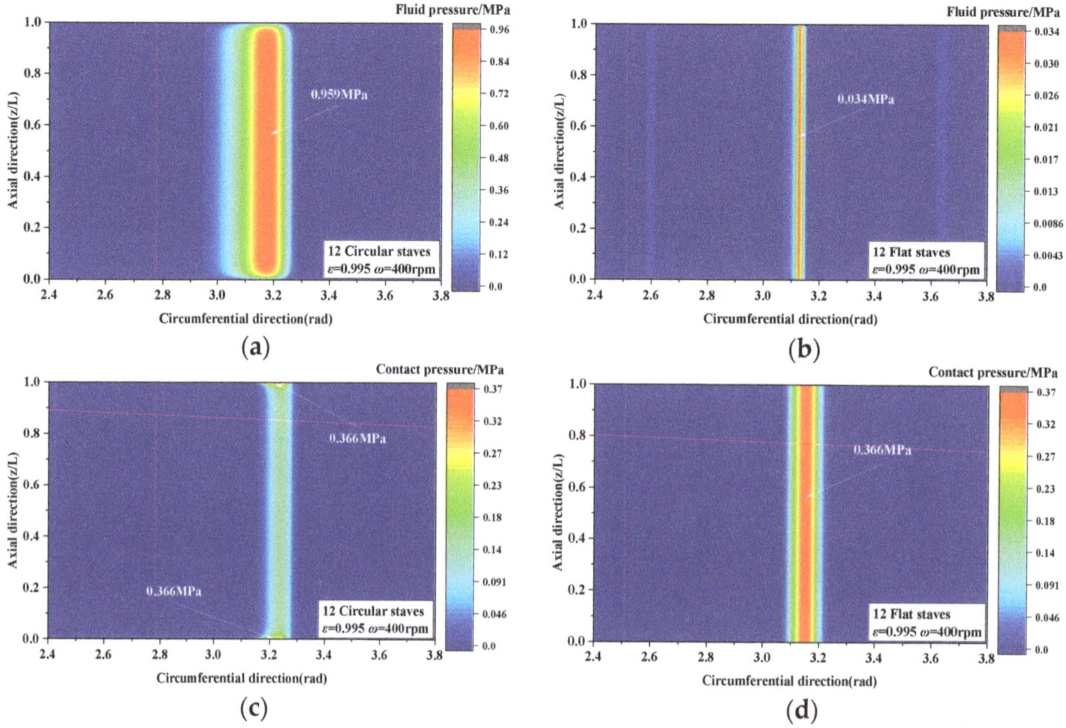

Figure 6. Mixed lubrication distributions of circular- and flat-staved bearings: (**a**,**b**) fluid pressure and (**c**,**d**) contact pressure.

3.3. *Effect of the Number of Staves on Mixed-Lubrication Performance*

3.3.1. Under Different Eccentricity Ratios

In this section, in order to investigate the effect of different numbers of staves on the mixed-lubrication performance of the circular- and flat-staved bearings and to reveal the influence mechanism, different numbers of staves from 8 to 30 were employed. As can be observed from Figure 7, with increasing numbers of staves, the load capacity of the circular-staved bearing decreases, while the contact load increases. This is because the grooves between adjacent staves break the continuity of the water film, and more staves means more grooves, thus causing a greater amount of destruction to the water film. Therefore, circular-staved bearings with smaller numbers of staves present a higher load capacity and a lower contact load.

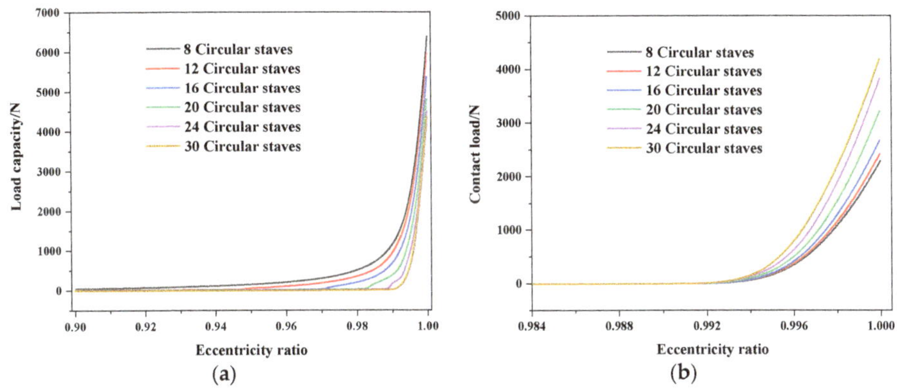

Figure 7. Variations in (**a**) load capacity and (**b**) contact load of circular-staved bearings with eccentricity ratio at ω = 400 rpm.

Figure 8 shows the variation in load capacity and contact load for flat-staved bearings with eccentricity ratio under different numbers of staves, where it can be seen that in macro terms, for load capacity and contact load, different numbers of staves make little differences for flat-staved bearings, but on a small scale, differences are still evident. As illustrated in Figure 8a, the load capacity of a bearing with 30 flat staves is lower than the other five cases at first, but with increasing eccentricity ratio, this gradually surpasses the other five cases, becoming the highest among the cases in the present study at high eccentricity ratios. As has been mentioned above, the presence of grooves breaks the continuity of the water film; this phenomenon also occurs in flat-staved bearing, similar to the case of circular-staved bearings. However, it is different from the circular-staved bearing in that every flat stave along the circumferential direction provides a wedge gap that generates fluid pressure. There is a greater number of grooves with a greater number of staves, as well as more wedge gaps, which makes the fluid pressure a trade-off between water film damage from multiple grooves and the hydrodynamic effect of multiple wedge gaps, which provides the possibility of optimizing the number of staves in order to obtain better mixed-lubrication performance of flat-staved bearings.

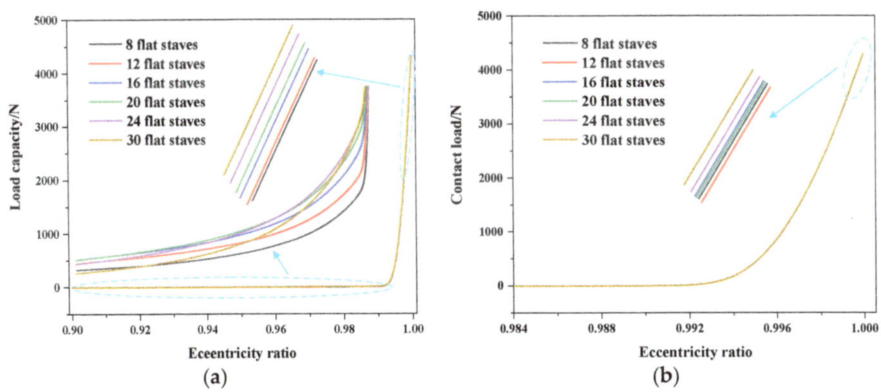

Figure 8. Variations in (**a**) load capacity and (**b**) contact load of flat-staved bearings with eccentricity ratio at ω = 400 rpm.

3.3.2. Under Different Rotation Speeds

As shown in Figures 9 and 10, with the increase in rotation speed, the load capacity rises, while the contact load decreases, for both the circular- and the flat-staved bearings. It can be seen in Figure 9 that under mixed-lubrication conditions, the bearing with fewer circular staves provides greater load capacity, and generates a lower contact load at $\varepsilon = 0.995$, indicating that it is able to achieve better mixed-lubrication performance, because of the presence of damage to the water film caused by the grooves.

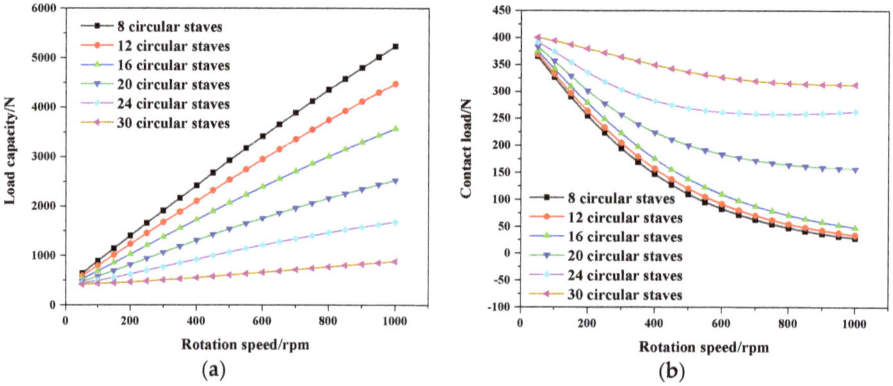

Figure 9. Variations in (**a**) load capacity and (**b**) contact load of circular-staved bearings with rotation speed at $\varepsilon = 0.995$.

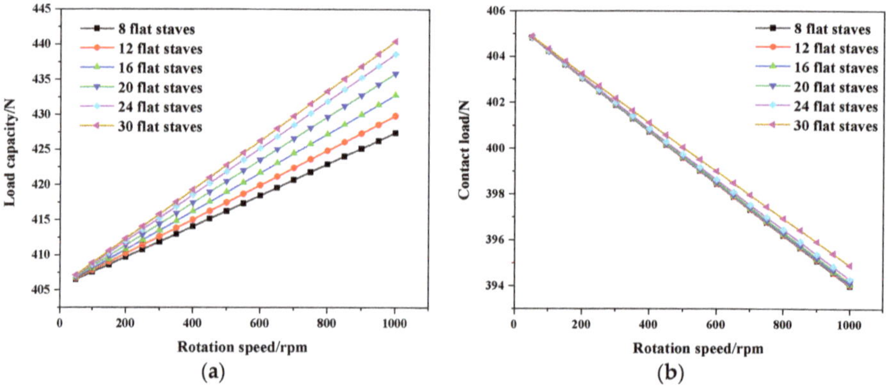

Figure 10. Variations in (**a**) load capacity and (**b**) contact load of flat-staved bearings with rotation speed at $\varepsilon = 0.995$.

As can be observed in Figure 10, under mixed-lubrication conditions, among the investigated numbers of staves, bearings with a greater number of flat staves generate greater load capacity and contact load at $\varepsilon = 0.995$. As shown in Figure 10a,b, under mixed-lubrication conditions, the load capacity of the flat-staved bearing is mainly provided by the contact load, but with increasing rotation speed, the hydrodynamic effect improves, causing the contact load ratio in the load capacity to gradually decrease. Moreover, the higher the rotation speed, the better the load capacity performance presented by the bearing with 30 flat staves.

3.3.3. Mechanism of Action of the Number of Staves on Lubrication Performance

In this section, to explain the mechanism of action by which the number of staves affects the mixed-lubrication performance of staved bearings, the eccentricity ratio and rotation speed were set to 0.995 and 400 rpm, respectively. As shown in Figure 11, the number of staves has a significant effect on the fluid pressure distribution, with this effect mainly being a result of the grooves between adjacent staves, as well as the structure of the staves. The grooves divide the continuous bearing surface into multiple discontinuous parts, making the water film discontinuous, as shown in Figure 2; thus, the fluid pressure distribution becomes discontinuous. Additionally, as illustrated in Figure 11a, with the increase in the number of staves, the width and peak value of the fluid pressure decrease. This is because a greater number of staves means a greater number of grooves, greater discontinuity in the parts, and a narrower width of individual staves. Greater numbers of grooves result in more serious breaking of the water film, and narrower staves provide a smaller wedge gap, both of which weaken the hydrodynamics. This phenomenon also occurs in flat-staved bearings, but as shown in Figure 11b, every single stave in the flat-staved bearing provides an additional fluid pressure peak, with more fluid pressure peaks counteracting the lower main fluid pressure peak value and providing a greater hydrodynamic effect. In fact, these two kinds of stave are both able to generate extra fluid pressure peaks in the convergence region of the bearing, while in the divergence region, due to the difference between the structure of circular and flat staves, only flat staves are able to generate extra fluid pressure peaks. Meanwhile, the extra fluid pressure peaks generated by circular staves are very low compared to the main pressure peak, and thus have little effect on the fluid load, while that generated by the flat staves are obvious, and have a significant effect on the fluid load of the bearing. This causes flat-staved bearings with a greater number of staves to generate higher load capacities in the current cases, as shown by the results in Figures 8a and 10a.

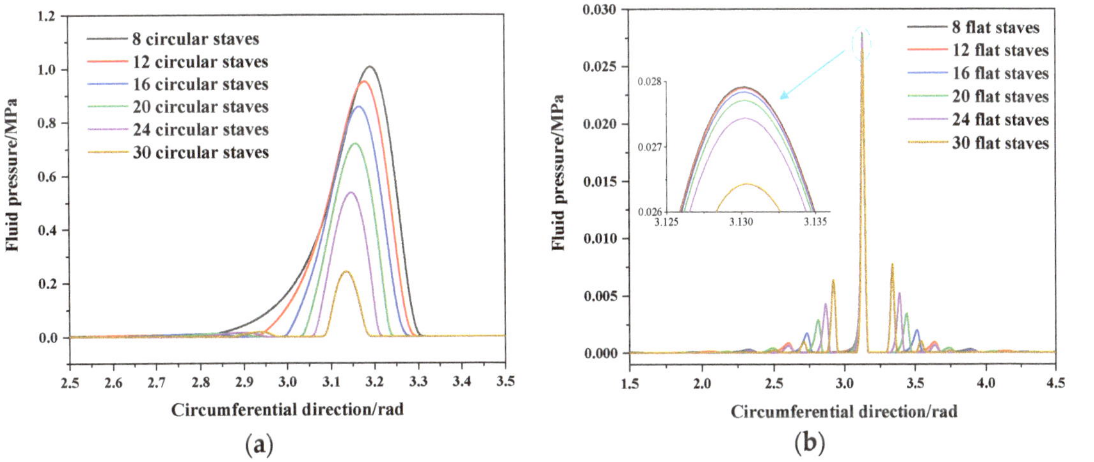

Figure 11. Central fluid pressure distribution of (**a**) circular-staved bearing and (**b**) flat-staved bearing under different number of staves.

3.4. Optimization of Number of Staves in Flat-Staved Bearings

According to the mechanism analysis results above, there may be an optimal value for the number of staves to use in order to improve the lubrication performance of flat-staved bearings. To determine the optimal number of staves, the load capacity as well as the fluid load of flat-staved bearings with an additional set of different numbers of staves were compared under different eccentricity and rotation speeds. As can be observed from Figure 12a, the flat-staved bearing with 30 staves shows better carrying capacity than the other bearings at $\varepsilon = 0.995$. This is a result of the trade-off between the pressure breakdown caused by multiple grooves and the extra pressure peaks generated by the use of multiple flat staves. As shown in Figure 12b, the fluid load significantly decreases when the number of staves exceeds 30, and when the number of staves reaches 48, almost no fluid load is generated. Meanwhile, it is noteworthy that there is a crossing point between the curve with 24 flat staves and that with 30 flat staves, indicating that the optimal number of staves is highly dependent on the operation conditions. In addition, as can be observed from

Figure 13, the flat-staved bearing with 30 flat staves presents better lubrication performance than the others at different rotation speeds, in particular at high rotation speeds. Moreover, consistent with Figure 12b, there is almost no fluid load generated with 48 flat staves at rotation speeds from 50 rpm to 1000 rpm compared with the other bearings. Hence, in the present case, 30 flat staves is the optimal number for maximizing the lubrication performance of flat-staved stern tube bearings lubricated with water.

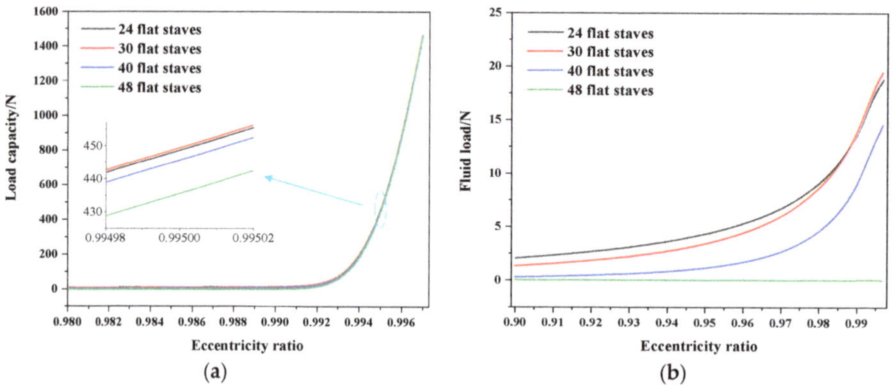

Figure 12. Variations in (**a**) load capacity and (**b**) fluid load of flat-staved bearings with eccentricity ratio at ω = 400 rpm.

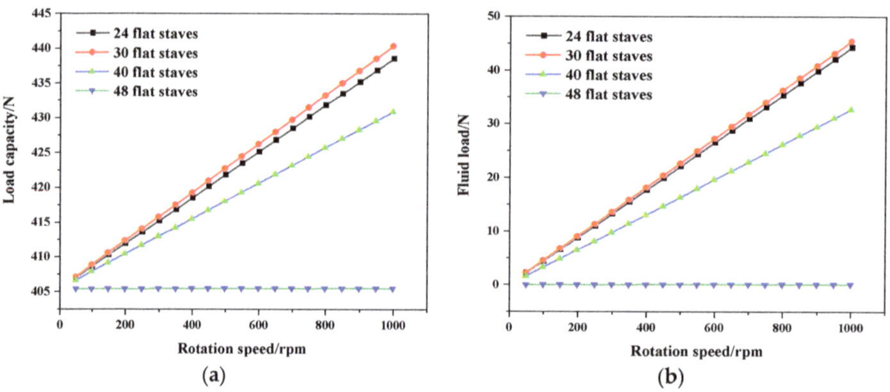

Figure 13. Variations of (**a**) load capacity and (**b**) fluid load of flat-staved bearings with rotation speed at ε = 0.995.

4. Conclusions

In this paper, a numerical model was developed to investigate the mixed-lubrication performance of staved stern tube bearings lubricated with water, and the mixed-lubrication and asperity contact behaviors were obtained by solving the average Reynolds equation and the KE model. Compared with general bearings, the particularity of staved bearings lies in their discontinuous lubrication gap. The appropriateness and the accuracy of the established model were verified by comparing the results obtained with data from the published literature. Subsequently, the effect of the number of staves was compared for circular- and flat-staved bearings using the established model. The comparative results showed that the number of staves has a significant influence on the mixed-lubrication performance of staved bearings. In addition, the mechanism of action by which number of staves affects the lubrication performance of circular- and flat-staved bearings was revealed, and the optimal number of staves in order to maximize the lubrication performance of flat-staved bearings was obtained. The conclusions are listed as follows.

(1) For circular-staved bearings, a smaller number of staves provides a better hydrodynamic effect as well as improved load capacity.

(2) Additionally, there is an optimal value, which highly depends on the operation conditions, for the number of staves in order to improve the mixed-lubrication performance of flat-staved stern tube bearings lubricated with water. In the current simulation, the optimal value was 30.

Author Contributions: D.T.: conceptualization, methodology, validation, investigation, formal analysis, writing—original draft. Y.H.: conceptualization, methodology, investigation. L.Y.: investigation, formal analysis, modification, check. Y.C.: language, modification, checking. All authors have read and agreed to the published version of the manuscript.

Funding: The author(s) disclose receipt of the following financial support for the research, authorship, and/or publication of this article: National Natural Science Foundation of China (Nos. 51975064 and 52105052).

Data Availability Statement: The data that support the results of this study are available from the authors, upon reasonable request.

Conflicts of Interest: The authors declare no conflict of interest.

Nomenclature

C	bearing clearance, mm
ε	eccentricity ratio
θ	circumferential direction
z	axial direction
ψ	attitude angle, rad
R_{J1}	inner radius of the journal bearing, mm
L	journal bearing width, mm
h_J	lubrication gap of the journal bearing, μm
N_{st}	number of staves
h_p	geometry clearance, mm
δ_J	journal bearing elastic deform, μm
p_{hJ}	hydrodynamic pressure of the Journal bearing, MPa
ρ	density of the water, kg/m^3
η	viscosity of the water, Pa·s
σ	composite roughness, μm
ϕ_θ	flow factor in the circumferential direction
ϕ_r	flow factor in the radial direction
ϕ_s	shear factor
ϕ_c	contact factor
ω	rotational speed, rad/s
p_c	contact pressure, MPa
β	asperity curvature
D	asperity density
E_J	elastic modulus of the journal bearing, GPa
E_s	elastic modulus of the shaft, GPa
ν_J	Poisson ratio of the journal bearing
ν_s	Poisson ratio of the shaft
γ	surface orientation

References

1. Xiang, G.; Han, Y.; He, T.; Wang, J.; Xiao, K. A Dynamic Wear Model for Micro-Grooved Water-Lubricated Bearings Under Transient Mixed Lubrication Condition. *J. Tribol.* **2020**, *142*, 7101. [CrossRef]
2. Prajapati, D.K.; Ramkumar, P. Surface topography effect on tribological performance of water-lubricated journal bearing under mixed-EHL regime. *Surf. Topogr. Metrol. Prop.* **2022**, *10*, 45022. [CrossRef]

3. He, T.; Zou, D.; Lu, X.; Guo, Y.; Wang, Z.; Li, W. Mixed-lubrication analysis of marine stern tube bearing considering bending deformation of stern shaft and cavitation. *Tribol. Int.* **2014**, *73*, 108–116. [CrossRef]
4. Lv, F.; Rao, Z.; Ta, N.; Jiao, C. Mixed-lubrication analysis of thin polymer film overplayed metallic marine stern bearing considering wall slip and journal misalignment. *Tribol. Int.* **2017**, *109*, 390–397. [CrossRef]
5. Patir, N.; Cheng, h.S. An Average Flow Model for Determining Effects of Three-Dimensional Roughness on Partial Hydrodynamic Lubrication. *J. Tribol. Trans. ASME* **1978**, *100*, 12–17. [CrossRef]
6. Yang, H.; Zhang, Y.; Lu, L. Numerical investigation of after stern tube bearing during ship turning maneuver. *J. Mar. Sci. Technol.* **2019**, *25*, 707–717. [CrossRef]
7. Huang, Q.; Yan, X. Impact factors on lubricant performance of stern bearing with misalignment angle induced by transverse vibration of shaft. *Ocean Eng.* **2020**, *216*, 108051. [CrossRef]
8. Rossopoulos, G.N.; Papadopoulos, C.I.; Leontopoulos, C. Tribological comparison of an optimum single and double slope design of the stern tube bearing, case study for a marine vessel. *Tribol. Int.* **2020**, *150*, 106343. [CrossRef]
9. Xie, Z.; Zhu, W. Theoretical and experimental exploration on the micro asperity contact load ratios and lubrication regimes transition for water-lubricated stern tube bearing. *Tribol. Int.* **2021**, *164*, 107105. [CrossRef]
10. Liu, Q.; Ouyang, W.; Cheng, Q.; Li, J.; Cheng, Q.; Li, R. Influences of bidirectional shaft inclination on lubrication and dynamic characteristics of the water-lubricated stern bearing. *Mech. Syst. Signal Process.* **2021**, *169*, 108623. [CrossRef]
11. He, T.; Xie, Z.; Ke, Z.; Dai, L.; Liu, Y.; Ma, C.; Jiao, J. Theoretical Study on the Dynamic Characteristics of Marine Stern Bearing Considering Cavitation and Bending Deformation Effects of the Shaft. *Lubricants* **2022**, *10*, 242. [CrossRef]
12. Zhu, J.; Wei, G.; Peng, Z.; Xia, Z.; Zheng, L.; Zhu., h. Analysis of underwater explosion shock on ship shaft stern bearing lubrication characteristics under different bearing working conditions. *Int. J. Nav. Arch. Ocean Eng.* **2022**, *14*, 100444. [CrossRef]
13. Wodtke, M.; Litwin, W. Water-lubricated stern tube bearing - experimental and theoretical investigations of thermal effects. *Tribol. Int.* **2020**, *153*, 106608. [CrossRef]
14. Zhang, H.; Yuan, C.; Tan, Z. A novel approach to investigate temperature field evolution of water lubricated stern bearings (WLSBs) under hydrodynamic lubrication. *Adv. Mech. Eng.* **2021**, *13*, 1–15. [CrossRef]
15. Zhou, G.; Wu, K.; Pu, W.; Li, P.; Han, Y. Tribological modification of hydrogenated nitrile rubber nanocomposites for water-lubricated bearing of ship stern shaft. *Wear* **2022**, *504*, 204432. [CrossRef]
16. Han, H.S.; Lee, K.H. Experimental verification of the mechanism on stick-slip nonlinear friction induced vibration and its evaluation method in water-lubricated stern tube bearing. *Ocean Eng.* **2019**, *182*, 147–161. [CrossRef]
17. Lin, C.-G.; Zou, M.-S.; Sima, C.; Liu, S.-X.; Jiang, L.-W. Friction-induced vibration and noise of marine stern tube bearings considering perturbations of the stochastic rough surface. *Tribol. Int.* **2018**, *131*, 661–671. [CrossRef]
18. Hongling, Q.; Chang, Y.; Hefa, Z.; Xufei, L.; Zhixiong, L.; Xiang, X. Experimental analysis on friction-induced vibration of water-lubricated bearings in a submarine propulsion system. *Ocean Eng.* **2020**, *203*, 107239. [CrossRef]
19. Jin, Y.; Liu, Z.; Zhou, X. Theoretical, numerical, and experimental studies on friction vibration of marine water-lubricated bearing coupled with lateral vibration. *J. Mar. Sci. Technol.* **2019**, *25*, 298–311. [CrossRef]
20. Jin, Y.; Deng, T.; Liu, Z.; Zhou, J. Research on the influence of the normal vibration on the friction-induced vibration of the water-lubricated stern bearing. *J. Vibroengineering* **2020**, *22*, 762–772. [CrossRef]
21. Lin, C.-G.; Zou, M.-S.; Zhang, H.C.; Qi, L.-B.; Liu, S.-X. Influence of different parameters on nonlinear friction-induced vibration characteristics of water lubricated stern bearings. *Int. J. Nav. Arch. Ocean Eng.* **2021**, *13*, 746–757. [CrossRef]
22. Ning, C.; Hu, F.; Ouyang, W.; Yan, X.; Xu, D. Wear monitoring method of water-lubricated polymer thrust bearing based on ultrasonic reflection coefficient amplitude spectrum. *Friction* **2022**, *11*, 685–703. [CrossRef]
23. Ouyang, W.; Liu, Q.; Xiao, J.; Huang, J.; Zhang, Z.; Wang, L. Experimental study on the distributed lubrication characteristics of full-size water-lubricated stern bearings under hull deformation. *Ocean Eng.* **2023**, *267*, 113226. [CrossRef]
24. Cai, J.; Han, Y.; Xiang, G.; Wang, J.; Wang, L. Effects of wear and shaft-shape error defects on the tribo-dynamic response of water-lubricated bearings under propeller disturbance. *Phys. Fluids* **2022**, *34*, 077118. [CrossRef]
25. Cai, J.; Han, Y.; Xiang, G.; Wang, C.; Wang, L.; Chen, S. Influence of the mass conservation cavitation boundary on the tribo-dynamic responses of the micro-groove water-lubricated bearing. *Surf. Topogr. Metrol. Prop.* **2022**, *10*, 045011. [CrossRef]
26. Cai, J.; Xiang, G.; Li, S.; Guo, J.; Wang, J.; Chen, S.; Yang, T. Mathematical modeling for nonlinear dynamic mixed friction behaviors of novel coupled bearing lubricated with low-viscosity fluid. *Phys. Fluids* **2022**, *34*, 093612. [CrossRef]
27. Xiang, G.; Yang, T.; Guo, J.; Wang, J.; Liu, B.; Chen, S. Optimization transient wear and contact performances of water-lubricated bearings under fluid-solid-thermal coupling condition using profile modification. *Wear* **2022**, *502–503*, 204379. [CrossRef]
28. Chen, S.; Xiang, G.; Fillon, M.; Guo, J.; Wang, J.; Cai, J. On the tribo-dynamic behaviors during start-up of water lubricated bearing considering imperfect journal. *Tribol. Int.* **2022**, *174*, 107685. [CrossRef]
29. Serdjuchenko, A.; Ursolov, A.; Batrak, Y. Asymptotic estimation of the fluid film pressure in non-metallic water-lubricated staved stern tube bearings. *Tribol. Int.* **2022**, *175*, 107798. [CrossRef]
30. Kogut, L.; Etsion, I. A Finite Element Based Elastic-Plastic Model for the Contact of Rough Surfaces. *Tribol. Trans.* **2003**, *46*, 383–390. [CrossRef]

31. Kogut, L.; Etsion, I. A Static Friction Model for Elastic-Plastic Contacting Rough Surfaces. *J. Tribol.* **2004**, *126*, 34–40. [CrossRef]
32. Beheshti, A.; Khonsari, M. An engineering approach for the prediction of wear in mixed lubricated contacts. *Wear* **2013**, *308*, 121–131. [CrossRef]

lubricants

MDPI

Article

A Fluid-Structure Interaction Method for the Elastohydrodynamic Lubrication Characteristics of Rubber-Plastic Double-Layer Water-Lubricated Journal Bearings

Yuankang Shen [1,2], Yao Zhang [1], Xiuli Zhang [1,2,*], Hongyu Zheng [1], Guorui Wei [1] and Mingyang Wang [1]

1 School of Mechanical Engineering, Shandong University of Technology, Zibo 255000, China; shenyuankang1@163.com (Y.S.); yaozhang0511@163.com (Y.Z.); zhenghongyu@sdut.edu.cn (H.Z.); weiguorui5@163.com (G.W.); 13176266935@163.com (M.W.)

2 Shandong Provincial Key Laboratory of Precision Manufacturing and Non-traditional Machining, Shandong University of Technology, Zibo 255000, China

* Correspondence: zhangxiulli@163.com

Abstract: This paper proposes a fluid-structure interaction (FSI) numerical calculation method for investigation of the elastohydrodynamic lubrication performance of the rubber-plastic double-layer water-lubricated journal bearings. The accuracy and rapidity of the FSI method are improved by studying the effect of mesh density and by comparing the calculation results with those in the literature. Based on the proposed method, a series of numerical simulations are carried out to reveal the influence of operating conditions and structural parameters on the lubrication performance of the rubber-plastic bearings. Numerical results show that the bush deformation of the rubber-plastic bearing is between that of the rubber bearing and the plastic bearing, and the deformation area is close to that of the rubber bearing. The bearing load carrying capacity increases significantly with the rotational speed, eccentricity ratio, bearing length, and decrease with the clearance. But the influences of the plastic layer elastic modulus and thickness on bearing load are unremarkable. The effect of bush deformation on bearing load is noticeable when the eccentricity ratio is more than 0.8. The results are expected to provide design references for the bearings.

Keywords: rubber-plastic double-layer bush; water lubrication; journal bearing; elastohydrodynamic lubrication; fluid–structure interaction

Citation: Shen, Y.; Zhang, Y.; Zhang, X.; Zheng, H.; Wei, G.; Wang, M. A Fluid-Structure Interaction Method for the Elastohydrodynamic Lubrication Characteristics of Rubber-Plastic Double-Layer Water-Lubricated Journal Bearings. *Lubricants* **2023**, *11*, 240. https://doi.org/10.3390/lubricants11060240

Received: 8 May 2023
Revised: 24 May 2023
Accepted: 26 May 2023
Published: 28 May 2023

1. Introduction

With the awareness of environmental protection becoming increasingly prominent, water-lubricated bearings are widely employed in huge ships and submarines because of their high reliability, easy maintenance and environmental protection. As the crucial component of the transmission system of marine vehicles, the performances of bearings directly impact the security and reliability of working ships [1,2]. Due to the rapid development of modern science and technology in recent years, both civil and military ships have put forward higher requirements for the performance of water-lubricated bearings to enhance service life and reduce vibration and noise during ship operation [3]. The lubrication performance of water-lubricated bearings is significantly influenced by their material composition and structural design. Therefore, it is crucial to study the lubrication performance and lubrication mechanism of water-lubricated bearings of different materials and structures [4,5].

Currently, the bush materials are mainly rubber, plastic and other elastic materials [6]. Rubber bush has high elasticity and good damping capacity; thus, it is widely used in stern bearings. However, rubber bearings have large starting torque and high friction noise at low speeds and heavy loads. Plastic bush has a low friction coefficient but poor compliance and damping performance. To achieve better bearing performance, rubber and plastic can be superposed to form rubber-plastic double-layer bush water-lubricated bearings. Yamajo et al. [7] studied the application characteristics of poly tetra fluoroethylene (PTFE) three-layer composite bearing and found that the bearing has a smaller friction coefficient, better wear resistance and longer service life than rubber bearing. Litwin [8,9] studied the lubrication characteristics of a three-layer PTFE-nitrile rubber (NBR)-bronze water-lubricated bearing with lubricating grooves by experiments. The results showed that the bearing works under a hydrodynamic lubrication state for most working conditions, and its starting and running resistance is less than those of the NBR bearing. Xie et al. [3] designed a double-liner bearing structure. The results show that the acoustic performances are closely related to the material combinations. Although studies have shown that rubber-plastic double-layer bush water-lubricated bearings have superior performance, their load-carrying mechanism and structural design method are not clear, which restricts their application.

At present, scholars have carried out extensive research on the lubrication and load-carrying mechanism of water-lubricated rubber bearings and plastic bearings. Shi et al. [10] studied the dynamic characteristics of water-lubricated rubber bearings and found that the eccentricity ratio, length-to-diameter ratio, radial clearance and rotational speed have a great influence on the bearing dynamic performance. Liu et al. [11] built a test rig to study the lubrication characteristics of water-lubricated rubber bearings at high rotating speeds. The experimental results showed that, different from the laminar flow, the bearing load increases faster as the rotating speed increases, indicating a turbulent flow at high rotating speeds. Qiao et al. [12] proposed a mixed lubrication model considering the effect of turbulence and studied the effects of axial misalignment, load, rotational speed and radial clearance on the lubrication performance of water-lubricated rubber bearings. Kuznetsov et al. [13] studied the effect of PTFE lining compliance on the bearing characteristics based on a thermohydrodynamic (THD) model including liner deformation, and the analysis showed increased load carrying capacity, significantly reduced peak pressure and thicker oil film in the loaded zone compared to a white metal bearing. Wang et al. [14] analyzed the lubrication performance of PTFE-based water-lubricated bearings using computational fluid dynamics (CFD) and fluid–structure interaction (FSI) method and studied the effects of elastic modulus and Poisson's ratio on bearing elastic deformation and load-carrying capacity. Yang et al. [15] analyzed the lubrication performance of UHMWPE water-lubricated bearings. Cheng et al. [16] studied the tribological properties of PEEK under seawater lubrication.

The above studies showed that when the elastic deformation of the bearing bush is close to the minimum water film thickness, the bush deformation must be considered in the calculation. Currently, the literature mainly adopts two methods to study elastohydrodynamic lubrication problems, namely the CFD-FSI method [17,18] and the programming method [19]. The CFD-FSI method solves 3-dimensional fluid and solid models. It is intuitive and convenient to post-processing, but it is prone to failure when the deformation is large and causes mesh distortion [20]. The programming method solves the fluid model based on the two-dimensional Reynolds equation and calculates the elastic deformation by approximate formula or the finite element method (FEM); thus, there is no mesh distortion, and it can calculate large deformation conditions. The Reynolds equation is usually solved by the finite difference method (FDM), and good results can be obtained for circular plain bearings. The FEM can also solve the Reynolds equation, but its program structure is more complex, and the calculation time is longer. To solve the bush elastic deformation, some researchers used approximate formulas. For example, Thomsen and Klit [21] calculated the deformation of the polymer liner using the Winkler/Column model. Xiang et al. [22–24] and Gong et al. [25] used an influence coefficient method to determine the elastic deformation. Du et al. [26] used an approximate formula to consider bush deformation. Other literature used the FEM to calculate the bearing deformation [27–33]. Relative to the approximate formula, the FEM can describe the elastic deformation more accurately and has higher adaptability. Therefore, in this work, the FDM is used to solve the Reynolds equation, and the FEM is used to calculate the elastic deformation, which is a tradeoff of accuracy, applicability and computing efficiency.

In summary, scholars have conducted extensive research on the lubrication mechanisms, frictional behavior and structural design of water-lubricated rubber bearings and plastic bearings. Rubber-plastic double-layer water-lubricated bearings may achieve better performance than rubber bearings and plastic bearings. However, the research on this type of bearing is limited, and the FSI lubrication characteristics of the bearings are not very clear. The previous work [20] intended to reveal the bearing performance by the CFD-FSI method, while the results only showed the lubrication performance for small eccentricity ratios, and mesh distortion occurs and causes calculation failure for large eccentricity ratios.

In order to reveal the elastohydrodynamic lubrication characteristics of the rubber-plastic double-layer water-lubricated journal bearings and obtain bearing performance under various operating conditions, especially heavy load conditions, a two-way FSI numerical calculation method is proposed and established by solving the water film pressure based on the Reynolds equation by the FDM and solving the bush deformation by the FEM. The validity of the numerical calculation method is verified by comparing the results with the published literature results. The water film pressure and bush deformation of the bearing are compared with those of the rubber bearing and the plastic bearing. The load-carrying performance of the bearings for different eccentricity ratios, rotational speed, length-to-diameter ratio, clearance ratio and elastic modulus and thickness of the plastic layer bush are investigated. The results will provide theoretical guidance for the structural design and application of this type of bearings.

2. Geometry

The rubber-plastic double-layer water-lubricated journal bearing geometry is shown in Figure 1. O_b is the bearing center; O_j is the journal center; e is the eccentricity between the bearing and journal centers; ϕ is the attitude angle; N is the rotational speed of the journal; h_{min} is the minimum film thickness; D is the bearing inner diameter; L is the bearing length; t_1 is the thickness of the plastic layer bush; and t_2 is the thickness of the rubber layer bush.

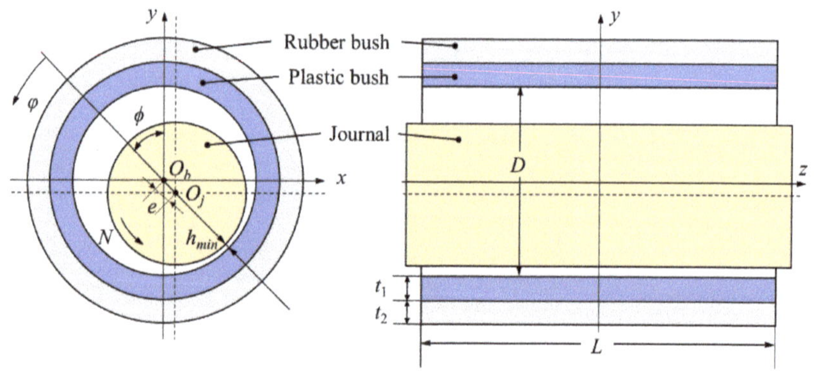

Figure 1. Rubber-plastic double-layer water-lubricated journal bearing geometry.

3. Numerical Analysis

3.1. Water Film Fluid Model

As the water viscosity is small, the temperature rise of water-lubricated bearing is low. The water film flow is assumed to be isothermal, steady and incompressible. The Reynolds equation can be written as [34]

$$\frac{1}{R^2}\frac{\partial}{\partial\varphi}\left(\frac{h^3}{12\mu}\frac{\partial p}{\partial\varphi}\right)+\frac{\partial}{\partial z}\left(\frac{h^3}{12\mu}\frac{\partial p}{\partial z}\right)=\frac{U}{2R}\frac{\partial h}{\partial\varphi} \tag{1}$$

where R ($R = D/2$) is the inner radius of the bearing bush; φ is the angular coordinate; h is the water film thickness; μ is the fluid dynamic viscosity; p is the hydrodynamic pressure; z is the axial coordinate; and U is the journal surface velocity.

Considering the elastic deformation of the bearing bush, the water film thickness h is expressed as

$$h = c(1 + \varepsilon \cos \varphi) + \delta \tag{2}$$

where c is the radial clearance, ε is the eccentricity ratio ($\varepsilon = e/c$), and δ indicates the radial elastic deformation of the bearing bush.

Reynolds boundary conditions [35] are used:

$$\begin{cases} p(\varphi, z = -\frac{L}{2}) = p(\varphi, z = +\frac{L}{2}) = 0 \\ p(\varphi_0, z) = 0 \\ \dfrac{\partial p(\varphi_0, z)}{\partial \varphi} = 0 \end{cases} \tag{3}$$

where φ_0 represents the water film rupture position.

The finite difference method is used to solve Equation (1). The mesh used for the calculations is set as $m \times n$ for the circumferential and axial directions, respectively. The influence of the mesh density is studied in Section 3.3 for accuracy and rapidity of the calculations. The super relaxation iteration (SOR) method [36] is used when calculating the pressure values, and all convergence residuals are taken as 10^{-4}.

The water film load-carrying capacity W in the eccentric direction and its vertical direction can be calculated by integrating the pressure over the journal surface:

$$W_e = \int_{-L/2}^{L/2} \int_0^{2\pi} p \cos \varphi R d\varphi dz \tag{4}$$

$$W_\varphi = \int_{-L/2}^{L/2} \int_0^{2\pi} p \sin \varphi R d\varphi dz \tag{5}$$

The water film load-carrying capacity is

$$W = \sqrt{W_e^2 + W_\varphi^2} \tag{6}$$

The attitude angle is

$$\phi = -\arctan \frac{W_\varphi}{W_e} \tag{7}$$

The friction torque M_f of the rotating journal is a composition of the shear flow friction torque M_{f1} and the pressure flow friction torque M_{f2}. The shear flow friction torque M_{f1} is the sum of the shear flow resistance of the water film intact area and ruptured are as follows.

$$M_{f1} = \frac{1}{2} \int_{-L/2}^{L/2} \int_0^{\varphi_b} \left(\frac{\mu U}{h} \right) R^2 d\varphi dz + \frac{1}{2} \int_{-L/2}^{L/2} \int_{\varphi_b}^{2\pi} \frac{\mu U h_b}{h^2} R^2 d\varphi dz \tag{8}$$

where φ_b is the angular position of the ruptured edge, and h_b is the film thickness on the edge.

The pressure flow friction torque M_{f2} is

$$M_{f2} = \frac{e \sin \phi}{2} W \tag{9}$$

The total friction torque M_f is

$$M_f = M_{f1} + M_{f2} \tag{10}$$

3.2. Bearing Bush Solid Model

The finite element method [37] is used to calculate the elastic deformation of the bearing bush. The finite element model of the rubber-plastic double-layer bearing bush is shown in Figure 2. The upper layer is the plastic bush and the lower layer is the rubber bush. The meshing method in the circumferential and axial directions is consistent with that of the fluid model. Eight-nodes hexahedron elements are used. Each node has three degrees of freedom, and the displacement field of the element can be written as

$$\underset{(3\times1)}{\boldsymbol{u}}(x,y,z)=\begin{bmatrix}u\\v\\w\end{bmatrix}=\begin{bmatrix}N_1&0&0&\vdots&N_2&0&0&\vdots&\ldots&\vdots&N_8&0&0\\0&N_1&0&\vdots&0&N_2&0&\vdots&\ldots&\vdots&0&N_8&0\\0&0&N_1&\vdots&0&0&N_2&\vdots&\ldots&\vdots&0&0&N_8\end{bmatrix}\cdot\boldsymbol{q}^e=\underset{(3\times24)}{\boldsymbol{N}}\cdot\underset{(24\times1)}{\boldsymbol{q}^e} \tag{11}$$

where $\boldsymbol{N}$ is the shape function matrix, and $\boldsymbol{q}^e$ is the node displacement array, $\boldsymbol{q}^e = [u_1\ v_1\ w_1\ u_2\ v_2\ w_2\ \ldots\ u_8\ v_8\ w_8]^T$.

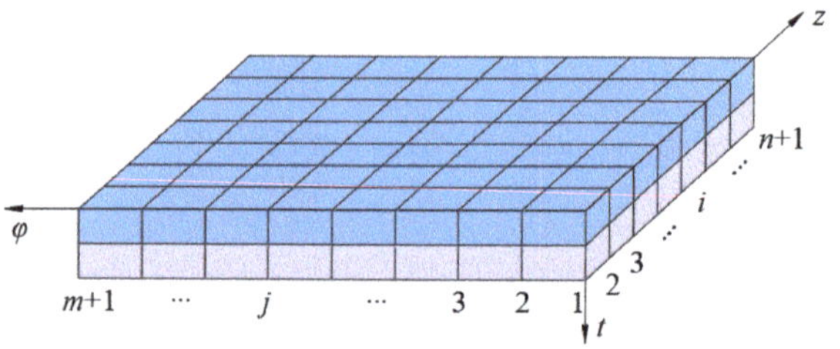

Figure 2. Finite element model of the rubber-plastic double-layer bearing bush.

According to the geometric equations of three-dimensional problem of elasticity mechanics, the strain field of the element can be expressed as

$$\underset{(6\times1)}{\varepsilon}(x,y,z)=\begin{bmatrix}\varepsilon_{xx}\\\varepsilon_{yy}\\\varepsilon_{zz}\\\gamma_{xy}\\\gamma_{yz}\\\gamma_{zx}\end{bmatrix}=\begin{bmatrix}\frac{\partial}{\partial x}&0&0\\0&\frac{\partial}{\partial y}&0\\0&0&\frac{\partial}{\partial z}\\\frac{\partial}{\partial y}&\frac{\partial}{\partial x}&0\\0&\frac{\partial}{\partial z}&\frac{\partial}{\partial y}\\\frac{\partial}{\partial z}&0&\frac{\partial}{\partial x}\end{bmatrix}\begin{bmatrix}u\\v\\w\end{bmatrix}=\underset{(6\times3)}{[\partial]}\underset{(3\times1)}{\boldsymbol{u}}=\underset{(6\times3)}{[\partial]}\underset{(3\times24)}{\boldsymbol{N}}\underset{(24\times1)}{\boldsymbol{q}^e}=\underset{(6\times24)}{\boldsymbol{B}}\underset{(24\times1)}{\boldsymbol{q}^e} \tag{12}$$

where $\boldsymbol{B} = [\partial]\boldsymbol{N}$.

As the elastic modulus and Poisson's ratio of the rubber bush and plastic bush are different, the stress field functions of the two-layer elements are different. Based on the linear elastic constitutive equation, the stress field function of the element in layer i (i = 1,2) can be expressed as

$$\underset{(6\times1)}{\sigma_i}=\underset{(6\times6)}{\boldsymbol{D}_i}\underset{(6\times1)}{\varepsilon} \tag{13}$$

where D_i is the elastic coefficient matrix of the element in layer i (i = 1,2), and

$$\boldsymbol{D}_i=\frac{E_i}{(1+v_i)(1-2v_i)}\begin{bmatrix}1-v_i&v_i&v_i&0&0&0\\v_i&1-v_i&v_i&0&0&0\\v_i&v_i&1-v_i&0&0&0\\0&0&0&\frac{1-2v_i}{2}&0&0\\0&0&0&0&\frac{1-2v_i}{2}&0\\0&0&0&0&0&\frac{1-2v_i}{2}\end{bmatrix}(i=1,2) \tag{14}$$

where v_i is the Poisson's ratio in layer i, and E_i is the elastic modulus in layer i.

The stiffness matrix of the element in layer i (i = 1,2) is

$$\underset{(24\times 24)}{\boldsymbol{K}_i^e} = \int_{\Omega^e} \underset{(24\times 6)}{\boldsymbol{B}^T} \underset{(6\times 6)}{\boldsymbol{D}_i} \underset{(6\times 24)}{\boldsymbol{B}} \, d\Omega \tag{15}$$

The equivalent node load matrix of the element is

$$\underset{(24\times 1)}{\boldsymbol{P}^e} = \int_{\Omega^e} \underset{(24\times 3)}{\boldsymbol{N}^T} \underset{(3\times 1)}{\overline{\boldsymbol{b}}} \, d\Omega + \int_{S_p^e} \underset{(24\times 3)}{\boldsymbol{N}^T} \underset{(3\times 1)}{\overline{\boldsymbol{p}}} \, dA \tag{16}$$

The element stiffness equation is

$$\underset{(24\times 24)}{\boldsymbol{K}^e} \underset{(24\times 1)}{\boldsymbol{q}^e} = \underset{(24\times 1)}{\boldsymbol{P}^e} \tag{17}$$

To solve the stiffness equation, it is necessary to introduce the boundary conditions according to the actual situation. The upper surface of the plastic layer is in direct contact with the water film, and the nodes are subjected to the hydrodynamic pressure. The lower surface of the rubber layer is in contact with the rigid bearing pedestal, and the node displacements are set as zero. The left and right sections of the bearing bush model are the same section, so the node displacements are the same. The two layers of the elastic bush fit tightly, so the node displacements are also the same.

3.3. Fluid–Solid Coupling Calculation

Figure 3 shows the flow chart of the fluid–solid coupling numerical calculation. Firstly, the initial water film thickness distribution is calculated according to the input parameters, and the steady-state Reynolds equation is solved using FDM and SOR algorithm to obtain the water film pressure distribution. Then, the elastic deformation distribution of the bearing bush is calculated using FEM, and the water film thickness distribution is updated by including the calculated deformation. The water film pressure distribution and the bush deformation distribution are recalculated until the convergence condition is reached. Finally, the load-carrying capacity, the attitude angle and the friction torque are calculated and output.

To obtain more accurate calculation results, the influence of circumferential and axial grid number, i.e., m and n, on the maximum hydrodynamic pressure and calculation time were studied. The results are shown in Figure 4. The bearing parameters and operating conditions are shown in Table 1. The circumferential grid number was set as 20 for Figure 4a, and the axial grid number was set as 160 for Figure 4b. The results showed that the axial grid number n has a large effect on the calculation result, and $p_{\max}$ approaches a constant when n is more than 160. The circumferential grid number has a small effect on $p_{\max}$, and the maximum difference is 5.25%. However, the calculation time increases exponentially as the grid number increases. Therefore, for the bearing with L/D = 2, the axial grid number 160 and the circumferential grid number 40 were used in this paper. To ensure the grid aspect ratio unchanged for models with different L/D, the circumferential grid number is always taken as 40, and the axial grid number is determined according to L/D in equal proportion.

To verify the accuracy of the calculation method, the water film pressure distribution and elastic deformation distribution were compared with the calculation results of Litwin [38] as shown in Figure 5. The bearing parameters are D = 100 mm, L = 200 mm, $t_1 = t_2$ = 6 mm, c = 0.15 mm, N = 660 r/min and $E_1 = E_2$ = 800 MPa. Figure 5 shows that the results of this work are in good agreement with Litwin's calculation results. The difference in the maximum water film pressure is 8.5%, and the difference in the maximum elastic deformation of the bearing bush is 2.1%. Since the mesh density we used is different from that used by Litwin, we believe that the reason for the difference may be attributed to the differences in calculation conditions and mesh density.

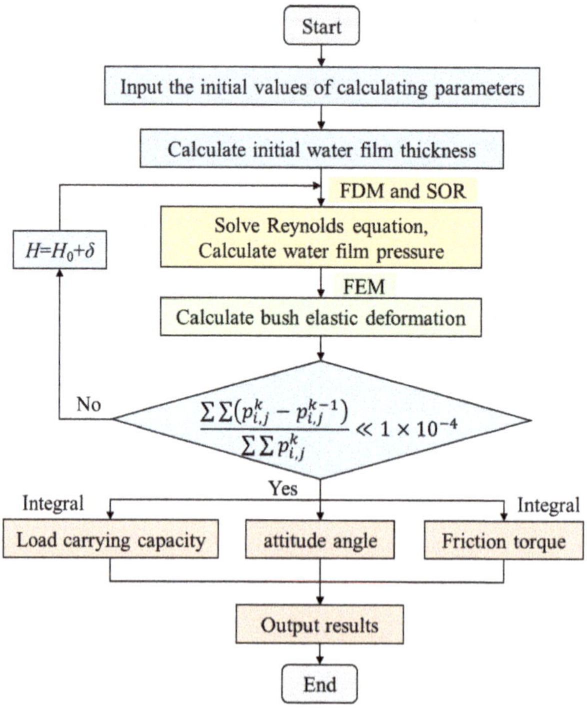

Figure 3. Flow chart of fluid–solid coupling calculation.

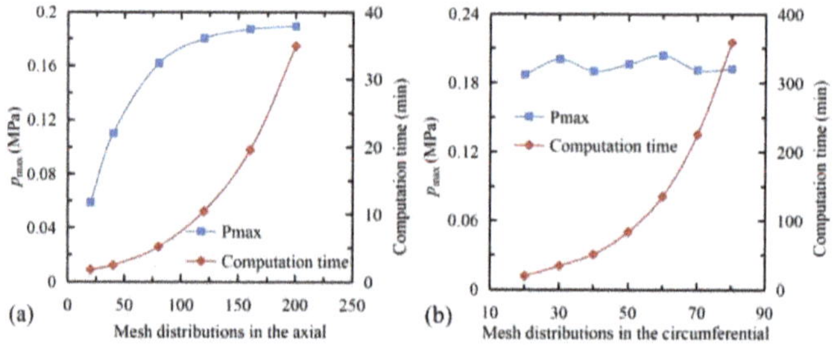

Figure 4. Influence of (**a**) axial grid number, (**b**) circumferential grid number on the result and computation time.

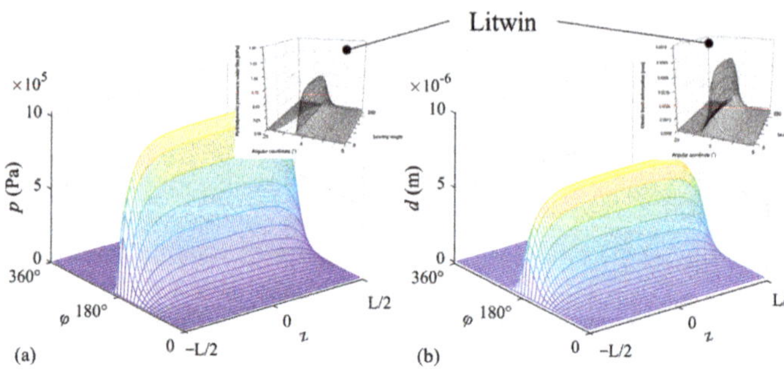

Figure 5. Comparison of (**a**) pressure distribution and (**b**) elastic deformation distribution with the results of Litwin [38].

Table 1. Bearing parameters and operating conditions.

Parameter	Symbol	Value
Bearing diameter (mm)	D	100
Relative clearance (%)	ψ	0.4
Length-to-diameter ratio	L/D	2
Total bush thickness (mm)	t	10
Plastic bush thickness (mm)	t_1	5
Rubber bush thickness (mm)	t_2	5
Plastic bush elastic modulus (MPa)	E_1	700
Rubber bush elastic modulus (MPa)	E_2	40
Plastic bush Poisson's ratio	ν_1	0.46
Rubber bush Poisson's ratio	ν_2	0.497
Eccentricity ratio	ε	0.9
Journal rotational speed (r/min)	N	1000
Dynamic viscosity (Pa·s)	μ	0.001

4. Results and Discussion

4.1. Performance Comparison

Figures 6 and 7 compare the water film pressure distribution, bush deformation distribution and water film thickness distribution of water-lubricated rubber bearings, rubber-plastic double-layer bearings and plastic bearings. The bearing parameters are D = 100 mm, L/D = 2, ψ = 0.2%, t = 10 mm, ε = 0.9, N = 1500 r/min and $t_1 = t_2$ = 5 mm for rubber-plastic double-layer bearing. The bush material properties are as shown in Table 1. The results showed that the maximum water film pressure of the three kinds of bearings differ greatly, and the values are 0.46 MPa, 0.67 MPa and 1.18 MPa, respectively. Compared with the rubber bearing, the pressure distributions of the rubber-plastic double-layer bearing and the plastic bearing are more concentrated. The bush deformation distribution is similar to the pressure distribution, and the maximum bush deformation values are 17.8 μm, 9.96 μm and 2.47 μm for the three kinds of bearings, respectively. Although the maximum pressure of the plastic bearing is the largest, its bush deformation is the smallest. The bush deformation of the rubber-plastic double-layer bearing is between those of the rubber bearing and the plastic bearing, and its distribution area is close to that of the rubber bearing. However, for φ = 220°~280°, the elastic deformation of the rubber-plastic double-layer bearing is larger than that of the rubber bearing. This is because the plastic layer bush is stiffer than the rubber layer, and the shape change of the rubber-plastic double-layer bearing is smaller than that of the rubber bearing. The bush deformation increases the minimum film thickness. The film thickness distribution of the three kinds of bearings is the same in the low-pressure region. In the high-pressure region, the water film thickness of the rubber bearing is the largest, followed by the rubber-plastic double-layer bearing and the plastic bearing.

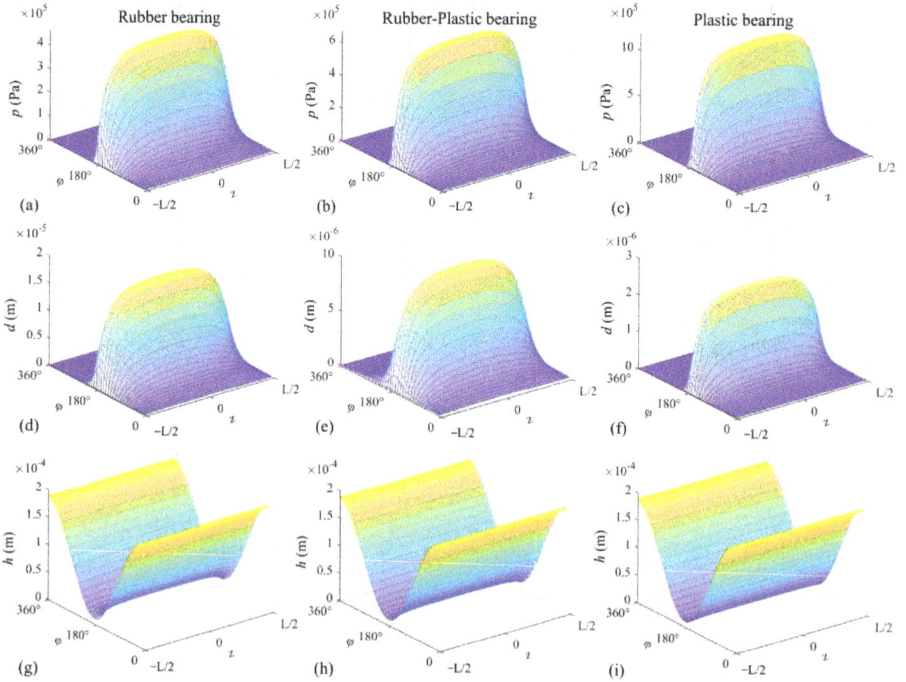

Figure 6. Performance comparison of water-lubricated rubber bearing, rubber-plastic double-layer bearing and plastic bearing (**a–c**) water film pressure distribution, (**d–f**) bush deformation distribution and (**g–i**) water film thickness distribution.

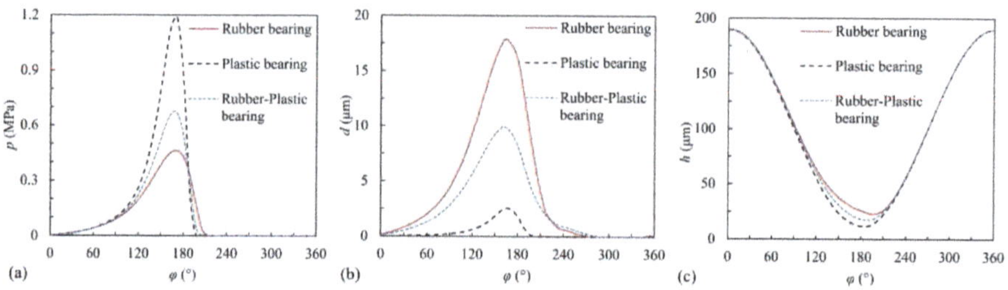

Figure 7. Distribution curves of (**a**) water film pressure, (**b**) bush deformation and (**c**) water film thickness on the mid-surface of the bearing ($z = 0$).

4.2. Effect of Rotational Speed on Bearing Performance

Figure 8 shows the water film pressure distribution, the bush deformation distribution and the water film thickness distribution of rubber-plastic double-layer water-lubricated bearing for different rotational speeds. The relative clearance is set to 0.2%, the rotational speed ranges from 300 r/min to 2100 r/min, and the other parameters are the same to those in Table 1. It shows that the rotational speed has a significant effect on the pressure distribution and bush deformation distribution. As the rotational speed increases, the water film pressure, the bush deformation and the minimum film thickness increase gradually. Figure 9 summarizes the effect of rotational speed on bearing lubrication and load-carrying properties for different eccentricity ratios. It shows that the maximum pressure, the maximum bush deformation, the load-carrying capacity and friction torque increase obviously with the rotational speed, and the variation increases with the eccentricity ratio. The increase in rotational speed enhances the hydrodynamic lubrication, thus the water film pressure and the load-carrying capacity increase, which make the increases of bush deformation and the minimum film thickness. The bush deformation is decided by the pressure and the bush material properties. The load-carrying capacity is obtained by integrating the pressure on the journal surface. Thus, the bush deformation and the load-carrying capacity change the same as the bearing pressure. When the eccentricity ratio is small, the bush deformation has little effect on the minimum water film thickness, and the load-carrying capacity increases almost linearly with the rotational speed. When the eccentricity ratio is large ($\varepsilon > 0.8$), the minimum water film thickness is small, and the influence of bush deformation on the minimum water film thickness becomes greater, the increase in load-carrying capacity with rotational speed thus slows down. In practical applications, the bearing usually works at high eccentricity ratios, so the bush deformation cannot be ignored. Figure 10 shows the variation curve of W to ε and the dimensionless journal center locus. The arc in Figure 10b indicates the clearance circle with $\varepsilon = 1$. Designers can obtain the eccentricity ratio according to the load and rotational speed from Figure 10a and then check the minimum water film thickness according to Figure 9c. Figure 10b shows that the journal center locus is basically the same for different rotational speeds.

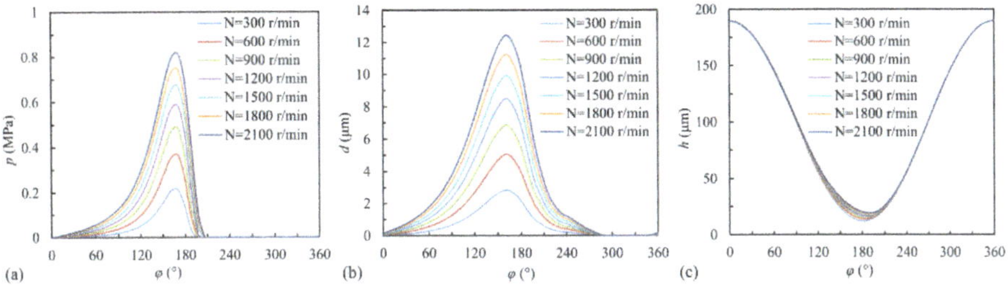

Figure 8. (**a**) Water film pressure distribution, (**b**) bush deformation distribution and (**c**) water film thickness distribution of rubber-plastic double-layer water-lubricated bearing for different rotational speeds ($\varepsilon = 0.9$).

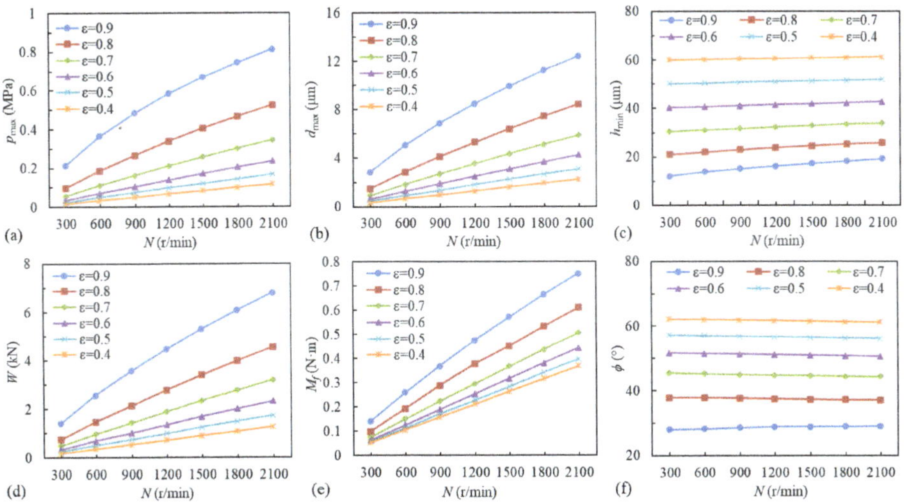

Figure 9. Variation of (**a**) maximum pressure, (**b**) maximum bush deformation, (**c**) minimum water film thickness, (**d**) load-carrying capacity, (**e**) friction torque and (**f**) attitude angle with rotational speed for different eccentricity ratios.

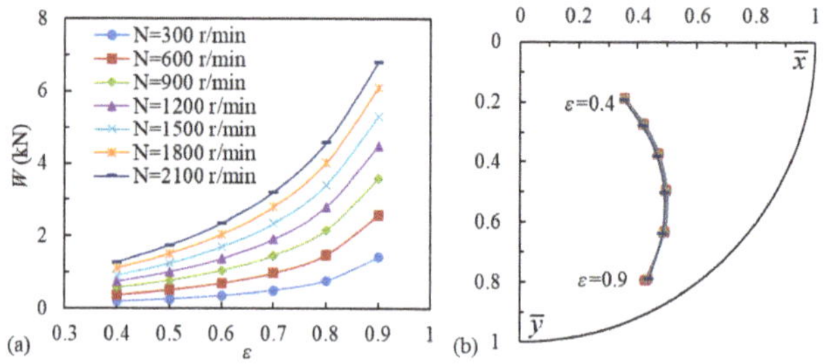

Figure 10. Variation of (**a**) load-carrying capacity and (**b**) journal center position with eccentricity ratio for different rotational speeds.

4.3. Effect of Length-to-Diameter Ratio on Bearing Performance

Figure 11 shows the water film pressure distribution, bush deformation distribution and water film thickness distribution of rubber-plastic double-layer bearing for different length-to-diameter ratios. The relative clearance is set to 0.2%; the rotational speed is 1500 r/min; L/D ranges from 0.5 to 2; and the other parameters are the same as those in Table 1. When L/D is small, the water film pressure distribution and bush deformation distribution are concentrated. The larger L/D is, the larger the distribution area of pressure and bush deformation is. However, when L/D is more than 1, the water film pressure and bush deformation change little. L/D has a small effect on the water film thickness distribution. Figure 12 shows the effect of L/D on bearing lubrication performance for different eccentricity ratios. It shows that for small L/D values, the maximum pressure increases significantly with L/D, but as L/D increases, the maximum pressure tends to a certain value. This is because the side leakage of short bearings has a large effect on the water film pressure. The longer the bearing is, the smaller the effect of side leakage is, and the maximum pressure tends to the value of the infinite length journal bearing. The changing rule of the maximum bush deformation is the same as that of the maximum pressure. As the maximum bush deformation increases only 1~3 μm when L/D increases from 0.5 to 2, the minimum water film thickness changes little. Due to the increase in bearing area, the load-carrying capacity and friction torque increase significantly with L/D. The attitude angle increases slightly. Figure 13 shows the variation curve of W to ε and the dimensionless journal center locus for different L/D. Designers can obtain the eccentricity ratio according to the load and length-to-diameter ratio from Figure 13a and check the minimum water film thickness according to Figure 12c. Figure 13b shows that the curvature radius of the journal center locus increases as L/D increases.

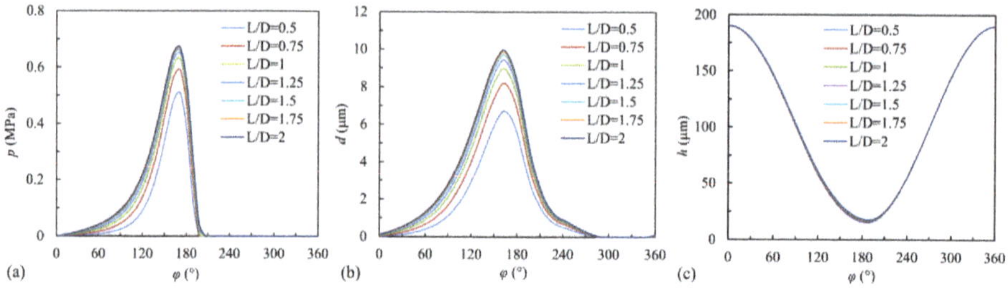

Figure 11. (**a**) Water film pressure distribution, (**b**) bush deformation distribution and (**c**) water film thickness distribution of rubber-plastic double-layer water-lubricated bearing for different length-to-diameter ratios (ε = 0.9).

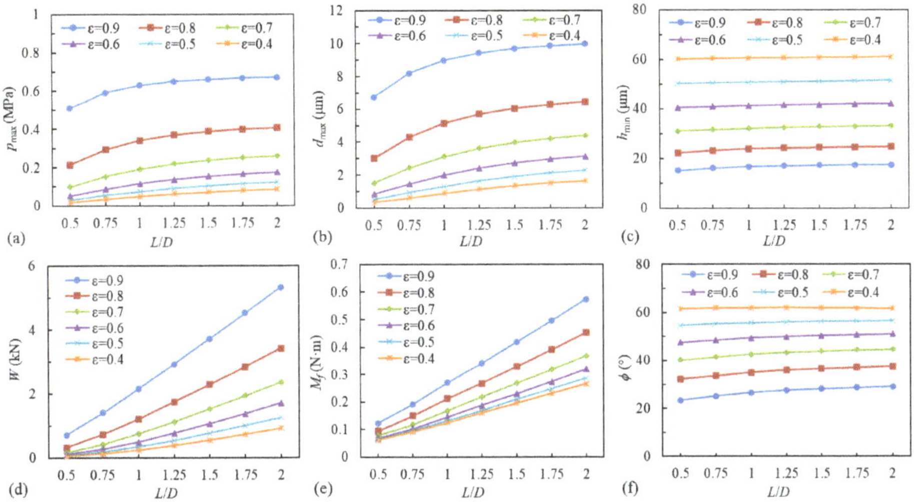

Figure 12. Variation of (**a**) maximum pressure, (**b**) maximum bush deformation, (**c**) minimum water film thickness, (**d**) load-carrying capacity, (**e**) friction torque and (**f**) attitude angle with length-to-diameter ratio for different eccentricity ratios.

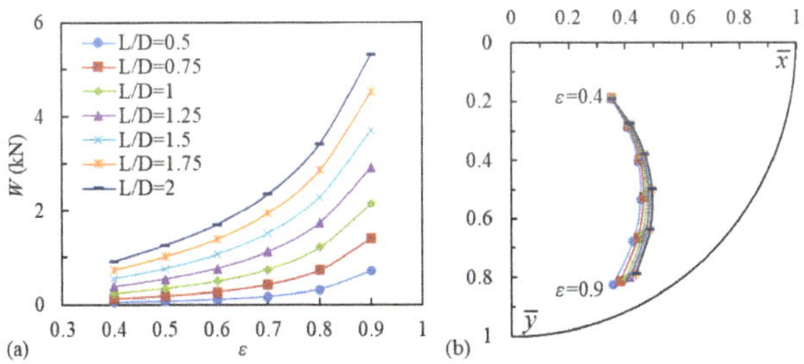

Figure 13. Variation of (**a**) load-carrying capacity and (**b**) journal center position with eccentricity ratio for different length-to-diameter ratios.

4.4. Effect of Relative Clearance on Bearing Performance

Figure 14 shows the water film pressure distribution, the bush deformation distribution and water film thickness distribution of rubber-plastic double-layer water-lubricated bearing for different relative clearances. The rotational speed is set to 1500 r/min, the relative clearance ranges from 0.1% to 0.4%, and the other parameters are the same as those in Table 1. Figure 15 shows the effect of the relative clearance on bearing lubrication performance for different eccentricity ratios. As the relative clearance increases, the water film thickness increases, the hydrodynamic pressure decreases, and the load-carrying capacity, friction torque and the maximum bush deformation are reduced. The actual minimum film thickness is the sum of $R\psi(1-\varepsilon)$ and bush deformation. When the relative clearance is small, the minimum water film thickness is small, the bush deformation is large, and the bearing lubrication performance is greatly affected by the bush deformation. As shown in Figure 15b, when the eccentricity ratio is 0.9 and the relative clearance is less than 0.2%, the bush deformation is much larger than the minimum film thickness value calculated by the formula $R\psi(1-\varepsilon)$. Figure 15c shows that the minimum film thickness decreases as the relative clearance increases from 0.1% to 0.15% for $\varepsilon = 0.9$. This is because the reduction of the bush deformation is greater than the increase in $R\psi(1-\varepsilon)$. The attitude angle changes little with the relative clearance. Figure 16 shows the variation curve of W to ε and the dimensionless journal center locus for different relative clearances. Designers can obtain the eccentricity ratio according to the load and relative clearance from Figure 16a and check the minimum water film thickness according to Figure 15c. Figure 16b shows that the curvature radius of the journal center locus decreases as the relative clearance increases.

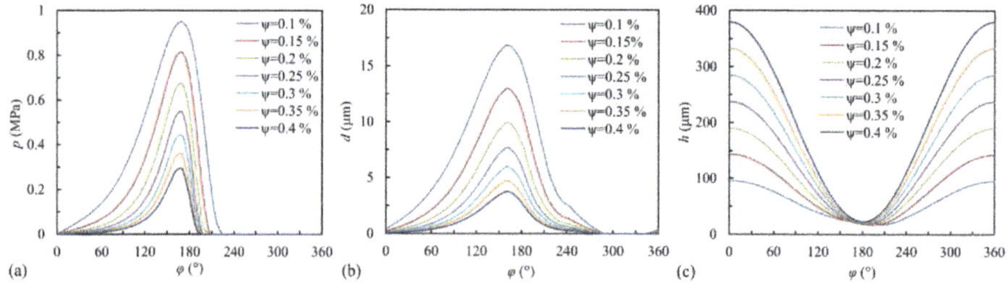

Figure 14. (**a**) Water film pressure distribution, (**b**) bush deformation distribution and (**c**) water film thickness distribution of rubber-plastic double-layer water-lubricated bearing for different relative clearances (ε = 0.9).

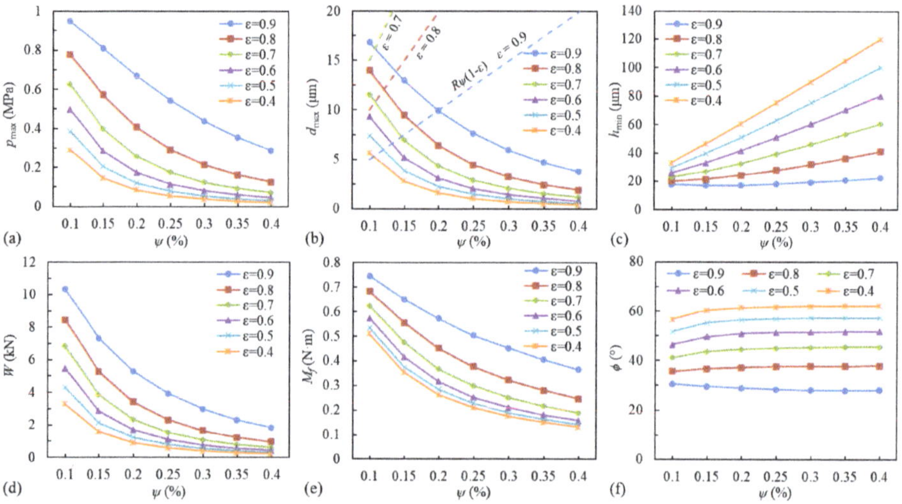

Figure 15. Variation of (**a**) maximum pressure, (**b**) maximum bush deformation, (**c**) minimum water film thickness, (**d**) load-carrying capacity, (**e**) friction torque and (**f**) attitude angle with relative clearance for different eccentricity ratios.

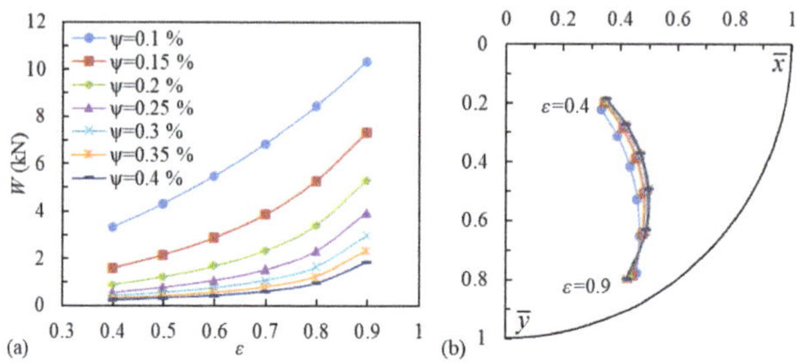

Figure 16. Variation of (**a**) load-carrying capacity and (**b**) journal center position with eccentricity ratio for different relative clearances.

4.5. Effect of Plastic Layer Elastic Modulus on Bearing Performance

Figure 17 shows the water film pressure distribution, bush deformation distribution and water film thickness distribution of rubber-plastic double-layer bearing for different elastic modulus of plastic layer bush, E_1. The relative clearance is set to 0.2%; the rotational speed is 1500 r/min; E_1 ranges from 100 MPa to 1300 MPa; and the other parameters are the same to those in Table 1. As E_1 increases, the bush deformation decreases, and the deformation distribution curve becomes smoother. The effect of E_1 on the water film thickness distribution and pressure distribution is limited, and the change mainly occurs at the loading area. Figure 18 shows the effect of E_1 on bearing lubrication performance for different eccentricity ratios. As E_1 increases from 100 MPa to 1300 MPa, the maximum bush deformation decreases by about 1 μm for ε = 0.4 and 6 μm for ε = 0.9. The decrease in bush deformation is small relative to the minimum film thickness; thus, E_1 has little effect on the minimum water film thickness. Only when ε is larger than 0.8, and E_1 is less than 500 MPa, the minimum water film thickness decreases significantly with the increase in E_1. The maximum pressure is affected by the minimum water film thickness. When the water film thickness decreases, the maximum pressure increases. The load-carrying capacity has the same changing rule with the maximum pressure. The friction torque and attitude angle increase slightly with E_1. Figure 19 shows the variation curve of W to ε and dimensionless journal center locus for different E_1. Figure 19a shows that when the eccentricity ratio is less than 0.8, E_1 has little influence on the load-carrying capacity. When the eccentricity ratio is larger than 0.8, and E_1 is larger than 300 MPa, E_1 has little influence on the load-carrying capacity. Figure 19b shows that the curvature radius of the journal center locus increases slightly as E_1 increases.

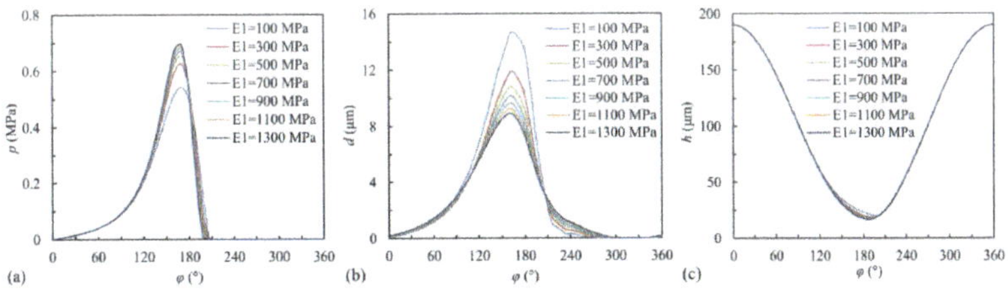

Figure 17. (**a**) Water film pressure distribution, (**b**) bush deformation distribution and (**c**) water film thickness distribution of rubber-plastic double-layer water-lubricated bearing for different plastic layer elastic modulus (ε = 0.9).

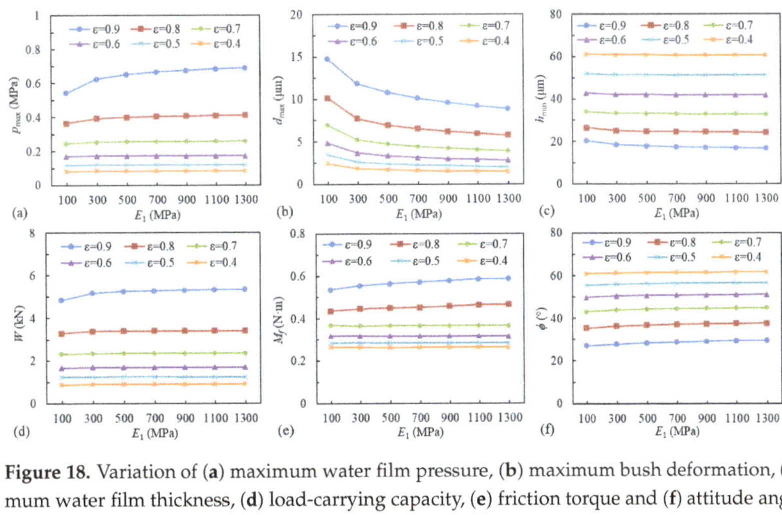

Figure 18. Variation of (**a**) maximum water film pressure, (**b**) maximum bush deformation, (**c**) minimum water film thickness, (**d**) load-carrying capacity, (**e**) friction torque and (**f**) attitude angle with the plastic layer elastic modulus for different eccentricity ratios.

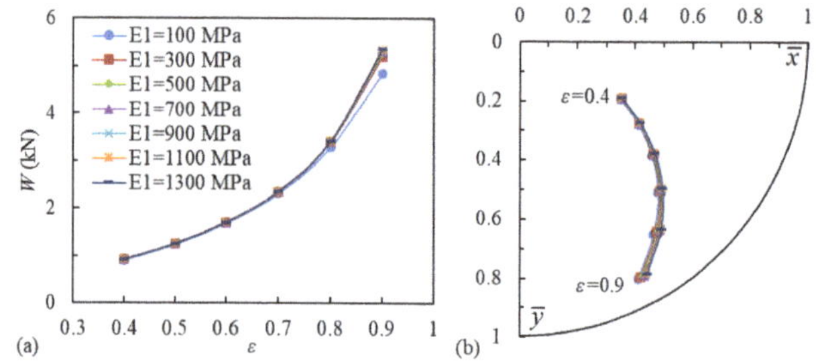

Figure 19. Variation of (**a**) load-carrying capacity and (**b**) journal center position with eccentricity ratio for different plastic layer elastic modulus.

4.6. Effect of Plastic Layer Thickness on Bearing Performance

Figure 20 shows the water film pressure distribution, the bush deformation distribution and the water film thickness distribution of rubber-plastic double-layer water-lubricated bearing for different plastic layer thicknesses. The relative clearance is set to 0.2%, the rotational speed is 1500 r/min, and the total bearing bush thickness t is 10 mm. The plastic layer thickness t_1 increases from 3 mm to 7 mm, and the rubber layer thickness t_2 decreases from 7 mm to 3 mm. The other parameters are the same as those in Table 1. As t_1 increases, the bush deformation decreases. The change of bush deformation mainly affects the peak pressure value, and it has little effect on the water film thickness and pressure distribution. Figure 21 shows the effect of plastic layer thickness on bearing lubrication performance for different eccentricity ratios. It shows that as t_1 increases from 3 mm to 7 mm, the maximum bush deformation decreases by about 3.5 μm for ε = 0.4 and 8.5 μm for ε = 0.9. The decrease in bush deformation is small relative to the minimum film thickness, thus the thickness distribution of the double-layer bush has a limited effect on the minimum water film thickness. Only when the eccentricity ratio is larger than 0.8 does the minimum water film thickness decrease significantly as t_1 increases. As the water film thickness decreases, the maximum pressure increases. The load-carrying capacity has the same changing rule with the maximum pressure. The friction torque increases slightly with t_1. The attitude angle changes little. Figure 22a shows that when the eccentricity ratio is less than 0.8, t_1 has little effect on the load-carrying capacity. When the eccentricity ratio is greater than 0.8, and t_1 is more than 6 mm, t_1 has little effect on the load-carrying capacity. Figure 22b shows that the plastic layer thickness has little effect on the journal center locus.

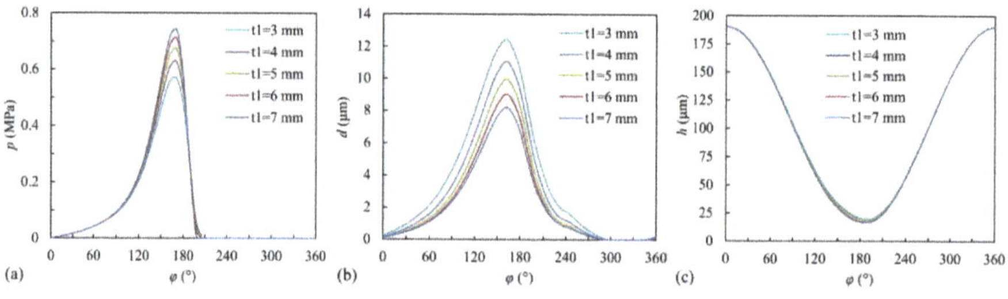

Figure 20. (**a**) Water film pressure distribution, (**b**) bush deformation distribution and (**c**) water film thickness distribution of rubber-plastic double-layer water-lubricated bearing for different plastic layer thicknesses (ε = 0.9).

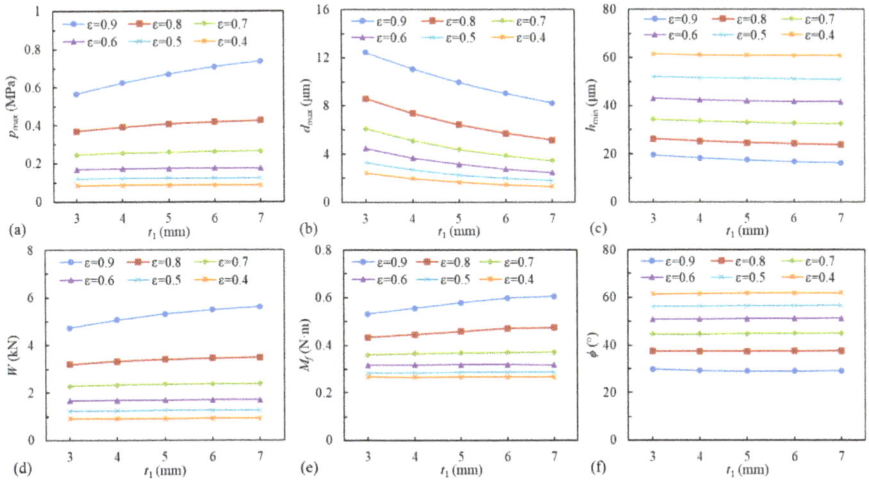

Figure 21. Variation of (**a**) maximum water film pressure, (**b**) maximum bush deformation, (**c**) minimum water film thickness, (**d**) load-carrying capacity, (**e**) friction torque and (**f**) attitude angle with the plastic layer thickness for different eccentricity ratios.

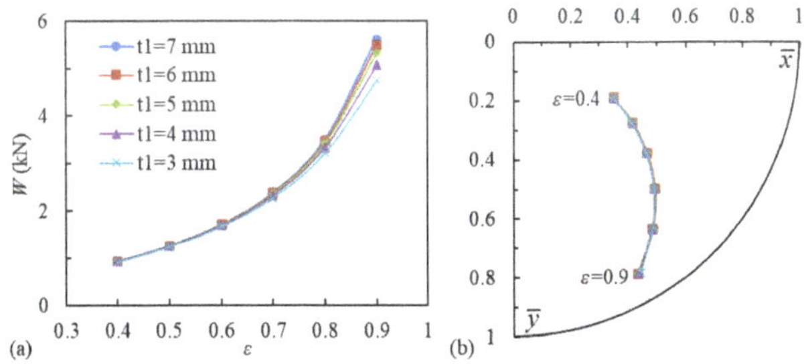

Figure 22. Variation of (**a**) load-carrying capacity and (**b**) journal center position with eccentricity ratio for different plastic layer thicknesses.

5. Conclusions

This study established a fluid-structure coupling numerical calculation model for investigation of lubrication performance of the rubber-plastic double-layer water-lubricated journal bearings. The influence of eccentricity ratio ε, rotational speed N, length-to-diameter ratio L/D, relative clearance ψ, elastic modulus E_1 and thickness t_1 of the plastic layer bush on the bearing lubrication performance were studied. From the results, the following conclusions are drawn:

(1) The bush deformation of the rubber-plastic double-layer bearing is between those of the rubber bearing and the plastic bearing, and its distribution area is close to that of the rubber bearing. However, the shape change of the double-layer bearing is smaller than that of the rubber bearing due to the larger stiffness of the plastic layer bush.

(2) The maximum pressure, the maximum bush deformation, the load-carrying capacity and friction torque increase noticeably with N, and the variation increases with ε. The load-carrying capacity increases almost linearly with N for small ε values. When ε is more than 0.8, the influence of bush deformation on the minimum water film thickness becomes more significant, and the increase in load-carrying capacity with N slows down.

(3) Under the influence of side leakage, the maximum water film pressure increases significantly with L/D for small L/D values, but as L/D increases, the maximum

pressure increases to a certain value. L/D affects the load-carrying capacity and friction torque significantly.

(4) With the increase in ψ, the load-carrying capacity and frictional torque decrease, the maximum bush deformation decreases, and the attitude angle has little changes. The bearing lubrication performance is greatly affected by the bush deformation for small ψ values.

(5) The elastic modulus and thickness of the plastic layer bush mainly affect the bearing lubrication performance under heavy load conditions ($\varepsilon > 0.8$). As E_1 and t_1 increase, the maximum bush deformation and the minimum water film thickness decrease, and the load-carrying capacity increase. The friction torque and the attitude angle change little.

(6) The data provided in Figures 9 and 10, Figures 12 and 13, Figures 15 and 16, Figures 18 and 19, Figures 21 and 22 can be used for rapid structure design of this type of bearing.

The research results provide a theoretical basis for the structural design and application of rubber-plastic double-layer water-lubricated journal bearings. The static and dynamic characteristics of the bearings will be studied experimentally in future work.

Author Contributions: Y.S.: conceptualization, methodology, data curation, writing—editing of the manuscript, writing—original draft preparation, software, validation. Y.Z.: methodology, software, validation. X.Z.: methodology, data curation, supervision, project administration, writing—editing of the manuscript. H.Z.: writing—editing of the manuscript, modification, check. G.W.: validation, software. M.W.: validation, software. All authors have read and agreed to the published version of the manuscript.

Funding: The financial support is from National Key Research and Development Program of China (Grant No. 2022YFE0199100) and National Natural Science Foundation of China (Grant No. 51905317). The authors would like to sincerely express their appreciation.

Data Availability Statement: Data will be made available on request.

Conflicts of Interest: The authors declare no conflict of interest.

References

1. Ouyang, W.; Liu, Q.; Xiao, J.; Huang, J.; Zhang, Z.; Wang, L. Experimental study on the distributed lubrication characteristics of full-size water-lubricated stern bearings under hull deformation. *Ocean Eng.* **2023**, *267*, 113226. [CrossRef]
2. Szlapczynski, R.; Krata, P. Determining and visualizing safe motion parameters of a ship navigating in severe weather conditions. *Ocean Eng.* **2018**, *158*, 263–274. [CrossRef]
3. Xie, Z.; Jiao, J.; Yang, K. Theoretical and experimental study on the fluid-structure-acoustic coupling dynamics of a new water lubricated bearing. *Tribol. Int.* **2023**, *177*, 107982. [CrossRef]
4. Michalec, M.; Svoboda, P.; Křupka, I.; Hartl, M. A review of the design and optimization of large-scale hydrostatic bearing systems. *Eng. Sci. Technol. Int. J.* **2021**, *24*, 936–958. [CrossRef]
5. Nijssen, J.; Faludi, J.; van Ostayen, R. An eco-impact design metric for water lubricated bearings based on anticipatory Life Cycle Assessment. *J. Clean. Prod.* **2021**, *321*, 128874. [CrossRef]
6. Xie, Z.; Jiao, J.; Yang, K.; Zhang, H. A State-of-Art Review on the Water-Lubricated Bearing. *Tribol. Int.* **2023**, *180*, 108276. [CrossRef]
7. Yamajo, S.; Kikkawa, F. PTFE Compound Bearing for Water Lubricated Shaft Systems. In Proceedings of the SNAME 10th Propeller and Shafting Symposium, Virginia Beach, VA, USA, 12–13 September 2003. [CrossRef]
8. Litwin, W. Experimental research on water lubricated three layer sliding bearing with lubrication grooves in the upper part of the bush and its comparison with a rubber bearing. *Tribol. Int.* **2015**, *82*, 153–161. [CrossRef]
9. Litwin, W. Properties comparison of rubber and three layer PTFE-NBR-bronze water lubricated bearings with lubricating grooves along entire bush circumference based on experimental tests. *Tribol. Int.* **2015**, *90*, 404–411. [CrossRef]
10. Shi, Y.; Li, M.; Zhu, G.; Yu, Y. Dynamics of a rotor system coupled with water-lubricated rubber bearings. *Proc. Inst. Mech. Eng. Part C J. Mech. Eng. Sci.* **2018**, *232*, 4263–4277. [CrossRef]
11. Liu, G.; Li, M. Experimental study on the lubrication characteristics of water-lubricated rubber bearings at high rotating speeds. *Tribol. Int.* **2021**, *157*, 106868. [CrossRef]
12. Qiao, J.; Zhou, G.; Pu, W.; Li, R.; He, M. Coupling analysis of turbulent and mixed lubrication of water-lubricated rubber bearings. *Tribol. Int.* **2022**, *172*, 107644. [CrossRef]
13. Kuznetsov, E.; Glavatskih, S.; Fillon, M. THD analysis of compliant journal bearings considering liner deformation. *Tribol. Int.* **2011**, *4*, 1629–1641. [CrossRef]

14. Wang, Y.; Yin, Z.; Jiang, D.; Gao, G.; Zhang, X. Study of the lubrication performance of water-lubricated journal bearings with CFD and FSI method. *Ind. Lubr. Tribol.* **2016**, *8*, 341–348. [CrossRef]
15. Yang, Z.; Guo, Z.; Yang, Z.; Wang, C.; Yuan, C. Study on tribological properties of a novel composite by filling microcapsules into UHMWPE matrix for water lubrication. *Tribol. Int.* **2021**, *153*, 106629. [CrossRef]
16. Cheng, B.; Duan, H.; Chen, Q.; Shang, H.; Zhang, Y.; Li, J.; Shao, T. Effect of laser treatment on the tribological performance of polyetheretherketone (PEEK) under seawater lubrication. *Appl. Surf. Sci.* **2021**, *566*, 150668. [CrossRef]
17. Balmus, M.; Massing, A.; Hoffman, J.; Razavi, R.; Nordsletten, D.A. A partition of unity approach to fluid mechanics and fluid–structure interaction. *Comput. Methods Appl. Mech. Eng.* **2020**, *362*, 112842. [CrossRef]
18. Burman, E.; Fernández, M.A.; Frei, S.; Gerosa, F.M. A mechanically consistent model for fluid–structure interactions with contact including seepage. *Comput. Methods Appl. Mech. Eng.* **2022**, *392*, 114637. [CrossRef]
19. Tang, D.; Han, Y.; Yin, L.; Chen, Y. Numerical Analysis of the Mixed-Lubrication Performance of Staved Stern Tube Bearings Lubricated with Water. *Lubricants* **2023**, *11*, 168. [CrossRef]
20. Yang, Z.; Zhang, X.; Wang, Y.; Zhao, Y. Study on elastohydrodynamic lubrication performance of double-layer composite water-lubricated bearings. *Mech. Ind.* **2023**, *24*, 3. [CrossRef]
21. Thomsen, K.; Klit, P. A study on compliant layers and its influence on dynamic response of a hydrodynamic journal bearing. *Tribol. Int.* **2011**, *44*, 1872–1877. [CrossRef]
22. Xiang, G.; Yang, T.; Guo, J.; Wang, J.; Liu, B.; Chen, S. Optimization transient wear and contact performances of water-lubricated bearings under fluid-solid-thermal coupling condition using profile modification. *Wear* **2022**, *502*, 204379. [CrossRef]
23. Chen, S.; Xiang, G.; Fillon, M.; Guo, J.; Wang, J.; Cai, J. On the tribo-dynamic behaviors during start-up of water lubricated bearing considering imperfect journal. *Tribol. Int.* **2022**, *174*, 107685. [CrossRef]
24. Cai, J.; Xiang, G.; Li, S.; Guo, J.; Wang, J.; Chen, S.; Yang, T. Mathematical modeling for nonlinear dynamic mixed friction behaviors of novel coupled bearing lubricated with low-viscosity fluid. *Phys. Fluids* **2022**, *34*, 093612. [CrossRef]
25. Gong, J.; Jin, Y.; Liu, Z.; Jiang, H.; Xiao, M. Study on influencing factors of lubrication performance of water-lubricated micro-groove bearing. *Tribol. Int.* **2019**, *129*, 390–397. [CrossRef]
26. Du, Y.; Lan, J.; Quan, H.; Sun, C.; Liu, X.; Yang, X. Effect of different turbulent lubrication models on the lubrication characteristics of water-lubricated rubber bearings at a high Reynolds number. *Phys. Fluids* **2021**, *33*, 065118. [CrossRef]
27. Xie, Z.; Song, P.; Hao, L.; Shen, N.; Zhu, W.; Liu, H.; Shi, J.; Wang, Y.; Tian, W. Investigation on effects of Fluid-Structure-Interaction (FSI) on the lubrication performances of water lubricated bearing in primary circuit loop system of nuclear power plant. *Ann. Nucl. Energy* **2020**, *141*, 107355. [CrossRef]
28. Xie, Z.; Wang, X.; Zhu, W. Theoretical and experimental exploration into the fluid structure coupling dynamic behaviors towards water-lubricated bearing with axial asymmetric grooves. *Mech. Syst. Signal Process.* **2022**, *168*, 108624. [CrossRef]
29. Wang, Y.; Yin, Z.; Gao, G.; Zhang, X. Analysis of the performance of worn hydrodynamic water-lubricated plain journal bearings considering cavitation and elastic deformation. *Mech. Ind.* **2017**, *18*, 508. [CrossRef]
30. Hili, M.A.; Bouaziz, S.; Maatar, M.; Fakhfakh, T.; Haddar, M. Hydrodynamic and Elastohydrodynamic Studies of a Cylindrical Journal Bearing. *J. Hydrodyn.* **2010**, *22*, 155–163. [CrossRef]
31. Rohilla, P.K.; Verma, R.; Verma, S. Performance analysis of couple stress fluid operated elastic hydrodynamic journal bearing. *Tribol. Online* **2019**, *14*, 143–154. [CrossRef]
32. Linjamaa, A.; Lehtovaara, A.; Larsson, R.; Kallio, M.; Söchting, S. Modelling and analysis of elastic and thermal deformations of a hybrid journal bearing. *Tribol. Int.* **2018**, *118*, 451–457. [CrossRef]
33. Huang, B.; Wang, L.-Q.; Guo, J. Performance comparison of circular, two-lobe and elliptical journal bearings based on TEHD analysis. *Ind. Lubr. Tribol.* **2014**, *66*, 184–193. [CrossRef]
34. Smith, E.H. On the Design and Lubrication of Water-Lubricated, Rubber, Cutlass Bearings Operating in the Soft EHL Regime. *Lubricants* **2020**, *8*, 75. [CrossRef]
35. Xiang, G.; Han, Y. Study on the tribo-dynamic performances of water-lubricated microgroove bearings during start-up. *Tribol. Int.* **2020**, *151*, 106395. [CrossRef]
36. Ma, J.; Fu, C.; Zhang, H.; Chu, F.; Shi, Z.; Gu, F.; Ball, A.D. Modelling non-Gaussian surfaces and misalignment for condition monitoring of journal bearings. *Measurement* **2021**, *174*, 108983. [CrossRef]
37. Zienkiewicz, O.C.; Taylor, R.L.; Zhu, J.Z. *The Finite Element Method: Its Basis and Fundamentals*, 7th ed.; Butterworth Heinemann: Oxford, UK, 2013. [CrossRef]
38. Litwin, W.; Olszewski, A. Water-Lubricated Sintered Bronze Journal Bearings—Theoretical and Experimental Research. *Tribol. Trans.* **2014**, *57*, 114–122. [CrossRef]

lubricants

MDPI

Article

Effect of Groove Structure on Lubrication Performance of Water-Lubricated Stern Tube Bearings

Shengdong Zhang

School of Industry, Jining University, Qufu 273199, China; sdzluck@163.com

Abstract: This study investigated the lubrication characteristics (i.e., the groove ratio and width) of water-lubricated stern tube bearings, based on the flexibility matrix method and lubrication theory. Considering the elastic deformation of the lining, a fluid structure interaction (FSI) model of the surface micro-groove texture of a water-lubricated stern tube bearing was established, and the correctness and rationality of the model were verified by experiments. Micro-grooved, surface-lubricated, water-lubricated stern tube bearings with three different cross-sectional shapes (rectangular, circular, and triangular) were designed. The influences of the groove area ratio and width on the bearing load-carrying capacity and friction coefficient were analyzed. At a groove area ratio of 0.31, the load-carrying capacity of the rectangular grooved stern tube bearing reached the maximum value and the friction coefficient reached the minimum value. It is recommended to design and use water-lubricated stern tube bearings, especially Thordon water-lubricated stern tube bearings, with rectangular micro-grooves, with a groove area ratio of 0.30–0.32, so that the best lubrication performance can be obtained. With the increase in the micro-groove width, the lubrication of water-lubricated stern tube bearings with partial rectangular micro-grooves is significantly better than that of others. Under the same conditions, the bearing load-carrying capacity and friction performance of local groove stern bearings is significantly better than that of global groove stern bearings.

Keywords: flexibility matrix method; micro-groove; FSI; water-lubricated bearing

Citation: Zhang, S. Effect of Groove Structure on Lubrication Performance of Water-Lubricated Stern Tube Bearings. *Lubricants* **2023**, *11*, 374. https://doi.org/10.3390/lubricants11090374

Received: 25 June 2023
Revised: 27 August 2023
Accepted: 28 August 2023
Published: 05 September 2023

1. Introduction

The water-lubricated stern tube bearing is a crucial supporting component in a propulsion system, with low-viscosity seawater or freshwater as the lubricant [1–4]. Considering the pollution caused by the leakage of lubricating fluid to the environment, this type of bearing has natural environmental advantages. In the operation of a water-lubricated stern tube bearing, water is both the lubricant and the coolant. Given that the specific heat of water is greater than that of lubricating oil, the cooling effect is good, and long-term operation will not result in the overheating phenomenon or shaft holding phenomenon. However, due to the low viscosity of water, water-lubricated stern tube bearings exhibit poor lubrication performance, resulting in the lubricating water film not being formed effectively. Specifically, the water-lubricated stern tube bearing's load-carrying capacity is insufficient, and the friction coefficient is high, which affects the reliability and safety of the whole dynamic system [5,6]. At present, the arrangement of the surface microtexture on the inner surface of the bearing is an effective means by which to improve the lubrication effect of a water-lubricated stern tube bearing [7,8]. The types of surface textures mainly include micro-grooves and micro-dimples. Scholars have studied the location, shape, and geometry of micro-dimples. Tala-Ighil et al. pointed out that the texture of the whole bearing surface does not necessarily improve the performance of the bearing, but the complete texture will have a positive effect on the low journal rotation speed, while the local texture at the outlet of the active pressure zone has a positive effect on the high rotation speed [9,10]. Choudhury et al. studied the dimple shape of an artificial hip joint, and the results show that a prosthesis with a square dimple array is the best in the formation of the lubricating film of the artificial hip joint [11]. Shinde et al. studied the performance of conical fluid dynamic journal bearings using numerical and experimental methods. On a smooth surface, the local texture of the bearing surface in the 90–180° area increased the maximum fluid film pressure by 42.08% [12]. Dong Ning et al. established a mathematical model of the elastohydrodynamic lubrication of surface texture sliding bearings, and numerically calculated and analyzed the pits in three forms: circular, rectangular, and isosceles triangular. It was found that the pits have a significant effect on the peak pressure fluctuation and film thickness of the bearing [13,14]. The parameters that have been studied include the shape of the texture, the depth of the texture, the number of textures, the percentage of textures, and the surface roughness. The findings demonstrate that specific parameter textures on the bearing surface can increase the load-carrying capacity by 36.32% and reduce frictional torque by 1.66% compared to bearings without texture. However, the numerical model has not been verified by experiments, and the effect of elastic deformation on the lining material has not been considered [15]. Blatter et al. [16] conducted experimental research on the effect of micro-grooves using a pin-on-disk tribometer. They highlighted that appropriately sized and shaped micro-grooves can reduce wear and significantly extend the sliding life. Yuan et al. [17] designed and executed experiments to study the influence of groove orientation on friction performance. Their findings revealed that micro-grooves aligned perpendicularly or parallel to the sliding direction strongly impact the friction performance of sliding surfaces. Wahab et al. [18] experimentally investigated the effects of micro-grooves on the tribological behavior of an Al_2O_3-13%TiO_2 coating. The results demonstrate that the textured coating effectively decreased the wear rate by entrapping wear debris within the grooves. Ali et al. [19] explored the behavior of transverse limited micro-grooves in EHL point contacts. They observed that surfaces with such micro-grooves exhibited lower friction and larger film thickness during reciprocation motion compared to smooth surfaces. Ehret et al. [20] studied the influence of waviness orientation in EHL point contacts on lubrication performance using the multigrid multi-integration method. Their results indicate that transverse waviness provide the best lubrication capability. Thakre et al. [21] conducted theoretical research on the impact of surface textures with various waveforms on the lubricating behavior of ionic liquids. The results highlight that the texture pattern and material combinations significantly affect micro-EHL contact behavior. Kaneta [22] and Shi [23] also investigated micro-grooves using numerical and experimental methods, respectively. Gong Jiayu et al. established a mathematical model of micro-groove water-lubricated stern bearings for fluid solid coupling, conducted strip specimen tests on a testing machine, and used orthogonal methods to study the influence of groove geometry and position on lubrication performance. The results show that the two types of water-lubricated stern bearings, namely, locally micro-grooved and globally micro-grooved, were significantly affected by the geometry and position, but the study only considered the effects of a single eccentricity and the distance between micro-grooves [24]. A numerical model based on the average flow Reynolds equation was developed and validated by Wu et al. [25]. The effects of three types of grooves (rectangular, triangular, and wedge-shaped), groove location, groove size, and number of grooves on the low-speed lubrication performance of a water-lubricated groove bearing were investigated. The water-lubricated bearing without grooves had poor lubrication performance at low speeds. The groove shape parameter has more influence on the lubrication performance. However, Wu's study did not specify which structure of micro-groove bearing had the best performance and did not consider the effects of groove ratio and groove width on the lubrication performance of the water-lubricated stern tube bearing.

Therefore, although scholars have researched the effect of surface texturing on the lubrication characteristics of stern bearings, there has been no in-depth research on the influence of groove proportion and width. The main purpose of this study was to research the effects of different groove parameters, especially the micro-groove structure type, micro-groove occupancy ratio, and micro-groove width, on the lubrication performance of water-lubricated stern tube bearings. In this case, based on the flexibility matrix method and lubrication theory, the fluid solid coupling model of a surface micro-groove weave structure water-lubricated stern tube bearing was established to account for the elastic deformation of the liner, and the correctness and reasonableness of the model were verified through experiments. Three different cross-sectional shapes of micro-groove surface woven water-lubricated stern tube bearing (rectangle, circle, and isosceles triangle) were designed, and the influences of the groove ratio and width on the load-carrying capacity and friction coefficient of the water-lubricated stern tube bearing were compared and analyzed. This study provides a theoretical basis for the selection of grooves in water-lubricated stern tube bearings.

2. Mathematical Model

The lubrication model of the micro-groove stern tube bearing is shown in Figure 1.

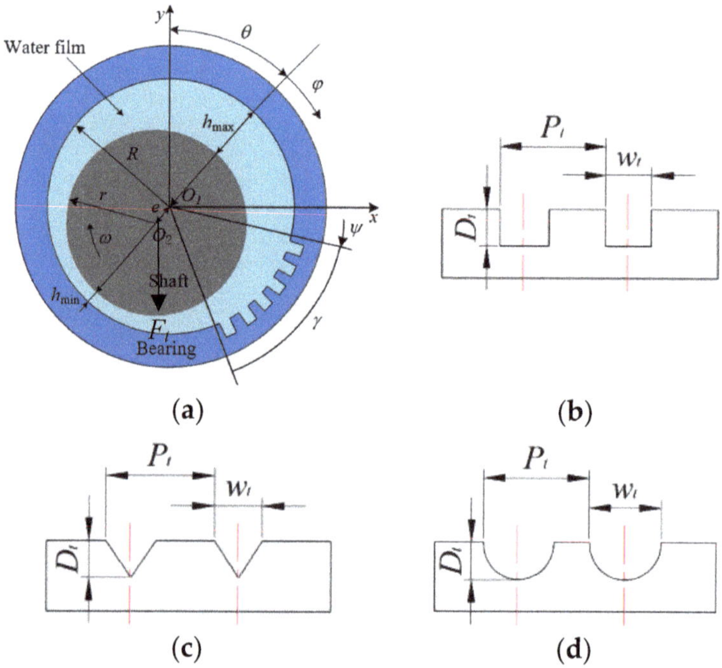

Figure 1. Lubrication model of micro-groove stern tube bearing. (**a**) Schematic diagram of micro-groove stern tube bearing structure; (**b**) rectangular structure of micro-grooves; (**c**) triangle structure of micro-grooves; (**d**) semi-circular structure of micro-grooves. Adapted from [15].

R and r denote the inner radius of the bearing and the radius of the shaft, respectively, m; O_1 and O_2 are the centers of the bearing and the shaft, respectively; e is the eccentricity distance, m; θ is the attitude angle, °; h_{min} and h_{max} are the minimum and maximum film thickness, respectively, m; φ is the circumferential coordinate starting from the maximum film thickness, °; Ψ and γ are the circumferential starting angle and the circumferential arrangement range of the micro-groove area, respectively, °; W_t and D_t are the width and depth of the micro-grooves, respectively, m; F_t is the radial load applied on the spindle, N; ω is the rotational speed of shaft, rad/s; and P_t is the span of the micro-grooves, m.

The flow solid coupling theory for hydrodynamic lubricated stern tube bearings makes the following assumptions:

(1) The inner diameter of the stern tube bearing, and the outer diameter of the shaft are much larger than the thickness of the water film.

(2) There is no relative sliding between the water film and the inner wall of the stern tube bearing or the outer wall of the shaft.
(3) Inertial forces and other volume forces of the water film are ignored.
(4) Due to the extremely thin water film, small changes in water film pressure in the direction of the thickness are not considered.
(5) Only the velocity gradient along the thickness direction of the water film is considered, and other directions are neglected.
(6) Only the laminar flow of water is considered.
(7) The variation of lubricant viscosity and density along the direction of water film thickness is not considered.
(8) The working condition of the bearing is the metastability state; that is, the balance state of the bearing. The starting and stopping conditions are not considered.
(9) The thermal effect of the shafting operation is ignored.

The Reynolds equation [26,27] for finite-width water-lubricated stern tube bearings in a steady state is

$$\frac{\partial}{\partial x}\left(\frac{h^3}{\eta}\frac{\partial p}{\partial x}\right)+\frac{\partial}{\partial y}\left(\frac{h^3}{\eta}\frac{\partial p}{\partial y}\right)=6\omega\frac{\partial(h)}{\partial x} \tag{1}$$

where x and y are the two coordinate directions of the water film, m; h is the water film thickness, m; p is the water film pressure, Pa; ω is the component of tangential velocity at the journal surface, rad/s; and η is the viscosity of the water, Pa·s.

The boundary conditions are the Reynolds boundary conditions [19,28]. In Figure 2, the pressures in columns 1 and (m + 1) are set to 0, which is used as the circumferential starting boundary. The circumferential rupture boundary is realized as follows: the pressure along the circumference is calculated point by point in rows from j = 2 to n, and in columns from i = 2 to m. When the pressure value is negative, it is determined that the water film rupture occurs at this point, and the pressure is set to zero at this point and the points after the row in which it is located. After iterative calculation, the rupture boundary gradually approaches the natural rupture boundary, and the pressure distribution satisfies the Reynolds boundary conditions.

$$\begin{cases} p=0 \text{ and } \frac{\partial p}{\partial \psi}=0, \text{ when } \psi=\theta_f \\ p=0, \text{ when } \lambda=0,l \end{cases} \tag{2}$$

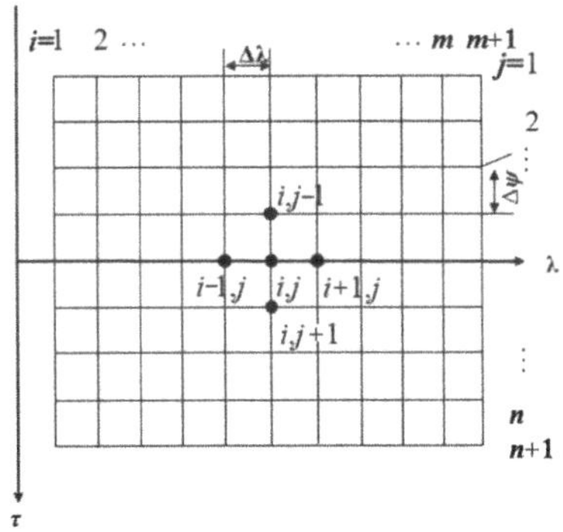

Figure 2. Grid division.

Here, λ is the axial direction coordinate of the bearing; l is the length of the bearing, m; and θ_f denotes the angular coordinate at the position of natural rupture of the water film, °.

The water film thickness equation [24,26,27,29,30] of micro-groove water-lubricated stern tube bearings is

$$h = \begin{cases} h_s + \delta & \text{Areas without texture} \\ h_s + \delta + D_t & \text{Areas with texture} \end{cases} \tag{3}$$

where δ is the radial deformation of the bearing inner surface nodes; h_s is the thickness of the water film when ignoring the bearing deformation and groove, $h_s = c + ecos(\Psi - \theta)$, c is the radius clearance, e is the eccentricity distance; and D_t is the depth of the micro-groove texture.

The flexibility matrix method [31–33] is used to input the meshing, cell numbering etc. into the finite element software ANSYS. The flexibility matrix method explicitly gives the relationship between elastic deformation and pressure distribution, which improves the coupling of the Reynolds equation and elastic deformation. The finite element model of the bearing liner is established in ANSYS, the finite element meshing of the bearing is adopted as a 6-plane cell, and the rectangular mesh of its inner surface is the mesh section required for the differential discrete Reynolds equation. For convenience, this is called the inner surface grid of the bearing and the inner surface cell. The solution of the elastic deformation equation involves calculating the radial deformation of the inner surface of the bearing under normal pressure distribution. For linear elastic materials, the elastic deformation equation can be expressed as

$$\{\delta\}_{node} = [G]_{node*elem}\{p\}_{node} \tag{4}$$

where the subscript "*node*" represents the number of nodes on the inner surface of the bearing; the subscript "*elem*" represents the number of elements on the inner surface of the bearing; $\{\delta\}$ is the radial deformation of bearing inner surface nodes; and $\{p\}$ is the average pressure of the inner surface elements of the bearing, and the pressure of the inner surface nodes is obtained by solving the Reynolds equation. The average pressure of the inner surface elements is the average of the pressures at the nodes of the constituent units; $[G]$ is the flexibility matrix. The physical significance of the jth column vector is the radial deformation of the inner surface node at the inner surface cell j under the action of element pressure distribution.

In generating the flexibility matrix using ANSYS, the mesh size is $3 \times 400 \times 40$, i.e., 3 nodes are taken in the radial direction, $n = 400$ nodes in the circumferential direction (τ direction), and $m = 40$ nodes in the axial direction (λ direction). Thus, the differential grid size for the discrete Reynolds equation is 400×40.

The equation for calculating the combined water film force is

$$\begin{cases} F_s = -\iint\limits_{\Omega} p\sin(\psi)dxdy \\ F_c = -\iint\limits_{\Omega} p\cos(\psi)dxdy \\ F = \sqrt{F_s^2 + F_c^2} \end{cases} \tag{5}$$

where F_s is horizontal component of the water film force; F_c is the vertical component of water film force; F is the combined force of the water film force; and Ω is the water film intact zone.

The friction coefficient equation is as follows:

$$\begin{cases} F_1 = u\int_{-l/2}^{l/2}\int_{\psi_1}^{\psi_2}\frac{\eta}{h}dxdy + u\int_{-l/2}^{l/2}h_b\int_{\psi_b}^{\psi_2}\frac{\eta}{h^2}dxdy \\ F_2 = \frac{e\sin\theta}{2r}F \\ F_T = F_1 + F_2 \\ f = \frac{F_t}{F} \end{cases} \tag{6}$$

where F_1 is the shear flow resistance, N; F_2 is the pressure flow resistance, N; F_T is the total resistance, N; f is the friction coefficient; Ψ_1, Ψ_2 are the starting and ending angular coordinates of the bearing, respectively; Ψ_b is the angular coordinate at the point of water film rupture; h_b is the water film thickness at the point of water film rupture; and l is the bearing length, m.

We dimensionalize Equations (1) to (6) and provide dimensionless relative units as follows:

$$x_0 = \frac{x}{\psi} = r;\ y_0 = \frac{y}{\lambda} = \frac{l}{2};\ \eta_0 = \frac{\eta}{\overline{\eta}};\ h_0 = \frac{h}{\overline{h}} = c;\ \varepsilon = \frac{e}{c};\ p_0 = \frac{p}{\overline{p}} = \frac{2\omega\eta_s}{\phi^2};\ \phi = \frac{c}{R}$$

c is the bearing radius clearance, m; R is the bearing radius, m; ϕ is the bearing clearance ratio; ε is eccentricity; η is the viscosity of the lubricant, and when the temperature is 20 °C, $\eta = \eta_0 = 0.001$ Pa·s; and l is the bearing length, m.

The dimensionless Reynolds equation for micro-groove water-lubricated stern tube bearings is

$$\frac{\partial}{\partial\psi}\left[\frac{\overline{h}^3}{\overline{\eta}}\frac{\partial\overline{p}}{\partial\psi}\right] + \left(\frac{d}{l}\right)^2\frac{\partial}{\partial\lambda}\left[\frac{\overline{h}^3}{\overline{\eta}}\frac{\partial\overline{p}}{\partial\lambda}\right] = 3\frac{\partial\left(\overline{h}\right)}{\partial\psi} \quad (7)$$

The dimensionless equation for the water film thickness $\overline{h}$ of micro-groove water-lubricated stern tube bearings is:

$$\overline{h} = \begin{cases} \overline{h}_s + \overline{\delta} & \text{Areas without texture} \\ \overline{h}_s + \overline{\delta} + \frac{D_t}{h_0} & \text{Areas with texture} \end{cases} \quad (8)$$

where $\overline{h}_s = 1 + \varepsilon\cos(\Psi - \theta)$, $\overline{\delta} = \frac{Gp_0\overline{p}}{h_0}$.

The equation for the dimensionless water film pressure in the horizontal direction is:

$$\overline{F}_s = -\iint_{\Omega} \overline{p}\sin(\psi)d\psi d\lambda \quad (9)$$

where Ω is the water film completion area.

The equation for the dimensionless water film pressure in the vertical direction is:

$$\overline{F}_c = -\iint_{\Omega} \overline{p}\cos(\psi)d\psi d\lambda \quad (10)$$

The dimensionless water film's load-carrying capacity equation obtained from Equations (9) and (10) is:

$$\overline{F} = \sqrt{\overline{F}_s^2 + \overline{F}_c^2} \quad (11)$$

The equation for dimensionless shear flow resistance is:

$$\overline{F}_1 = \frac{1}{2}\int_{-1}^{1}\int_{\psi_1}^{\psi_b}\frac{\overline{\eta}}{\overline{h}}d\psi d\lambda + \frac{1}{2}\int_{-1}^{1}\overline{h}_b\int_{\psi_1}^{\psi_b}\frac{\overline{\eta}}{\overline{h}^2}d\psi d\lambda \quad (12)$$

The equation for dimensionless pressure flow resistance is:

$$\overline{F}_2 = \frac{1}{2}\overline{F}\varepsilon\sin\theta \quad (13)$$

The total resistance equation obtained from Equations (12) and (13) is

$$\overline{F}_f = \overline{F}_1 + \overline{F}_2 \quad (14)$$

The friction coefficient equation obtained from Equations (11) to (14) is

$$f = \frac{\overline{F}_f}{\overline{F}}\phi \tag{15}$$

2.1. Discretization of the Reynolds Equation (7)

The finite difference method [34–37] is used to solve the Reynolds Equation (7) by dividing m mesh cell along the axial direction (λ direction) of the bearing and n mesh cell along the circumferential direction (τ direction) of the bearing; the mesh division is shown in Figure 2.

According to the idea of finite difference method, both $\frac{\partial p}{\partial \tau}$ and $\frac{\partial p}{\partial \lambda}$ can be expressed by the difference quotient formed by the pressure values on adjacent nodes.

$$\left(\frac{\partial \overline{p}}{\partial \psi}\right)_{i,j} \approx \frac{\overline{p}_{i+1,j} - \overline{p}_{i-1,j}}{2\Delta\psi};\ \left(\frac{\partial \overline{p}}{\partial \lambda}\right)_{i,j} \approx \frac{\overline{p}_{i,j-1} - \overline{p}_{i,j+1}}{2\Delta\lambda} \tag{16}$$

If the values on half a step size are used to form the difference quotient, the results will be more accurate.

$$\left(\frac{\partial \overline{p}}{\partial \psi}\right)_{i,j} \approx \frac{\overline{p}_{i+0.5,j} - \overline{p}_{i-0.5,j}}{\Delta\psi};\ \left(\frac{\partial \overline{p}}{\partial \lambda}\right)_{i,j} \approx \frac{\overline{p}_{i,j-0.5} - \overline{p}_{i,j+0.5}}{\Delta\lambda} \tag{17}$$

The second derivative at any node can be represented by the difference of the first derivative at the insertion point of adjacent half-steps.

$$\left(\frac{\partial}{\partial \psi}\left(\overline{h}^3\frac{\partial \overline{p}}{\partial \psi}\right)\right)_{i,j} \approx \frac{\left(\overline{h}^3\frac{\partial \overline{p}}{\partial \psi}\right)_{i+0.5,j} - \left(\overline{h}^3\frac{\partial \overline{p}}{\partial \psi}\right)_{i-0.5,j}}{\Delta\psi} \tag{18a}$$

$$\left(\frac{\partial}{\partial \lambda}\left(\overline{h}^3\frac{\partial \overline{p}}{\partial \lambda}\right)\right)_{i,j} \approx \frac{\left(\overline{h}^3\frac{\partial \overline{p}}{\partial \lambda}\right)_{i,j+0.5} - \left(\overline{h}^3\frac{\partial \overline{p}}{\partial \lambda}\right)_{i,j-0.5}}{\Delta\lambda} \tag{18b}$$

Furthermore, the first derivative in Equation (18) is represented by the difference quotient of the corresponding node.

$$\left(\overline{h}^3\frac{\partial \overline{p}}{\partial \psi}\right)_{i+0.5,j} \approx \overline{h}^3_{i+0.5,j}\frac{\overline{p}_{i+1,j} - \overline{p}_{i,j}}{\Delta\psi} \tag{19a}$$

$$\left(\overline{h}^3\frac{\partial \overline{p}}{\partial \psi}\right)_{i-0.5,j} \approx \overline{h}^3_{i-0.5,j}\frac{\overline{p}_{i,j} - \overline{p}_{i-1,j}}{\Delta\psi} \tag{19b}$$

$$\left(\overline{h}^3\frac{\partial \overline{p}}{\partial \lambda}\right)_{i,j+0.5} \approx \frac{\overline{h}^3_{i,j+0.5}\overline{p}_{i,j+1} - \overline{h}^3_{i,j+0.5}\overline{p}_{i,j}}{\Delta\lambda} \tag{19c}$$

$$\left(\overline{h}^3\frac{\partial \overline{p}}{\partial \lambda}\right)_{i,j-0.5} \approx \frac{\overline{h}^3_{i,j-0.5}\overline{p}_{i,j} - \overline{h}^3_{i,j-0.5}\overline{p}_{i,j-1}}{\Delta\lambda} \tag{19d}$$

We can obtain the following equation:

$$\left(\frac{\partial}{\partial \psi}\left(\overline{h}^3\frac{\partial \overline{p}}{\partial \psi}\right)\right)_{i,j} \approx \frac{\overline{h}^3_{i+0.5,j}\overline{p}_{i+1,j} + \overline{h}^3_{i-0.5,j}\overline{p}_{i-1,j} - \left(\overline{h}^3_{i+0.5,j} + \overline{h}^3_{i-0.5,j}\right)\overline{p}_{i,j}}{(\Delta\psi)^2} \tag{20a}$$

$$\left(\frac{\partial}{\partial\lambda}\left(\overline{h}^3\frac{\partial\overline{p}}{\partial\lambda}\right)\right)_{i,j}\approx\frac{\overline{h}^3_{i,j+0.5}\overline{p}_{i,j+1}+\overline{h}^3_{i,j-0.5}\overline{p}_{i,j-1}-\left(\overline{h}^3_{i,j+0.5}+\overline{h}^3_{i,j-0.5}\right)\overline{p}_{i,j}}{(\Delta\lambda)^2} \tag{20b}$$

Similarly, $\frac{\partial\overline{h}}{\partial\overline{p}}$ can also be expressed as

$$\left(\frac{\partial\overline{h}}{\partial\overline{p}}\right)_{i,j}\approx\frac{\overline{h}_{i+0.5,j}-\overline{h}_{i-0.5,j}}{\Delta\psi} \tag{21}$$

Through the above simplification, the dimensionless Reynolds equation of an incompressible water film with constant temperature, constant viscosity, and normal working conditions can be obtained:

$$\begin{array}{c}\frac{\overline{h}^3_{i+0.5,j}\overline{p}_{i+1,j}+\overline{h}^3_{i-0.5,j}\overline{p}_{i-1,j}-\left(\overline{h}^3_{i+0.5,j}+\overline{h}^3_{i-0.5,j}\right)\overline{p}_{i,j}}{(\Delta\psi)^2}+\\ \left(\frac{d}{l}\right)^2\frac{\overline{h}^3_{i,j+0.5}\overline{p}_{i,j+1}+\overline{h}^3_{i,j-0.5}\overline{p}_{i,j-1}-\left(\overline{h}^3_{i,j+0.5}+\overline{h}^3_{i,j-0.5}\right)\overline{p}_{i,j}}{(\Delta\lambda)^2}=\\ 3\frac{\overline{h}_{i+0.5,j}-\overline{h}_{i-0.5,j}}{\Delta\psi}\end{array} \tag{22}$$

After simplifying, we can obtain the discretized form of the continuous Reynolds Equation (7), as shown below:

$$B_{1i,j}\overline{p}_{i+1,j}+B_{2i,j}\overline{p}_{i-1,j}+B_{3i,j}\overline{p}_{i,j+1}+B_{4i,j}\overline{p}_{i,j-1}-B_{5i,j}\overline{p}_{i+1,j}=B_{6i,j} \tag{23}$$

The coefficients in Equation (23) are

$$\begin{cases}B_{1i,j}=\overline{h}^3_{i+0.5,j}\\ B_{2i,j}=\overline{h}^3_{i-0.5,j}\\ B_{3i,j}=\left(\frac{d}{l}\right)^2\left(\frac{\Delta\psi}{\Delta\lambda}\right)^2\overline{h}^3_{i,j+0.5}\\ B_{4i,j}=\left(\frac{d}{l}\right)^2\left(\frac{\Delta\psi}{\Delta\lambda}\right)^2\overline{h}^3_{i,j-0.5}\\ B_{5i,j}=B_{1i,j}+B_{2i,j}+B_{3i,j}+B_{4i,j}\\ B_{6i,j}=3\Delta\psi(h_{i+0.5,j}-h_{i-0.5,j})\end{cases} \tag{24}$$

2.2. Discretization of Elastic Deformation Equation $\overline{\delta}$, $\overline{\delta}$ in Equation (8)

Discretizing the continuous elastic deformation equation yields the following elastic deformation formulation:

$$\overline{\delta}_{i,j}=\frac{p_0}{h_0}\sum_i\sum_j G_{i,j}\overline{p}_{i,j} \tag{25}$$

2.3. Ultra-Relaxation Iterative Method for Solving Pressure Distribution and Pressure Convergence Conditions

Equation (23) obtained from the discretization is a system of linear algebraic equations, which is solved using the super-relaxed iterative method, calculated as:

$$(\overline{p})^k_{i,j}=\beta\left(\frac{B_{1i,j}(\overline{p})^{k-1}_{i+1,j}+B_{2i,j}(\overline{p})^{k-1}_{i-1,j}+B_{3i,j}(\overline{p})^{k-1}_{i,j+1}+B_{4i,j}(\overline{p})^{k-1}_{i,j-1}-B_{6i,j}}{B_{5i,j}}\right)+(\overline{p})^{k-1}_{i,j} \tag{26}$$

where β is the super relaxation factor, $1 < \beta < 2$; k is the number of stress iterations.

Combined with the boundary conditions, the pressure calculation formula can end the iterative calculation after a certain number of iterations to satisfy the following convergence conditions.

$$\frac{\sum\limits_{j=2}^{n}\sum\limits_{i=2}^{m}\left|(\overline{p})_{i,j}^{k}-(\overline{p})_{i,j}^{k-1}\right|}{\sum\limits_{j=2}^{n}\sum\limits_{i=2}^{m}\left|(\overline{p})_{i,j}^{k}\right|}\leq\zeta \quad (27)$$

ζ is the pressure convergence accuracy, $\zeta = 1.0 \times 10^{-4}$.

The diagram of calculation process is shown in Figure 3.

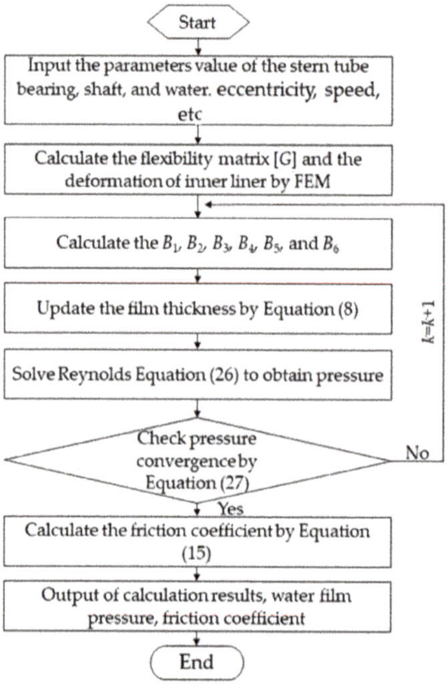

Figure 3. Diagram of calculation process.

3. Experiment

The test was carried out on a JXH-C liquid dynamic sliding bearing test bench, and the schematic diagram of the test bench is shown in Figure 4.

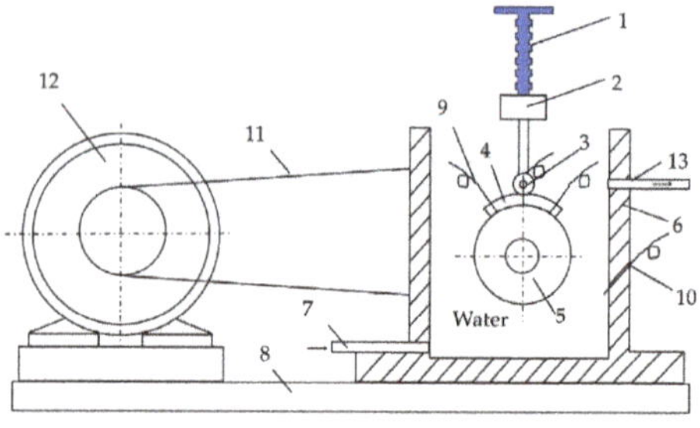

Figure 4. Test bench schematic: 1. screw for loading; 2. sensor for measuring loading force; 3. roller for loading; 4. test plate material to be tested; 5. shaft; 6. water tank; 7, 13. water hole; 8. base; 9. pressure sensor; 10. temperature sensor; 11. belt; 12. motor.

Experimental technique: The power part of the left side of the test bench is a small DC motor, for which the power is 600 W, the rated voltage is 220 V, and the maximum is 3000 r/min. The right side of the test part mainly consists of the test spindle and the two high-precision single-row centripetal ball bearings. The main spindle is equipped with a large pulley, and the pulley side of the belt wheel is equipped with an infrared velocimeter, which is used to measure the rotational speed of the journal. The diameter of the spindle is 60 mm, the length of the spindle is 100 mm, and the speed range of the spindle is 0~1500 r/min. Pressure sensor 9, as shown in Figure 4, uses an HM91A-H1-2-V2-F1-W1 sensor, with a maximum range of 2 MPa and an accuracy of 0.025%. A complete loading device (1, as shown in Figure 4) consisting of a screw loading rod, loading force sensor (2, as shown in Figure 4), and loading rollers (3, as shown in Figure 4) is mounted directly above the spindle. The device changes the size of the loading force through the screw, uses the loading roller to apply the loading force on the slat, and collects the size of the loading force through the loading force sensor. The loading force sensor 2 is a BHR-4 type load cell with a maximum range of 2000 N and an accuracy of 0.1%. The slats are arranged directly above the spindle for ease of loading. In order to simulate the dynamic pressure lubrication of the bearing slats more realistically, the slats should be completely submerged by water. The water in the tank is in a flowing state to ensure a constant water temperature, which is set at 20 °C.

Before the formal test, to ensure that the inner surface of the slat and the surface of the spindle had a better fit, the slat was soaked in water for 24 h, and then the slat was mounted on the test bench to break in for 30 min. After the break-in was completed, the formal test began. During the test, the water film pressure was measured by three pressure sensors on the slat, which were installed in the three round holes shown in Figure 5a.

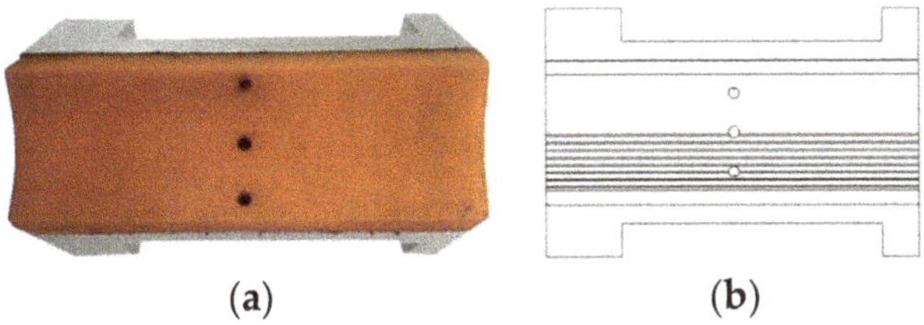

Figure 5. Test block with local micro-grooves: (**a**) image of the object; (**b**) schematic diagram.

To verify the correctness of the fluid solid coupling mathematical model of the water-lubricated micro-groove texture stern tube bearing, the lining material of the test strip was Thordon, the elastic modulus was 490 MPa, the density was 3220 kg/m^3, and Poisson's ratio was 0.45. The slab shell was 6061 aluminum alloy. The lining and shell were pasted with epoxy resin. There were two groups of slabs in this experiment: no micro-grooves and local micro-grooves. The shaft diameter was 60 mm, the bearing inner diameter was 60.25 mm, the bearing length was 100 mm, the rectangular groove was a through-type, the span was 0.4 mm, the groove spacing was 2 mm, and the groove depth was 50 μm. The load was 12 N and the speed was 1000 r/min. The local micro-groove lath was a micro-groove with a rectangular cross-section shape uniformly processed from the inlet edge to the circumferential 30° range. The test block with local micro-grooves is shown in Figure 5.

The water film pressure values were measured at 10°, 30°, and 50° positions of the test block, along the circumferential angle direction, three times at each position, and then the average value was taken. The tested average and simulated values are listed in Table 1.

Table 1. The water film pressure values of blocks.

	Without Grooves			Local Grooves		
Angle/°	**Simulation/kPa**	**Test/kPa**	**Relative Error**	**Simulation/kPa**	**Test/kPa**	**Relative Error**
10	0.02401	0.02423	0.91%	0.02466	0.02489	0.92%
30	0.06388	0.06423	5.44%	0.06411	0.06492	1.25%
50	0.01839	0.01856	0.92%	0.01872	0.01889	0.89%

The water film pressure test values of two types of micro-grooved stern tube bearing test blocks were obtained and compared with the simulation values, as shown in Figure 6.

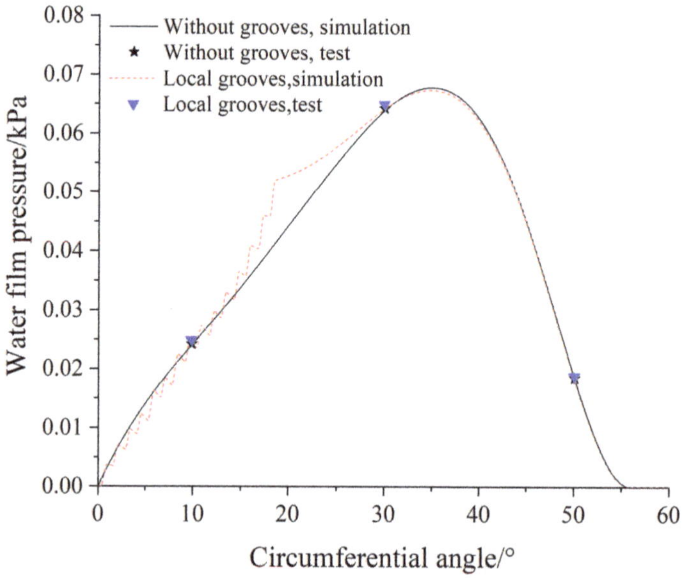

Figure 6. Tested and simulated pressure values of water film.

It can be seen from Figure 6 that the simulations of the two types of test blocks without micro-grooves and with local micro-grooves are in good agreement with the experimental water film pressure, thus verifying the correctness of the numerical results.

4. Results and Discussion

4.1. Effect of Groove Ratio

The groove ratio is the ratio of the area of the grooves to the area of the bearing lining. From the above experiments and simulations, the size of the groove area has a certain effect on the bearing load-carrying capacity. Therefore, we designed three kinds of through-type grooves with different cross-sectional shapes: rectangle, circle, and isosceles triangle. The shapes of the designed groove sections are shown in Figure 1. The geometric parameters of the groove were W_t = 0.4 mm, D_t = 0.05 mm, and P_t = 2 mm. The speed was 1000 r/min, and the eccentricity ε was 0.3. The groove ratio was set to 0, 0.25, 0.50, 0.75, and 1.00. The dimensionless bearing load-carrying capacity $\overline{F}$, Equation (11), and friction coefficient f, Equation (15), of the stern tube bearings were calculated, as shown in Figure 7.

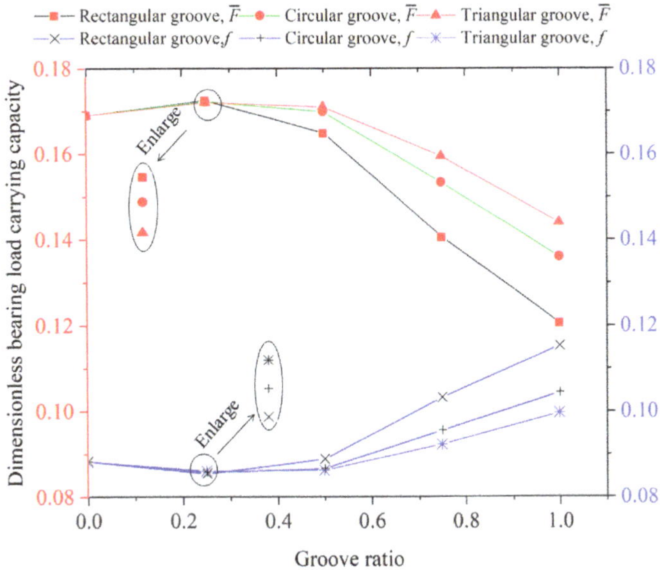

Figure 7. Effect of groove ratio on dimensional bearing load-carrying capacity $\overline{F}$, and friction coefficient f of stern tube bearing.

Figure 7 illustrates that the bearing load-carrying capacity of rectangular, circular, and isosceles triangular groove stern tube bearings increases nonlinearly with the groove ratio and reaches its maximum at around 0.25 before decreasing. At a groove ratio of 0.25, micro-grooves are positioned on the water inlet side of the minimum water film thickness of the stern tube bearing. During shaft rotation, water is drawn into the bearing gap, and the micro-grooves undergo micro deformation under load. This results in the formation of the water sac effect in the internal space of the micro-grooves, further increasing the hydrodynamic force and enhancing the bearing load-carrying capacity of the stern tube bearing. However, if the groove ratio exceeds 0.25, as the number of micro-grooves increases, the grooves will act badly on the water film, resulting in a thicker water film thickness in the groove region, which in turn causes a reduction in the water film pressure, leading to a reduction in the load-bearing capacity. The friction coefficient is inversely proportional to the bearing load-carrying capacity of the stern tube bearing, as shown in Equation (11). As the groove ratio increases, the friction coefficient decreases first and then increases nonlinearly. Comparing the load-carrying capacity and friction coefficient curves of the three different cross-sectional groove stern tube bearings, the rectangular groove texture stern tube bearing exhibits superior bearing load-carrying capacity and friction performance when the groove ratio is less than or equal to 0.25. When the groove ratio is greater than 0.25, the isosceles triangular groove texture stern tube bearing demonstrates better bearing load-carrying capacity and friction performance than the other two types.

To determine the optimal value for the local micro groove ratio, stern tube bearings with rectangular, circular, and isosceles triangular groove textures were selected with the target of achieving a maximum bearing load-carrying capacity and minimum friction coefficient. The speed was 1000 r/min and the eccentricity ε was 0.3. The calculation was carried out within the groove ratio range 0.20 to 0.35. Figure 8 shows the results obtained.

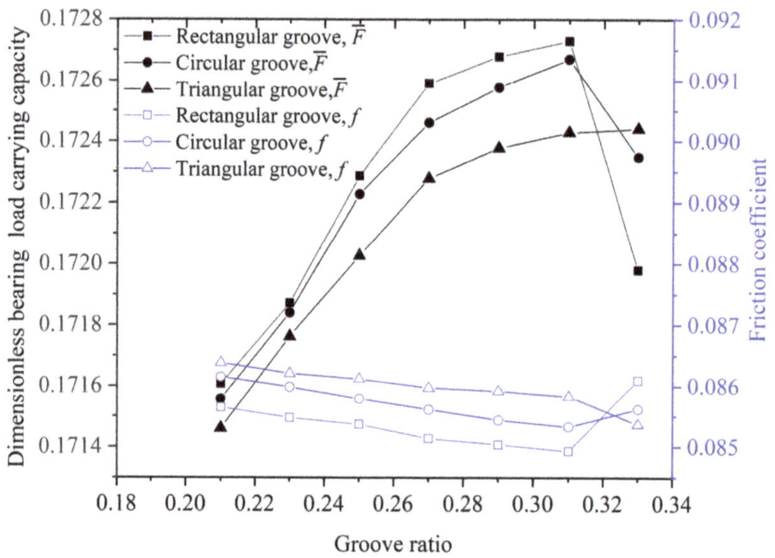

Figure 8. The dimensionless bearing load-carrying capacity and friction coefficient of the stern tube bearing as the groove ratio ranges from 0.20 to 0.35.

At a groove ratio of around 0.31, the rectangular groove stern tube bearing achieves its maximum bearing load-carrying capacity and minimum friction coefficient. To obtain optimal lubrication performance, especially with Thordon-lined water-lubricated stern tube bearings, a groove ratio of 0.30 to 0.32 is recommended. Isosceles triangular groove stern tube bearings outperform circular and rectangular groove stern tube bearings in terms of the bearing load-carrying capacity and friction coefficient at groove ratios above 0.31.

4.2. *Effect of Groove Width*

To investigate the effect of the groove width on the bearing load-carrying capacity and friction coefficient of stern tube bearings, we selected rectangular, circular, and isosceles triangular micro-groove stern tube bearings. The local groove ratio was fixed at 0.31, and the global groove ratio was set to 1.00. The groove dimensions were D_t = 0.05 mm and P_t = 2 mm, and the bearings were tested at a speed of 1000 r/min with eccentricity ε of 0.3. We varied the value of W_t from 200 to 1600 μm in intervals of 200 μm. The resulting calculations are presented in Figure 9.

As can be seen from Figure 9, under the same conditions, the bearing load-carrying capacity of the local micro-groove stern tube bearing test block is greater than that of the global micro-groove stern tube bearing test block. The reason for this is that the location of the local micro-groove is on the inlet side of the minimum water film thickness, which makes it easy to form the water sac effect and helps to increase the bearing load-carrying capacity. The global micro-groove stern tube bearing's groove position across the domain of the minimum water film thickness on both sides (that is, the inlet side and the outlet side; although a water bladder effect is formed on the inlet side, the thickness of the water film on the outlet side is thickened, resulting in a lower pressure on the water film on the outlet side) results in the bearing load-carrying capacity being less than that of the local micro-groove stern tube bearing.

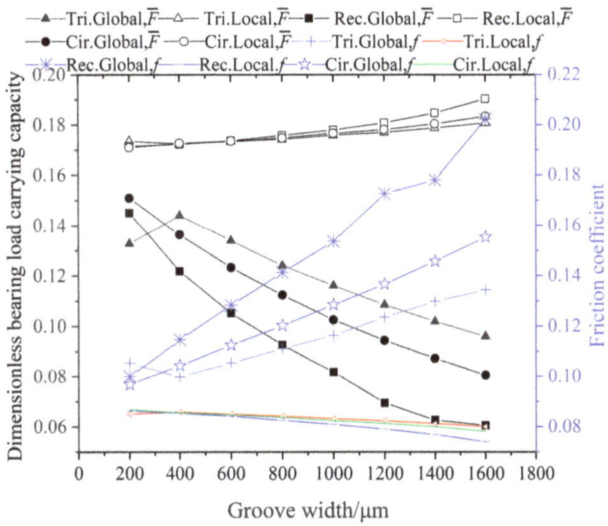

Figure 9. Effect of groove width on bearing load-carrying capacity and friction coefficient of stern tube Bearing. Tri. represents triangular; Rec. represents rectangular; Cir. represents circular; Global represents stern tube bearing with a triangular micro-groove ratio of 1.0; Local represents stern tube bearing with a triangular micro-groove ratio of 0.31; $\overline{F}$ represents the dimensionless bearing load-carrying capacity; and f represents the friction coefficient.

Comparing the water-lubricated stern tube bearing blocks with local micro-groove and global micro-grooves, the bearing load-carrying capacity of the local micro-groove and global micro-groove water-lubricated stern tube bearing blocks showed a nonlinear increase and a fractional linear decrease with the increase in the width of the micro-groove, respectively. This phenomenon is due to the location of the micro-groove. The groove of the local micro-groove water-lubricated stern tube bearing is on the water inlet side of the minimum water film thickness, and as the width of the micro-groove increases, a larger water sac is formed in the groove, and the larger water sac helps to increase the bearing load-carrying capacity of the water-lubricated stern tube bearing. The grooves of the global micro-groove water-lubricated stern tube bearing are not only distributed on the inlet side of the minimum water film thickness but also on the outlet side. As the width of the groove increases, the load-carrying capacity of the global micro-groove water-lubricated stern tube bearing increases on the inlet side, but the load-carrying capacity decreases due to the thickening of the water film thickness of the global micro-groove water lubricated stern tube bearings on the outlet side results in a lower water film pressure, and the decrease in the load-carrying capacity on the outlet side is greater than the increase in the load-carrying capacity on the inlet side as the width of the groove increases.

The dimensionless bearing load-carrying capacity of both global and local isosceles triangular micro-groove stern tube bearings first increases and then decreases nonlinearly with an increase in groove width. Similarly, the friction coefficient decreases first and then increases nonlinearly. For global and local rectangular groove stern tube bearings, the dimensionless bearing load-carrying capacity exhibits a nonlinear decreasing and increasing trend with the increase in groove width, while the friction coefficient shows a nonlinear increasing and decreasing trend. Moreover, the dimensionless bearing load-carrying capacity of global and local circular groove stern tube bearings shows a non-linear decreasing and non-linear increasing trend with an increase in groove width, and the friction coefficient exhibits a non-linear increasing and non-linear decreasing trend. Overall, the lubrication performance of the local rectangular groove stern tube bearing is superior to that of the isosceles triangle and circular groove stern tube bearings.

Under identical conditions, the bearing load-carrying capacity and friction performance of local groove stern tube bearings are significantly superior to those of global groove stern tube bearings. At a groove width of 1600 μm, the local rectangular groove stern tube bearing exhibits a 217% increase in dimensionless bearing load-carrying capacity compared to the global rectangular groove stern tube bearing (i.e., $|0.19 - 0.06|/0.06 \times 100\%$), while the friction coefficient decreases by 65% (i.e., $|0.20 - 0.07|/0.20 \times 100\%$). With the increase in groove width, the lubrication effect of the local rectangular groove water-lubricated stern tube bearing outperforms the other two types.

5. Conclusions

(1) Reasonably arranging the groove ratio is beneficial for improving the lubrication effect of water-lubricated stern tube bearings. Rectangular groove stern tube bearings exhibit superior bearing load-carrying capacity and friction coefficient performance compared to circular and isosceles triangular micro-groove stern tube bearings when the groove ratio is between 0.30 and 0.32, resulting in the best lubrication performance. Isosceles triangular groove stern tube bearings exhibit better bearing load-carrying capacity and friction coefficient performance than the other two types when the groove ratio is above 0.31.

(2) Increasing the width of the groove results in a significantly better lubrication effect of the local rectangular groove stern tube bearing compared to other stern bearings. The dimensionless bearing load-carrying capacity can be increased by 217%, and the friction coefficient can be reduced by 65%.

6. Further Research

In this study, we focused on investigating the water-lubricated stern tube bearing made of Thordon material as the research subject. We aimed to explore the impact of microscopic surface groove structures on the lubrication performance of water-lubricated stern tube bearings. The selection of lining materials for water-lubricated stern tube bearings is not restricted to Thordon material alone. Other options, such as NBR, phenol, ACM, SF, and iron pearwood, can also be utilized as linings for water-lubricated stern tube bearings.

Materials such as Thordon, phenol, ACM, SF, and iron pearwood exhibit high hardness. Thus, when they are employed as liners for water-lubricated stern tube bearings, it becomes more convenient to incorporate micro-grooves into the inner surface of the liner. However, the task of creating micro-grooves in rubber-based linings poses a significant challenge due to the low hardness and susceptibility to deformation under force. This characteristic greatly complicates the process of machining micro-grooves. In practical operations, the geometry of micro-grooves becomes difficult to maintain due to the extrusion and deformation of the rubber-based liner, resulting in substantial distortions of the micro-groove shape. Consequently, researchers are interested in investigating the lubrication characteristics of water-lubricated stern tube bearings equipped with micro-grooves. To facilitate their study, we recommend that scholars opt for materials with higher hardness.

Funding: This research was funded by Jiangxi Provincial Department of Science and Technology, grant number No. 20192BBEL50028, and the Jining University's "Hundred Outstanding Talents" Support Program Cultivation Project, and the APC was funded by No. 20192BBEL50028.

Data Availability Statement: Data sharing is not applicable to this article.

Acknowledgments: This work is supported by the Jiangxi Provincial Department of Science and Technology (No. 20192BBEL50028) and the Jining University's "Hundred Outstanding Talents" Support Program Cultivation Project.

Conflicts of Interest: The author declares no conflict of interest. The funders had no role in the design of the study, in the collection, analyses, or interpretation of data, in the writing of the manuscript, or in the decision to publish the results.

Nomenclature

c	bearing radius clearance
D_t	depth of the micro-grooves
e	eccentricity distance
"*elem*"	the number of elements on the inner surface of the bearing
f	friction coefficient
F	combined force of the water film force
$\overline{F}$	dimensionless bearing load-carrying capacity
F_1	shear flow resistance
F_2	pressure flow resistance
F_c	vertical component of water film force
F_s	horizontal component of the water film force
F_t	radial load applied on the spindle
F_T	total resistance
$[G]$	flexibility matrix
h	water film thickness
h_b	water film thickness at the point of water film rupture
h_{min}	minimum film thickness
h_{max}	maximum film thickness
h_s	thickness of the water film when ignoring the bearing deformation and groove
k	number of stress iterations
l	bearing length
"*node*"	the number of nodes on the inner surface of the bearing
O_1	centers of the bearing
O_2	centers of the shaft
p	water film pressure
P_t	span of the micro-grooves
$\{p\}$	average pressure of the inner surface elements of the bearing
r	radius of the shaft
R	inner radius of the bearing
W_t	width of the micro-grooves
β	super relaxation factor
γ	circumferential arrangement range of the micro-groove area
δ	radial deformation of the bearing inner surface nodes
$\{\delta\}$	radial deformation of bearing inner surface nodes
ε	eccentricity ratio
ζ	pressure convergence accuracy
η	viscosity of the water
θ	attitude angle
θ_f	angular coordinate at the position of natural rupture of the water film
λ	axial direction coordinate of bearing
φ	circumferential coordinate starting from the maximum film thickness
ω	rotational speed of shaft

References

1. Prajapati, D.K.; Ramkumar, P. Surface topography effect on tribological performance of water-lubricated journal bearing under mixed-EHL regime. *Surf. Topogr. Metrol. Prop.* **2022**, *10*, 045022. [CrossRef]
2. Chen, J.; Guo, Z.; Li, X.; Yuan, C. Development of gradient structural composite for improving tribological performance of PU material in water-lubricated bearings. *Tribol. Int.* **2022**, *176*, 107876. [CrossRef]
3. Lv, F.; Jiao, C.; Xu, J. Theoretical and experimental investigation on effective length-to-diameter ratio of misaligned water-lubricated bearings. *Tribol. Int.* **2023**, *187*, 108712. [CrossRef]
4. Xiang, G.; Wang, J.; Han, Y.; Yang, T.; Dai, H.; Yao, B.; Zhou, C.; Wang, L. Investigation on the nonlinear dynamic behaviors of water-lubricated bearings considering mixed thermoelastohydrodynamic performances. *Mech. Syst. Signal Process.* **2021**, *169*, 108627. [CrossRef]
5. Hanawa, N.; Kuniyoshi, M.; Miyatake, M.; Yoshimoto, S. Static characteristics of a water-lubricated hydrostatic thrust bearing with a porous land region and a capillary restrictor. *Precis. Eng.* **2017**, *50*, 293–307. [CrossRef]

6. Wu, Z.; Guo, Z.; Yuan, C. Insight into water lubrication performance of polyetheretherketone. *J. Appl. Polym. Sci.* **2020**, *138*, 49701. [CrossRef]
7. Hamilton, D.B.; Walowit, J.A.; Allen, C.M. A Theory of Lubrication by Microirregularities. *J. Basic Eng.* **1966**, *88*, 177–185. [CrossRef]
8. Etsion, I.; Burstein, L. A Model for Mechanical Seals with Regular Microsurface Structure. *Tribol. Trans.* **1996**, *39*, 677–683. [CrossRef]
9. Tala-Ighil, N.; Fillon, M. A numerical investigation of both thermal and texturing surface effects on the journal bearings static characteristics. *Tribol. Int.* **2015**, *90*, 228–239. [CrossRef]
10. Tala-Ighil, N.; Fillon, M.; Maspeyrot, P. Effect of textured area on the performances of a hydrodynamic journal bearing. *Tribol. Int.* **2011**, *44*, 211–219. [CrossRef]
11. Choudhury, D.; Rebenda, D.; Sasaki, S.; Hekrle, P.; Vrbka, M.; Zou, M. Enhanced lubricant film formation through micro-dimpled hard-on-hard artificial hip joint: An in-situ observation of dimple shape effects. *J. Mech. Behav. Biomed. Mater.* **2018**, *81*, 120–129. [CrossRef]
12. Shinde, A.; Pawar, P.; Shaikh, P.; Wangikar, S.; Salunkhe, S.; Dhamgaye, V. Experimental and numerical analysis of conical shape hydrodynamic journal bering with partial texturing. *Procedia Manuf.* **2018**, *20*, 300–310. [CrossRef]
13. Dong, N.; Wang, Y.Q.; Liu, Q.; Huang, X.B. Influence of Micro-Morphology on Thermal Elatohydrodynamic Lubrication of Water-Lubricated Tenmat Bearing. *Appl. Mech. Mater.* **2015**, *743*, 85–90. [CrossRef]
14. Wang, Y. Thermal Elastohydrodynamic Lubrication Property Analysis of Water-lubricated Tenmat Bearing Considering Debris and Surface Roughness. *J. Mech. Eng.* **2017**, *53*, 121–129. [CrossRef]
15. He, Z.; Song, Q.; Liu, Q.; Xin, J.; Yang, C.; Liu, M.; Li, B.; Yan, F. Analysis of the effect of texturing parameters on the static characteristics of radial rigid bore aerodynamic journal bearings. *Surf. Topogr. Metrol. Prop.* **2022**, *10*, 035025. [CrossRef]
16. Blatter, A.; Maillat, M.; Pimenov, S.; Shafeev, G.; Simakin, A.; Loubnin, E. Lubricated sliding performance of laser-patterned sapphire. *Wear* **1999**, *232*, 226–230. [CrossRef]
17. Yuan, S.; Huang, W.; Wang, X. Orientation effects of micro-grooves on sliding surfaces. *Tribol. Int.* **2011**, *44*, 1047–1054. [CrossRef]
18. A Wahab, J.; Ghazali, M.; Sajuri, Z.; Otsuka, Y.; Jayaprakash, M.; Nakamura, S.; Baharin, A. Effects of micro-grooves on tribological behaviour of plasma-sprayed alumina-13%titania coatings. *Ceram. Int.* **2017**, *43*, 6410–6416. [CrossRef]
19. Ali, F.; Kaneta, M.; Křupka, I.; Hartl, M. Experimental and numerical investigation on the behavior of transverse limited micro-grooves in EHL point contacts. *Tribol. Int.* **2015**, *84*, 81–89. [CrossRef]
20. Ehret, P.; Dowson, D.; Taylor, C. Waviness Orientation in EHL Point Contact. In *Tribology Series*; Elsevier: Amsterdam, The Netherlands, 1996; Volume 31, pp. 235–244. [CrossRef]
21. Thakre, G.D.; Sharma, S.C.; Harsha, S.; Tyagi, M. A theoretical study of ionic liquid lubricated μ-EHL line contacts considering surface texture. *Tribol. Int.* **2016**, *94*, 39–51. [CrossRef]
22. Kaneta, M.; Guo, F.; Wang, J.; Krupka, I.; Hartl, M. Formation of micro-grooves under impact loading in elliptical contacts with surface ridges. *Tribol. Int.* **2013**, *65*, 336–345. [CrossRef]
23. Shi, X.; Ni, T. Effects of groove textures on fully lubricated sliding with cavitation. *Tribol. Int.* **2011**, *44*, 2022–2028. [CrossRef]
24. Gong, J.; Jin, Y.; Liu, Z.; Jiang, H.; Xiao, M. Study on influencing factors of lubrication performance of water-lubricated micro-groove bearing. *Tribol. Int.* **2018**, *129*, 390–397. [CrossRef]
25. Wu, Z.; Yuan, C.; Guo, Z.; Huang, Q. Effect of the groove parameters on the lubricating performance of the water-lubricated bearing under low speed. *Wear* **2023**, *522*, 204708. [CrossRef]
26. Meng, F.; Chen, Y. Analysis of elasto-hydrodynamic lubrication of journal bearing based on different numerical methods. *Ind. Lubr. Tribol.* **2015**, *67*, 486–497. [CrossRef]
27. Su, B.; Huang, L.; Huang, W.; Wang, X. The load carrying capacity of textured sliding bearings with elastic deformation. *Tribol. Int.* **2017**, *109*, 86–96. [CrossRef]
28. Liu, S. On boundary conditions in lubrication with one dimensional analytical solutions. *Tribol. Int.* **2012**, *48*, 182–190. [CrossRef]
29. Du, Y.; Li, M. Effects on lubrication characteristics of water-lubricated rubber bearings with journal tilting and surface roughness. *Proc. Inst. Mech. Eng. Part J J. Eng. Tribol.* **2019**, *234*, 161–171. [CrossRef]
30. Xie, Z.; Liu, H. Experimental research on the interface lubrication regimes transition of water lubricated bearing. *Mech. Syst. Signal Process.* **2020**, *136*, 106522. [CrossRef]
31. Chen, C.; Lin, G.; Hu, Z. An innovative and efficient solution for axisymmetric contact problem between structure and half-space. *Eng. Anal. Bound. Elements* **2022**, *142*, 10–27. [CrossRef]
32. Liu, H.; Wu, B.; Li, Z. The generalized flexibility matrix method for structural damage detection with incomplete mode shape data. *Inverse Probl. Sci. Eng.* **2021**, *29*, 2019–2039. [CrossRef]
33. Mukhopadhyay, M.; Sheikh, A.H. *The Flexibility Matrix Method*; Springer International Publishing: Cham, Switzerland, 2022; pp. 133–148. [CrossRef]
34. Rekha, J.; Suma, S.P.; Shilpa, B.; Khan, U.; Hussain, S.M.; Zaib, A.; Galal, A.M. Solute transport exponentially varies with time in an unsaturated zone using finite element and finite difference method. *Int. J. Mod. Phys. B* **2022**, *37*, 2350089. [CrossRef]
35. Liu, G.; Li, M. Lubrication Characteristics of Water-Lubricated Rubber Bearings Based on an Improved Algorithm of the Mass Conservation Boundary Condition. *J. Tribol.* **2022**, *144*, 081804. [CrossRef]

36. Du, Y.; Quan, H. Characteristics of water-lubricated rubber bearings in mixed-flow lubrication state. *Lubr. Sci.* **2022**, *34*, 224–234. [CrossRef]
37. Prajapati, D.K.; Katiyar, J.K.; Prakash, C. Determination of friction coefficient for water-lubricated journal bearing considering rough surface EHL contacts. *Int. J. Interact. Des. Manuf.* **2023**, 1–9. [CrossRef]

lubricants

MDPI

Article

The Influence of Scratches on the Tribological Performance of Friction Pairs Made of Different Materials under Water-Lubrication Conditions

Qingchen Liang [1,2], Peng Liang [1,2,*], Feng Guo [1,2], Shuyi Li [1], Xiaohan Zhang [1] and Fulin Jiang [1]

1 School of Mechanical and Automotive Engineering, Qingdao University of Technology, Qingdao 266520, China; liangqingchen1@126.com (Q.L.)

2 Key Lab of Industrial Fluid Energy Conservation and Pollution Control, Qingdao University of Technology, Ministry of Education, Qingdao 266520, China

* Correspondence: liangpeng2009@126.com; Tel.: +86-532-88392179; Fax: +86-532-88395625

Abstract: Water-lubricated bearings are widely used in marine equipment, and the lubricating water often contains hard particles. Once these particles enter the gap between the bearing and the shaft, they can scratch the smooth surfaces of the shaft and bearing, influencing the working performance of the bearing system. To investigate the effect of scratch parameters on tribological performance, this paper conducts multiple block-on-ring experiments and constructs a mixed-lubrication model under water-lubrication conditions. The results show that among the three commonly used bearing materials, the tribological performance of graphite block is the most sensitive to scratches on the test ring surface. Under the condition of one scratch ($N = 1$), the loading area of water film pressure is divided into two separate zones (a trapezoidal pressure zone and an extremely low-pressure zone). In addition, the variation of maximum water film pressure is determined by the positive effect (hydrodynamic pressure effect of fluid) and negative effect ("piercing effect" of the asperities). Compared with the scratch depth and scratch location, the scratch width has the most significant effect on the tribological performance of the block-on-ring system. The maximum contact pressure is located at both edges of the scratch due to the formation of a water sac structure. The scratch has a great influence on the transition of the lubrication state of the block-on-ring system. The existence of scratches increases the critical speed at which the lubrication state transits from mixed-lubrication to elastohydrodynamic lubrication, and the critical speed is directly proportional to the scratch width.

Keywords: surface scratch; tribological performance; block-on-ring; elastohydrodynamic lubrication; scratch width

Citation: Liang, Q.; Liang, P.; Guo, F.; Li, S.; Zhang, X.; Jiang, F. The Influence of Scratches on the Tribological Performance of Friction Pairs Made of Different Materials under Water-Lubrication Conditions. *Lubricants* **2023**, *11*, 449. https://doi.org/10.3390/lubricants11100449

Received: 11 August 2023
Revised: 6 September 2023
Accepted: 11 October 2023
Published: 17 October 2023

1. Introduction

Due to the advantages of a simple structure, no pollution, and a good cooling effect, water-lubricated bearings are used as a key transmission component [1–3] in ship propulsion systems. However, because the lubricating water has low viscosity and poor load capacity, the bearing and shaft often operate in a mixed-lubrication state, resulting in rapid wear [4–6] of the bearing. The friction behavior under water-lubrication conditions has attracted great interest [7–10] from many scholars. Jia et al. [11] investigated the friction properties between PI-based composites and stainless steel using a block-on-ring test rig. The results showed that the plastic deformation and micro-cracking phenomena of the composite material were reduced under water-lubrication conditions, and the anti-wear performance of the composite material was enhanced. Chen et al. [12] studied the tribological properties of carbon fiber-reinforced polyetheretherketone composites under different lubricant conditions. The friction coefficient under the seawater lubrication condition was lower than that under the pure water-lubrication condition. This was because seawater helps to form compounds and deposit them on friction surfaces. Zhao et al. [13] discovered that increasing the content of potassium titanate whiskers helped reduce the wear rate of polyurethane composites under water-lubrication conditions. Xiao et al. [14] conducted block-on-ring friction experiments under different loads and speeds. Among the four lubrication conditions (dry friction, pure water lubrication, seawater lubrication, and sand water lubrication), the friction coefficient and wear rate under seawater lubrication were the lowest. Huang et al. [15] investigated the frictional characteristics of stainless-steel blocks and alloy steel rings under water-lubrication conditions. They found that the friction pairs entered a stable friction state (friction coefficient $\mu \geq 0.5$) after a running-in distance of 3 m. Wang et al. [16] studied the friction behavior between a rubber block and a coated ring whose surface had Cr/CrN/GLC coatings. Compared with the uncoated ring, the coated ring exhibited a markedly lower friction coefficient and wear rate under water-lubrication conditions. Xiong [17] pointed out that carbon fiber could improve the friction performance of a test block made of ultra-high molecular weight polyethylene under water-lubrication conditions. Nobili et al. [18] deposited DLC coatings on different substrate surfaces to obtain multiple groups of test blocks, and experimental results showed that DLC coatings deposited on high-hardness substrate surfaces had better wear resistance. Xu et al. [19] discovered that the friction coefficient (0.12) of a test block coated with carbon fiber-reinforced polyphenylene sulfide (PPS) under water lubrication was much lower than that (0.40) under dry friction. All the above studies focus on using the block-on-ring test rig to evaluate the friction performance under water-lubrication conditions, but there is currently a lack of research on using the block-on-ring theoretical model to explain the experimental phenomenon. Therefore, it is necessary to construct a finite-width mixed-lubrication model and research the friction behavior and the lubrication performance of the block-on-ring system.

If the lubricant contains hard particles, once particles enter the gap between the bearing and the shaft, the surfaces of the friction pairs may be scratched, and the scratched surfaces will affect the lubrication performance of the bearings. Currently, scholars have gradually started to pay attention to the relationship between surface scratches and the working performance of the bearing systems [20,21]. Dobrica et al. [22,23] used the numerical simulation method to study the influence of surface scratches on the performance of journal bearings. Shallow scratches had almost no effect on bearing performance, while deep scratches had a destructive effect. There was almost no fluid hydrodynamic pressure effect in the deep scratches, resulting in zero local pressure. Chatterton et al. [24] found that the dynamic performance of tilting pad bearings decreased with an increase in scratch depth and scratch number. Vo [25] discovered that scratches had a strong influence on bearings, with a significant decrease in film pressure in the scratched area and a reduction in the load capacity of the bearing. Deeper scratches increased the shaft center eccentricity and decreased the minimum film thickness. Chasalevris et al. [26] studied the influence of bearing scratch on system response and found subharmonics and superharmonics in the response signals of the worn bearing through continuous wavelet transform. Nicodemus et al. [27] researched the working performance of the scratched bearings under micro-polar lubricant conditions. The micro-polar effect made polar lubricant more sensitive to bearing scratches than Newtonian lubricant. Cai et al. [28] discussed the influence of the scratch of water-lubricated bearing on system dynamics response. The interesting thing is that the dynamic performance of water-lubricated bearings is improved under certain reasonable wear parameters. The above studies are all about the effects of bearing scratches on the static and dynamic performance of the system.

In fact, scratches may also appear on shaft surfaces, which is similar to the phenomenon whereby wear may occur both on the tool surface and the workpiece surface when the tool cuts the workpiece [29]. The scratches on the shaft surfaces will also significantly affect the bearing system, so scholars have started studying the influence of shaft scratches on bearing lubrication performance [30]. Jean et al. [31,32] found that different scratch locations could change the fluid pressure distribution. Branagan [33,34] simplified the long-bearing model into a short-bearing model to study the effects of scratch locations on the performance of oil-lubricated bearings. When the shaft scratch was located at the bearing center, it had a much greater effect on the bearing lubrication performance compared with the condition that the scratch was located at the bearing edge. Giraudeau et al. [35] studied the influence of the shaft scratch on the performance of double-leaf oil-lubricated bearings with experiments. When the order of magnitude of scratch depth and the minimum film thickness were the same, the existence of scratch dramatically affected the distribution of film pressure. To sum up, the above studies focus on the oil-lubricated bearings, and the structure parameters and location parameters of shaft scratches can markedly influence the pressure and thickness of the lubricating oil film.

Compared to oil-lubricated bearings, water-lubricated bearings often work in harsh environments where the lubricating water may contain hard particles (such as sand). When these hard particles enter the bearing clearance, although most of them will flow out from the bearing groove accompanying the flowing water, a small number of large particles may remain in the clearance between the shaft and the bearing. These large particles will scratch the smooth surfaces [36] of the shaft and the bearing when the shaft rotates, which is proved by the literature [37]. Regardless of whether scratches appear on the surface of the shaft or bearing, scratches will affect the lubrication performance and working quality of the bearing system. Because all the previous studies focus on the effects of shaft scratches on the oil-lubricated bearings, research on the influences of shaft scratches on the water-lubricated bearings is still in a blank state. In our previous work [38,39], we have already investigated the influence of shaft scratches on the lubrication performance of water-lubricated bearings using a theoretical model. However, further research is needed to study the effect of scratches on friction and wear performance of friction pairs by experiment. This paper experimentally studies the tribological performance under different scratch parameters and friction pair materials. We also established a theoretical model for a block-on-ring system to explain the change law of performance parameters observed in the experiments.

2. Theoretical Model

The model of the block-on-ring system is shown in Figure 1. The variable R represents the outer radius of the test ring, O represents the center of the test ring, ω represents the angular velocity of the test ring, Ω_1 represents the unscratched area, and Ω_2 represents the scratched area.

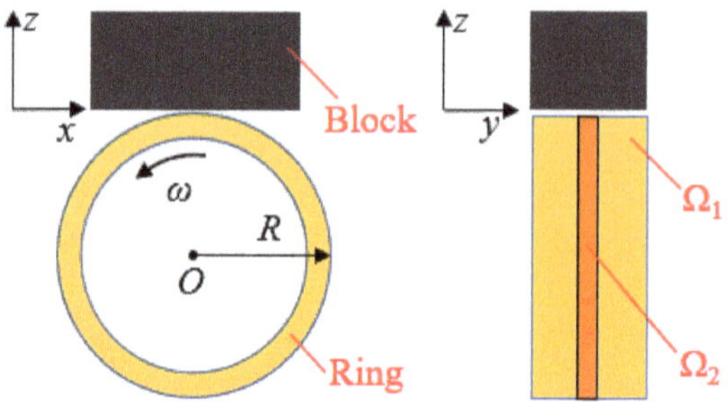

Figure 1. Block-on-ring friction pairs.

2.1. *Average Flow Reynolds Equation*

The average flow Reynolds equation [40,41] considering surface roughness and fluid flow state is used to calculate the fluid pressure, i.e.,

$$\begin{cases} \frac{\partial}{\partial x}\left(\phi_x \frac{h^3}{\eta}\frac{\partial p}{\partial x}\right) + \frac{\partial}{\partial y}\left(\phi_y \frac{h^3}{\eta}\frac{\partial p}{\partial y}\right) = 6u_s\phi_c\frac{\partial h}{\partial x} + 6u_s\sigma\frac{\partial \phi_s}{\partial x} & , Re < Re_c \\ \frac{\partial}{\partial x}\left(\phi_x \frac{h^3}{k_x\eta}\frac{\partial p}{\partial x}\right) + \frac{\partial}{\partial y}\left(\phi_y \frac{h^3}{k_y\eta}\frac{\partial p}{\partial y}\right) = 6u_s\phi_c\frac{\partial h}{\partial x} + 6u_s\sigma\frac{\partial \phi_s}{\partial x} & , Re \geq Re_c \end{cases} \quad (1)$$

where η is the dynamic viscosity of water, p is the water film pressure, h is the water film thickness, u_s is the linear velocity of the outer surface of the test ring, and (x, y) are the coordinates of points along block length and width directions. $\sigma = \sqrt{\sigma_b^2 + \sigma_r^2}$ is the combined surface roughness of test ring and test block, σ_b is the roughness of test block, and σ_r is the roughness of the test ring. $Re = \rho u_s h_{00}/\eta$ is the Reynolds number [42], ρ is the density of water, and h_{00} is the rigid central film thickness [43]. Re_c is the critical Reynolds number. $k_x = 1 + (0.0136/12)Re^{0.9}$ and $k_y = 1 + (0.0136/12)Re^{0.98}$ are both turbulence factors [44]. In this paper, the Reynolds number Re is 216.3, and the Reynolds number is significantly smaller than the critical Reynolds number [45] Re_c (2000), and the flow state of the lubricant is laminar [46]. (ϕ_x, ϕ_y) are the pressure flow factors [40] along x and y directions, ϕ_s is the shear flow factor [41], and ϕ_c is the contact factor [47]. The calculation formulas of pressure flow factors are

$$\begin{cases} \phi_x = 1 - Ce^{-r\left(\frac{h}{\sigma}\right)} & ,\gamma \le 1 \\ \phi_x = 1 + C\left(\frac{h}{\sigma}\right)^{-r} & ,\gamma > 1 \end{cases} \tag{2}$$

$$\phi_y\left(\frac{h}{\sigma}, \gamma\right) = \phi_x\left(\frac{h}{\sigma}, \frac{1}{\gamma}\right) \tag{3}$$

where the values of three variables (γ, C, r) in the pressure flow factor formulas can be selected according to reference [40].

The shear flow factor can be gained as

$$\phi_s = \left(\frac{\sigma_r}{\sigma}\right)^2 \Phi_s\left(\frac{h}{\sigma}, \gamma_1\right) - \left(\frac{\sigma_b}{\sigma}\right)^2 \Phi_s\left(\frac{h}{\sigma}, \gamma_2\right) \tag{4}$$

$$\begin{cases} \Phi_s = A_1\left(\frac{h}{\sigma}\right)^{\alpha_1} e^{-\alpha_2\left(\frac{h}{\sigma}\right) + \alpha_3\left(\frac{h}{\sigma}\right)^2} & ,\frac{h}{\sigma} \le 5 \\ \Phi_s = A_2 e^{-0.25\left(\frac{h}{\sigma}\right)} & ,\frac{h}{\sigma} > 5 \end{cases} \tag{5}$$

where the value selection of five variables $(A_1, A_2, \alpha_1, \alpha_2, \alpha_3)$ in Equations (4) and (5) can be found in reference [41].

The equation of contact factor is

$$\begin{cases} \phi_c = e^{-0.6912 + 0.782\frac{h}{\sigma} - 0.304\left(\frac{h}{\sigma}\right)^2 + 0.0401\left(\frac{h}{\sigma}\right)^3} & ,0 \le \frac{h}{\sigma} < 3 \\ \phi_c = 1 & ,\frac{h}{\sigma} \ge 3 \end{cases} \tag{6}$$

where h/σ is the film thickness ratio.

2.2. Film Thickness Equation

According to the geometric relationship and elastic mechanics theory of the block-on-ring system, the calculation formula of the film thickness is

$$\begin{cases} h(x,y) = h_{00} + \frac{x^2}{2R} + \frac{2}{\pi E'} \iint \frac{p(x',y')}{\sqrt{(x-x')^2 + (y-y')^2}} dx'dy' & ,\Omega \in \Omega_1 \\ h(x,y) = h_{00} + \frac{x^2}{2R} + h_c + \frac{2}{\pi E'} \iint \frac{p(x',y')}{\sqrt{(x-x')^2 + (y-y')^2}} dx'dy' & ,\Omega \in \Omega_2 \end{cases} \tag{7}$$

where h_c is the scratch depth, and Ω is the computational domain. $E' = \left[\frac{(1-v_r^2)}{E_r} + \frac{(1-v_b^2)}{E_b}\right]^{-1}$ is the composite elastic modulus, (v_r, v_b) are the Poisson ratio of ring and block, respectively, and (E_r, E_b) are the elastic modulus of ring and block, respectively.

2.3. Asperity Contact Model

To calculate the solid contact pressure and contact area between the ring surface and block surface, this paper adopts the GW model, i.e.,

$$\begin{cases} p_{asp} = \frac{16\sqrt{2}\pi}{15}(\sigma\beta D)^2\sqrt{\frac{\sigma}{\beta}}E'F_{2.5}\left(\frac{h}{\sigma}\right) \\ A_c = \pi^2(\sigma\beta D)^2 A_0 F_{2.0}\left(\frac{h}{\sigma}\right) \end{cases} \tag{8}$$

where p_{asp} is the contact pressure, β is the curvature radius of asperities on either surface, and D is the number of asperities per unit contact area. In this paper, the value of β is 2.0×10^{-6} m, and that of D is $5.0 \times 10^{11}/\text{m}^2$. A_c and A_0 are the real contact area and nominal contact area, respectively. $F_{2.5}$ and $F_{2.0}$ are both Gaussian probability density functions, and the calculation formulas are [48,49]

$$F_{2.5}\left(\frac{h}{\sigma}\right) = \begin{cases} 2.134 \cdot 10^{-4} e^{3.804ln\left(4-\frac{h}{\sigma}\right)+1.34\left[ln\left(4-\frac{h}{\sigma}\right)\right]^2} & ,\frac{h}{\sigma} \le 3.5 \\ 1.12 \cdot 10^{-4}\left(4-\frac{h}{\sigma}\right)^{1.9474} & ,3.5 < \frac{h}{\sigma} \le 4.0 \\ 0 & ,\frac{h}{\sigma} > 4.0 \end{cases} \tag{9}$$

$$F_{2.0}\left(\frac{h}{\sigma}\right) = \begin{cases} 1.705 \cdot 10^{-5} e^{4.504ln\left(4-\frac{h}{\sigma}\right)+1.37\left[ln\left(4-\frac{h}{\sigma}\right)\right]^2} & ,\frac{h}{\sigma} \le 3.5 \\ 8.8123 \cdot 10^{-5}\left(4-\frac{h}{\sigma}\right)^{2.15} & ,3.5 < \frac{h}{\sigma} \le 4.0 \\ 0 & ,\frac{h}{\sigma} > 4.0 \end{cases} \tag{10}$$

2.4. Load Capacity Equation

The synthetic load capacity can be obtained as

$$W = F_h + F_c = \iint\limits_{\Omega_1+\Omega_2} p(x,y)dxdy + \iint\limits_{\Omega_1+\Omega_2} p_{asp}(x,y)dxdy \tag{11}$$

where W is the synthetic load capacity, F_h is the component of load capacity generated by water film, F_c is the component of load capacity generated by asperity contact, Ω_1 represents the unscratched area, and Ω_2 represents the scratched area.

2.5. Friction Coefficient Equation

Friction is generated by fluid shear force and asperity shear force, so the calculation formulas [50] of friction f and friction coefficient μ are given as

$$f = \iint\limits_{\Omega_1+\Omega_2}\left(\frac{\eta\omega R^2}{h} + \frac{h}{2}\frac{\partial p}{\partial x}\right)dxdy + \sum_{i=1}^{N}\iint\limits_{\Omega_1+\Omega_2}\mu_c p_{asp}dxdy \tag{12}$$

$$\mu = \frac{f}{W} \tag{13}$$

where μ_c is the dry friction coefficient between the ring surface and block surface, and N is the mesh number of the computational domain. The value of the dry friction coefficient μ_c is 0.15.

2.6. Boundary Conditions

The boundary conditions used in this paper are

$$\begin{cases} p(x_{in},y) = p(x_{out},y) = p(x,y_{in}) = p(x,y_{out}) = 0 \\ p(x,y) \ge 0, (x_{in} < x < x_{out}, y_{in} < y < y_{out}) \end{cases} \tag{14}$$

where (x_{in}, y_{in}) are the entry boundary coordinates of the computational domain in the x and y directions, respectively. (x_{out}, y_{out}) are the exit boundary coordinates of the computational domain in the x and y directions, respectively.

2.7. Numeric Calculation Scheme

In this paper, the finite difference method is used to solve the block-on-ring mixed-lubrication model, which includes the iterative processes of water film pressure and asperity contact pressure. The specific numerical calculation steps of the lubrication model are presented in Figure 2.

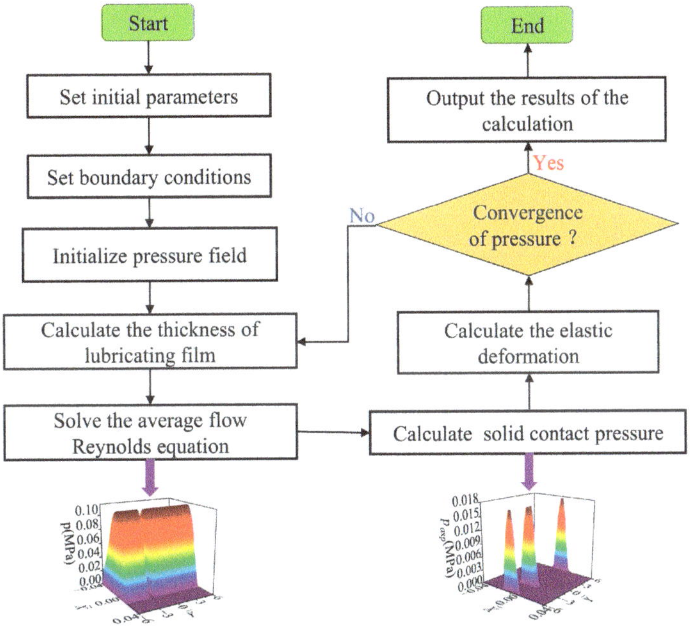

Figure 2. Flow chart of numerical simulation.

3. Experimental Equipment and Model Validation

3.1. Experimental Equipment

With the use of block-on-ring friction and wear tester (model number is MRH-3) shown in Figure 3, the friction coefficient between the test ring and the test block can be measured [51]. Table 1 gives the acquisition system and sensor parameters of the tester. The tester mainly includes a spring-loaded device, a control cabinet, a water chamber, and a friction sensor. The control cabinet makes the output shaft of the variable frequency motor rotate, therefore driving the test ring in the water chamber to rotate. The sensor records friction force and friction coefficient in real time. Figure 3 also provides an enlarged image of the water chamber; water is supplied from the inlet, and there is also a water outlet to achieve the circulation flow of lubricating water. Because the impurities in the water may affect the experimental results, a filter is installed at the water inlet to remove all contaminants.

Figure 4 gives three types of test blocks made of different materials (i.e., rubber, thordon, and graphite), which are commonly used in water-lubricated bearings [52–54]. The test ring is made of stainless steel [55], Figure 5a provides a test ring without scratches, and Figure 5b presents another test ring whose surface has one scratch. The structure of this scratch is described in Figure 5c, W_c is the scratch width, h_c is the scratch depth, L_c is the scratch location, L is the width of the stainless-steel ring, and the scratch is located in the middle of the test ring. The structure and material parameters of the friction pairs are shown in Table 2.

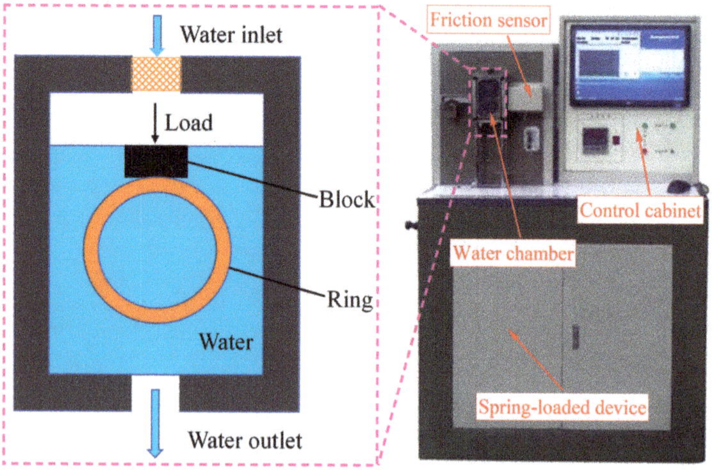

Figure 3. Block-on-ring friction and wear tester.

Table 1. Parameters of tester.

Parameters	Value
Maximum load (N)	3000
Speed range (r/min)	10~3000
Accuracy of pressure sensor	0.03%FS
Accuracy of friction torque sensor	0.2%FS
Sampling frequency of data acquisition card (KS/s)	500

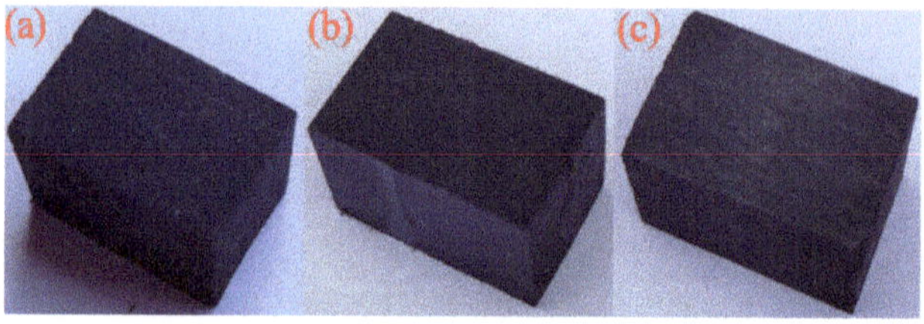

Figure 4. Test blocks. (**a**) Rubber block, (**b**) Thordon block, (**c**) Graphite block.

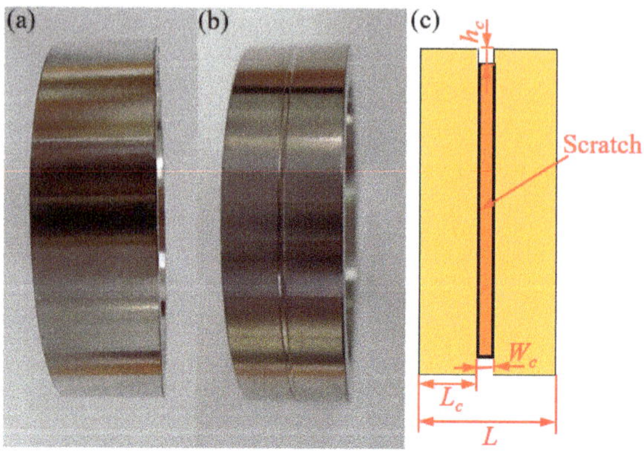

Figure 5. Test rings. (**a**)Test ring without scratch, (**b**) Test ring with one scratch, (**c**) Structure of scratch.

Table 2. Parameters of friction pairs.

	Test Ring	Test Block		
Material	316 Stainless steel	Nitrile rubber	Thordon	Graphite
Size (mm)	Outside diameterΦ49.00 Width 13.65	Length 19.05, width 12.32, thickness 12.32		
Roughness (μm)	Ra0.4	Ra0.6	Ra0.6	Ra0.6
Elastic modulus	210 GPa	6.1 MPa	490 MPa	7800 MPa
Poisson ratio	0.30	0.49	0.36	0.19

3.2. Model Validation

When the load is 20 N, the friction coefficients between the stainless-steel ring and the graphite block are measured, and the black curve in Figure 6a represents the experimental results of friction coefficients. The theoretical friction coefficients are also calculated by the block-on-ring model, and the blue curve in Figure 6a represents the theoretical friction coefficients. In addition, Figure 6b shows the simulation results (purple curve) and experimental results (orange curve) of the friction coefficients between the graphite block and scratched ring under a load of 30 N. As described in Figure 6, the theoretical simulation results are in good agreement with the experimental results, which validates the accuracy of the theoretical model proposed in this paper. It can also be seen from Figure 6 that (1) At low speeds, the experimental results are lower than the theoretical results. This is due to the surface wear of the test block during the experimental process, and the graphite powder enters the lubrication area and plays a lubricating role. Therefore, the friction coefficients under experimental conditions are lower than those under simulation conditions. (2) At high speeds, the experimental values of the friction coefficient are higher than the simulation values. This is because the high speeds cause vibration of the stainless-steel ring, enhancing the asperity contact between the stainless-steel ring and the graphite block. Therefore, the friction coefficient results under experimental conditions are greater.

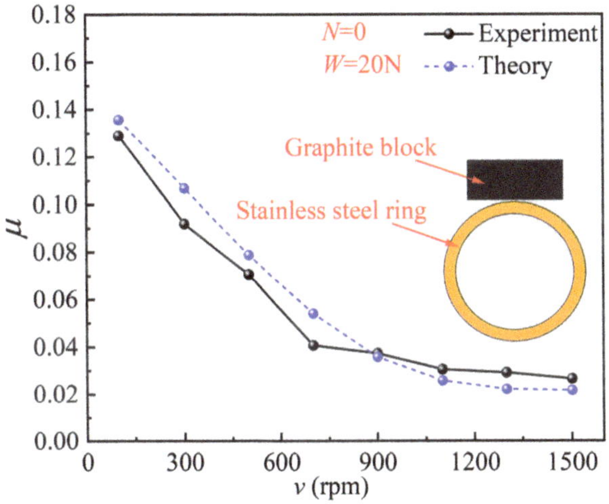

(a)

Figure 6. *Cont.*

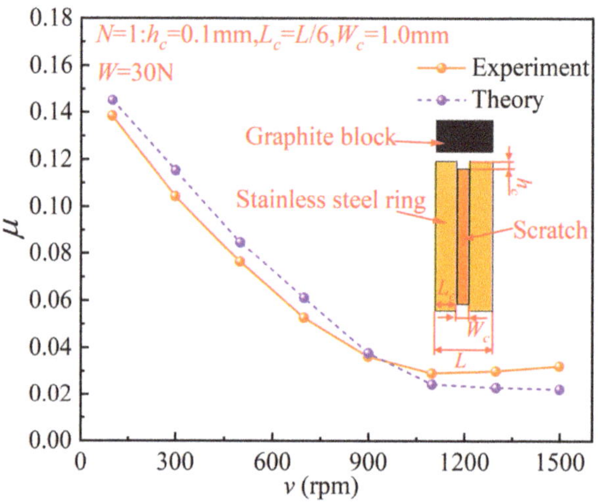

(b)

Figure 6. Experimental results and theoretical results. (**a**) Graphite block—Stainless-steel ring without scratches. (**b**) Graphite block—Stainless-steel ring with one scratch.

4. Results and Discussion

4.1. Sensitivity of Tribological Performance of Test Blocks Made of Different Materials to Scratches

Figure 7 plots the variations of the friction coefficients with time under water-lubrication conditions. During the experiment process, the load is 30 N, and the rotation speed of the test ring is 100 r/min. In the figure, "N = 0" indicates that there are no scratches on the surface of the test ring, while "N = 1" indicates that the test ring surface has one scratch. After the test ring starts rotating, the friction coefficient instantly rises to a peak value and then slightly decreases to a stable value. The material of the test block has a significant influence on the friction coefficient. When there are no scratches (N = 0) on the test ring surface, the stable friction coefficient of the rubber block and stainless-steel ring is approximately 0.101, that of the thordon block and stainless-steel ring is approximately 0.114, that of the graphite block and stainless-steel ring is approximately 0.127, and the friction coefficients of three types of test blocks have the following relationship: $\mu_{graphite} > \mu_{thordon} > \mu_{rubber}$. This indicates that the bigger the elastic modulus of the test block is, the higher the friction coefficient of frictional pairs becomes. The friction coefficient of the scratched ring (N = 1) is greater than that of the unscratched ring (N = 0). In addition, the differences in friction coefficient between the two cases vary with test block materials. The above results indicate that the existence of scratches can affect the friction performance of the block-on-ring system, and the degree of influence is related to the material of the test block.

The rotation speed of the test ring in Figure 7 is only 100 r/min, so the sensitivity of the friction coefficient of the block-on-ring system to scratches at other speeds is unclear. Figure 8 indicates the variation curves of the friction coefficients within the speed range of 100–1000 r/min, and the experiment load is still 30 N. It can be seen from the figures that the variation trend of the friction coefficient is similar to the Stribeck curve [56]. In the initial stage of the friction coefficient curve, the block-on-ring system is in the mixed-lubrication state, and the friction coefficients under different test blocks differ significantly. The relationship between friction coefficients of the three types of blocks is $\mu_{graphite} > \mu_{thordon} > \mu_{rubber}$. As the rotation speed goes up, the friction pairs gradually transit from the mixed-lubrication (ML) state to the elastohydrodynamic lubrication (EHL) state, and the differences in friction coefficient among the three test blocks become smaller. The critical speed is defined as the speed threshold at which the friction pairs transit from the ML state to the EHL state and is directly proportional to the elastic modulus of the test block. The comparison in friction coefficient between a scratched ring and an unscratched ring shows that the existence of scratches greatly increases the friction coefficient in the mixed-lubrication state but has very little effect on the friction coefficient in the elastohydrodynamic lubrication state. An increase in scratch number (from 0 to 1) raises the critical speed at which the block-on-ring system enters the EHL state, and there is no critical turning point (i.e., critical speed) in the friction coefficient curve of the graphite block–scratched ring. Figure 8 also presents the difference $|\Delta\mu|$ in friction coefficient between the scratched ring and unscratched ring and gives the average value $|\Delta\mu|_{av}$ of difference $|\Delta\mu|$ within the speed range of 100–1000 r/min. The difference $|\Delta\mu|$ is big when the rotation speed is slower than 600 r/min. However, when the rotation speed exceeds 600 r/min, the rubber block (or thordon block) and stainless-steel ring enter the elastohydrodynamic lubrication state, and the difference $|\Delta\mu|$ of friction coefficient becomes small. The graphite block and scratched ring are always in the mixed-lubrication state, so the friction coefficient of the scratched ring is markedly higher than that of the unscratched ring throughout the entire speed range. The average value $|\Delta\mu|_{av}$ for the rubber block is 0.0056, for the thordon block is 0.0047, and for the graphite block is 0.0112. In other words, the average values $|\Delta\mu|_{av}$ under different test blocks have the following relationship $|\Delta\mu_{\text{graphite}}|_{av} > |\Delta\mu_{\text{rubber}}|_{av} > |\Delta\mu_{\text{thordon}}|_{av}$.

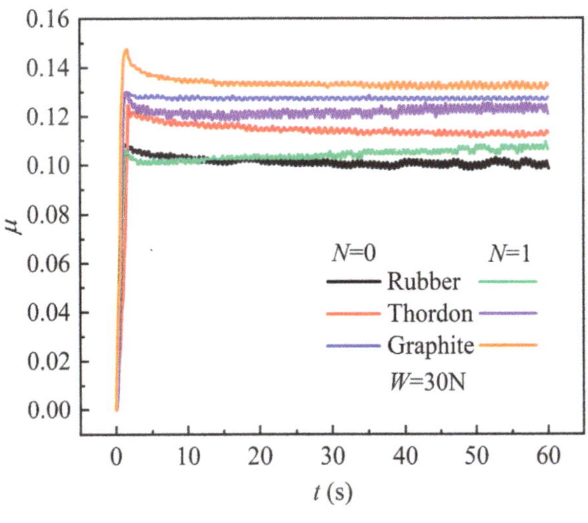

Figure 7. Variation of friction coefficient with time.

To analyze the relationship between the solid contact state and the friction coefficient, Figure 9a provides the simulation results of the contact pressure at a speed of 400 r/min and a load of 30 N. As the elastic modulus of the test block rises, the block-on-ring contact area (i.e., the area between two pink dashed lines in Figure 9a) reduces along the x direction, but the maximum contact pressure becomes bigger and bigger. For example, for the unscratched test ring (N = 0), the maximum contact pressure of the rubber block is only 0.075 MPa, while that of the graphite block grows to 0.187 MPa. Due to the small elastic modulus of the rubber block, a large deformation occurs in the middle area (black curve in Figure 10a) of the rubber block along the x direction, forming a water sac structure. This deformation makes the test ring only come into contact with the edge of the rubber block (see the upper left corner image in Figure 9a), avoiding large-scale solid contact. Because the graphite block hardly deforms, the test ring almost comes into contact with the entire graphite block along the y direction (see the bottom left corner image in Figure 9a). The maximum contact pressure of the rubber block–test ring and rubber block–scratched test ring is 0.075 MPa and 0.081 MPa, respectively. The maximum contact pressure of the thordon block–test ring and thordon block–scratched test ring is 0.107 MPa and 0.133 MPa, respectively. The maximum contact pressure of the graphite block–test ring and graphite block–scratched test ring is 0.187 MPa and 0.222 MPa, respectively. From the above results, it can be seen that (1) When there are scratches on the test ring surface, the maximum contact pressure of the test block is raised. (2) The larger the elastic modulus of block is, the greater the increment of the maximum contact pressure becomes. The above two aspects are the reasons why the friction coefficient of the graphite block–scratched ring is the highest in Figure 8.

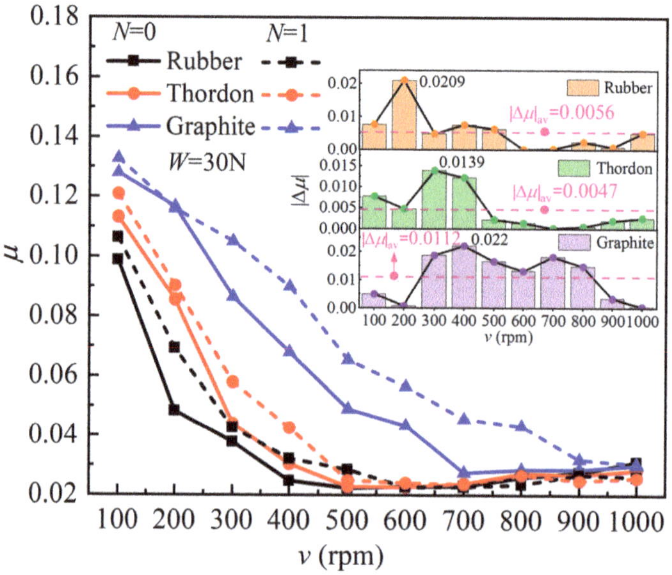

Figure 8. Variation of friction coefficient with speed.

Figure 9b presents the simulation results of water film pressure at a speed of 400 r/min. As the elastic modulus increases, the loading area (the area between two black dashed lines in Figure 9b) of water film pressure decreases along the *x* direction. When there are no scratches ($N = 0$) on the test ring surface, the maximum water film pressure of the thordon block is the highest (0.1635 MPa), while the maximum water film pressure (0.1255 MPa) of the graphite block is almost equal to that (0.1205 MPa) of the rubber block. There are different effects scratches exert on the maximum water film pressure of three test blocks when the test ring surface has scratches: (1) The scratches increase the maximum water film pressure of the rubber block. This is because the load on the rubber block is mainly supported by the water film (see the pressure distribution in Figure 9). If the test ring surface has scratches, then the water film thickness will be reduced, therefore increasing the maximum water film pressure of the rubber block. (2) Surface scratches decrease the maximum water film pressure of the graphite block (or thordon block). This is because the elastic modulus of the graphite block (or thordon block) is large, the block hardly deforms, and the water film is thin. Therefore, the load is mainly supported by asperity contact for the graphite block. When there are scratches on the test ring surface, although the decrease in film thickness tends to increase the water film pressure, this will also increase the number of asperities in contact state. The thin water film will be pierced by the microasperities, destroying the water film continuity and reducing the water film pressure. Therefore, the existence of scratches will decrease the maximum water film pressure of the graphite block due to the "piercing effect" of the asperities. To sum up, the variation of maximum water film pressure is determined by the positive effect (hydrodynamic pressure effect of fluid) and negative effect ("piercing effect" of the asperities).

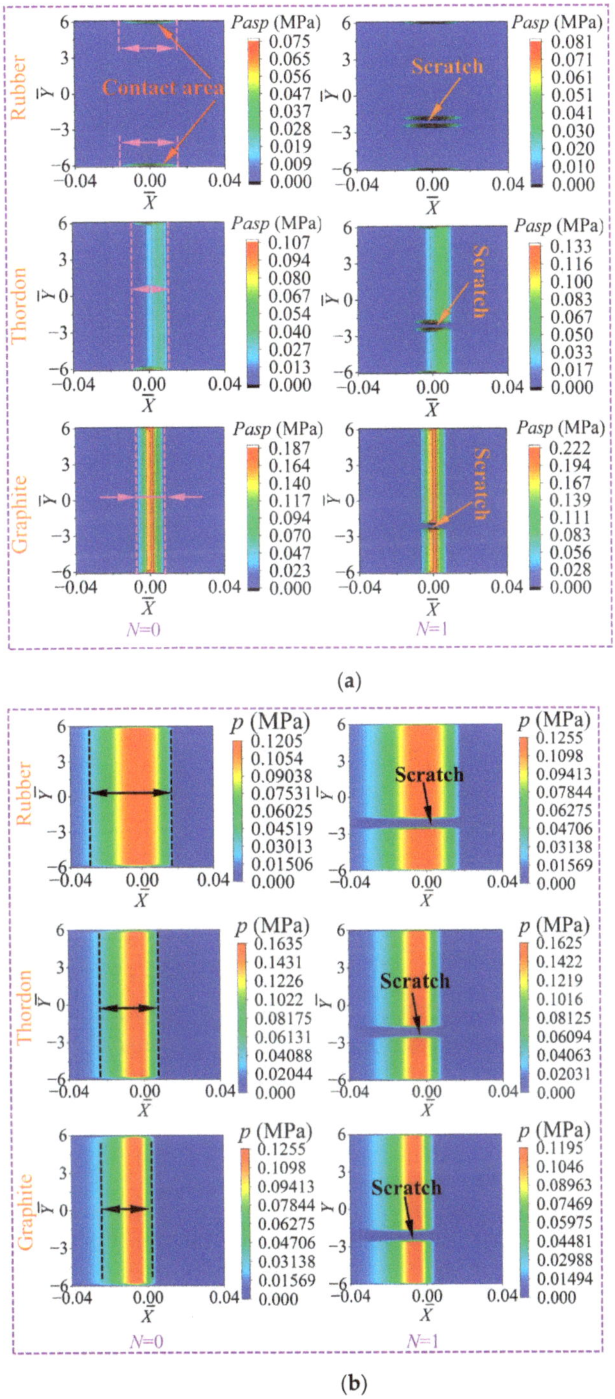

Figure 9. Simulation results of contact pressure and water film pressure. (**a**) Contact pressure. (**b**) Water film pressure.

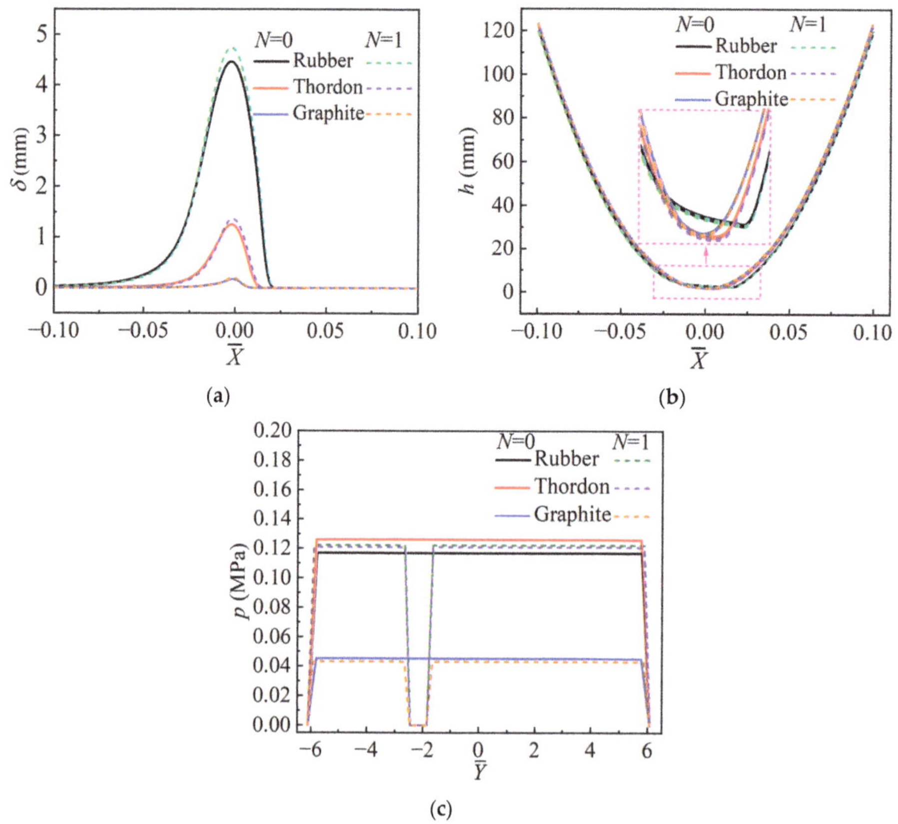

Figure 10. Simulation results of performance parameters in the middle section. (**a**) Deformation ($\overline{Y} = 0$). (**b**) Film thickness ($\overline{Y} = 0$). (**c**) Water pressure distribution ($\overline{X} = 0$).

Figure 10 shows the simulation results of lubrication parameters at the middle section of the test block. The rotation speed of the test ring is 400 r/min, and the experiment load is 30 N. As can be seen from Figure 10a, the deformation of the test block increases as the material elastic modulus decreases. In other words, the rubber block has the largest deformation, which causes the phenomenon that there is no asperity contact in the central area of the rubber block along the *y* direction (the contact pressure in this area is zero in the upper left corner image in Figure 9a). From Figures 9a and 10a, it can be seen that the asperity contact of the graphite block is the most severe. As shown in Figure 10a, when the test ring surface has a scratch, the deformation of the test block grows slightly. In addition, the smaller the elastic modulus of the test block is, the greater the increment of deformation becomes. Because the deformation of the rubber block is the largest, the formed wedge gap is also the biggest (see Figure 10b), and the loading area of the rubber block is wider than that of the other two blocks along the *x* direction (see Figure 9b). Figure 10c shows the pressure distribution of the water film at the middle section ($\overline{X} = 0$) along the *y* direction. When there are no scratches on the test ring surface, the water film pressure exhibits a trapezoidal variation law. However, under the condition of one scratch ($N = 1$), the loading area of water film pressure is divided into two separate zones, i.e., trapezoidal pressure zone and extremely low-pressure zone. The water film pressure of the graphite block is much lower than that of the other two types of blocks.

Based on the analysis of Figures 9 and 10, it can be concluded that the existence of scratch can reduce the loading area of asperity contact and water film, so the water film thickness should decrease, which results in a stronger contact pressure and a larger water film pressure to support the load. This is why the friction coefficient of a scratched ring is larger than that of an unscratched ring (see Figure 8).

4.2. The Effect of Scratch Parameters on the Tribological Performance of Block-on-Ring System

Because the friction pairs often operate in the mixed-lubrication state under water-lubrication conditions, the asperity contact has a great effect on the friction and wear of the friction pairs [57,58]. Among the three types of test blocks, graphite block has the most severe asperity contact. It can be seen from Figures 7 and 8 that the friction parameters ($|\Delta\mu|$ and $|\Delta\mu|_{av}$) of the graphite block are the largest, i.e., the tribological performance of the graphite block is the most sensitive to the scratches on the test ring surface. Therefore, the friction pairs composed of graphite block and stainless-steel ring are chosen as the object of further study. In the experiment, the scratch depth, scratch width, and scratch location are selected as the influencing factors on the tribological performance of the block-on-ring system. As shown in Table 3, each factor has three level values. According to the knowledge of permutation and combination, the total number of experimental groups can be calculated as

$$M = \left(C_n^1\right)^m = \left(C_3^1\right)^3 = 27 \tag{15}$$

where M is the number of experimental groups, n is the number of levels, m is the number of influence factors, and C is the combinatorial number. According to Equation (15), there are a total of 27 groups of experiments.

Table 3. Factors and levels of experiment.

Influence Factor	**Level Value**		
	1	**2**	**3**
Scratch depth h_c (mm)	0.1	0.2	0.3
Scratch width W_c (mm)	0.2	0.4	0.6
Scratch location L_c	1/6L	1/3L	1/2L

To improve experimental efficiency, this paper adopts the orthogonal method to design the experiment scheme. Based on the *L9* orthogonal design table [59], only 9 groups of experiments (instead of 27 groups) need to be carried out, and these 9 groups of experimental friction coefficients can be seen in Table 4. This table also gives the adjoint probabilities P of the three influence factors. For example, the adjoint probability P of scratch width W_c is 0.030, which is less than the significance level (0.050). The adjoint probabilities of scratch depth h_c and scratch location L_c are both greater than 0.050. The above results indicate that scratch width W_c has the most significant effect on the friction coefficient of the block-on-ring system, followed by the scratch location L_c, and lastly, the scratch depth h_c.

Table 4. Orthogonal experiment scheme and results.

Group Number	h_c **(mm)**	W_c **(mm)**	L_c	**Friction Coefficient**
1	0.3	0.6	1/6L	0.078
2	0.2	0.2	1/2L	0.059
3	0.1	0.6	1/2L	0.076
4	0.3	0.2	1/3L	0.057
5	0.3	0.4	1/2L	0.070
6	0.2	0.6	1/3L	0.074
7	0.2	0.4	1/6L	0.065
8	0.1	0.4	1/3L	0.069
9	0.1	0.2	1/6L	0.052
Adjoint probability (P)	0.596	0.030	0.528	
Importance of influence		$W_c > L_c > h_c$		

Figure 11a shows the friction coefficients of 9 groups of experiments. The rotation speed range of the experiments is between 100 r/min and 1500 r/min. As the rotation speed increases, the friction coefficient shows a trend of first decreasing and then increasing. The friction coefficient varies significantly with different scratch parameters at low speeds, but the differences in friction coefficient among the 9 groups of experiments are small when the speed is higher than 1100 r/min. As can be seen from Figure 8, the graphite block and scratched ring have not yet entered the elastohydrodynamic lubrication (EHL) state at a speed of 1000 r/min, but they will enter the EHL state (see Figure 11a) when v = 1100 r/min.

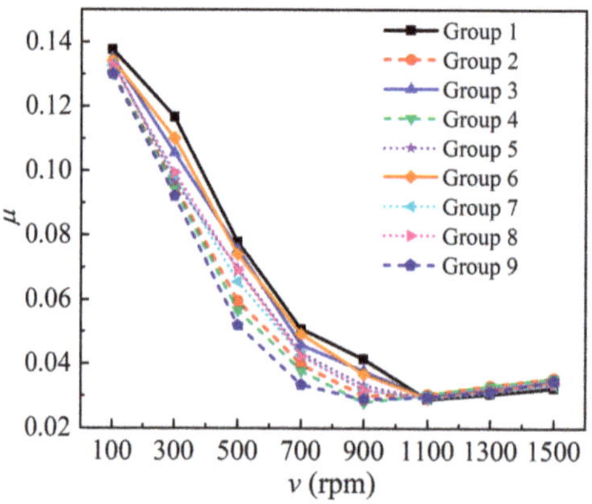

(a)

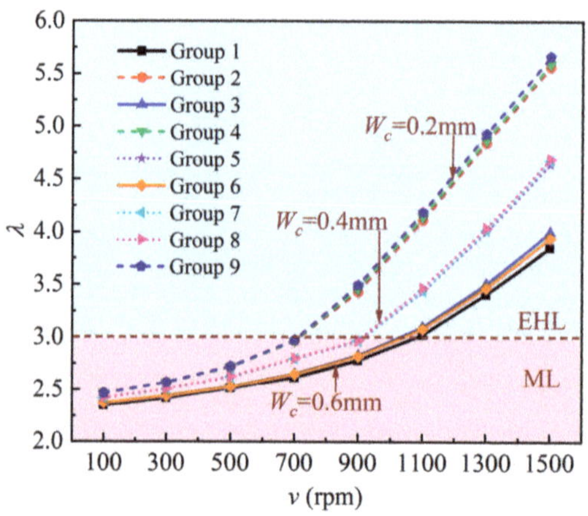

(b)

Figure 11. Friction coefficient and film thickness ratio of 9 groups of scratch parameters. (**a**) Experimental results of friction coefficient. (**b**) Simulation results of film thickness ratio.

Figure 11b gives the simulation results of the film thickness ratio λ, and the increase in rotation speed will lead to an increase in the film thickness ratio. The film thickness ratio curves can be easily divided into three major groups according to scratch width, which indicates that the scratch width has the largest influence on the lubrication performance of the block-on-ring system among the three scratch parameters. This is because the scratch width affects the loading area. In other words, widening scratch can reduce the loading area of water film and asperity contact. A thinner water film and a stronger solid contact are needed to support the load, and thus, the friction coefficient will significantly go up (see Figure 11a). The difference in film thickness curves among the three major groups is not significant in the mixed-lubrication (ML) stage [60], but it becomes more pronounced in the elastohydrodynamic lubrication (EHL) stage when the rotation speed increases. This is because the asperity contact pressure is the main support body of the load in the mixed-lubrication stage, then the test block and the test ring gradually separate as the speed increases, and the fluid film pressure becomes the main body to support the load in the elastohydrodynamic lubrication stage. On the one hand, low rotation speed has little effect on asperity contact, and high speed has a big effect on the hydrodynamic pressure effect of water film. On the other hand, wide scratches will strongly damage the continuity and hydrodynamic pressure effect of the water film in the EHL state, significantly reducing the lubricating film thickness h and film thickness ratio λ. The scratch width also influences the critical speed at which the friction pairs transit from the ML state to the EHL state. For example, when the scratch width is 0.2 mm, the critical speed is 700 r/min, but the critical speed is 1100 r/min under the scratch width of 0.6 mm. This indicates that an increase in scratch width will raise the speed threshold for the block-on-ring system to enter the EHL state.

Figure 12 gives the simulation results of water film pressure and asperity contact pressure. As can be seen from Figure 12a, the water film pressure in the scratched area is almost zero [61], regardless of the value of scratch width. Additionally, the maximum water film pressure is determined by the scratch width, while the scratch depth and scratch location have little effect on the maximum water film pressure. For example, the 2nd, 4th, and 9th groups have the same scratch width (W_c = 0.2 mm) and different scratch depths (h_c = 0.1 mm, 0.2 mm, 0.3 mm), and the maximum water film pressure of these three groups is equal (0.150 MPa). From Figure 12b, it can be noticed that the maximum contact pressure of the 1st group is the highest, and the scratch width is 0.6 mm, which indicates that increasing the scratch width leads to more severe asperity contact between the test ring and test block. Under the same scratch width, the influence of scratch depth and scratch location on the maximum contact pressure is very weak. The scratch widths of the 4th, 5th, and 6th groups in Figure 12 are 0.2 mm, 0.4 mm, and 0.6 mm, respectively. As can be seen from the simulation results of these three groups, the maximum water film pressure is inversely proportional to the scratch width, while the maximum contact pressure is directly proportional to the scratch width. The maximum contact pressure is located at both edges of the scratch. This is because there is no pressure or deformation in the scratched area, while deformation occurs in the other areas of the test block. This nonuniform deformation results in a water sac structure between the scratch and the end face of the test block. Figure 13 illustrates the water sac structure in the water-lubricated bearing. If the water film pressure at a certain point is high, the elastic deformation at the point will be large. In addition, the elastic deformation at surrounding points is relatively small because of low water film pressure. Then, the water sac structure [62,63] forms, and it is full of water.

Figure 14 shows the simulation results of the minimum film thickness h_{min} and maximum deformation δ_{max} of the graphite block. From the figure, it can be seen that the scratch depth h_c and the scratch location L_c have little influence on the minimum film thickness and the maximum deformation. For example, the scratch depths of 1st group, 3rd group, and 6th group are 0.3 mm, 0.1 mm, and 0.2 mm, respectively, but the minimum film thickness h_{min} of these three groups is almost equal to 1.927 μm, and the maximum deformation δ_{max} is around 0.184 μm. Among the three scratch parameters, the scratch width determines the minimum film thickness and maximum deformation, i.e., under the same scratch width, the maximum deformation (or minimum film thickness) is almost equal. As can be seen from Figure 14, the maximum deformation is clearly divided into three major groups according to the scratch width. Therefore, although the scratch depths of 1st group, 3rd group, and 6th group are different, the minimum film thickness (or maximum deformation) of these groups is almost the same because the scratch widths are all 0.6 mm. As the scratch width increases, the minimum film thickness decreases while the maximum deformation increases. For example, when the scratch width W_c is 0.2 mm, the maximum deformation of the test block is 0.166 μm, while the maximum deformation rises to 0.184 μm under the condition of W_c = 0.6 mm.

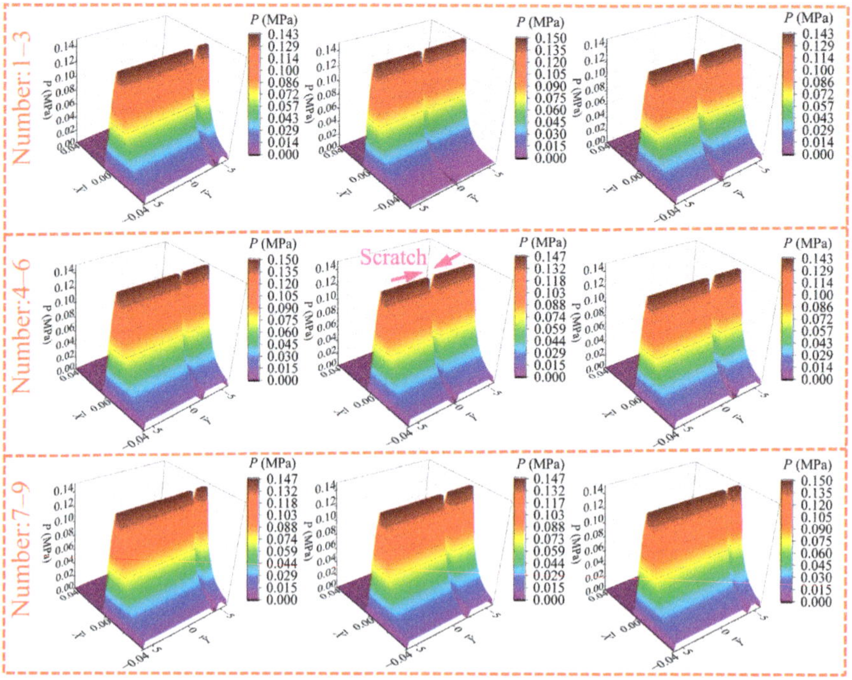

(**a**)

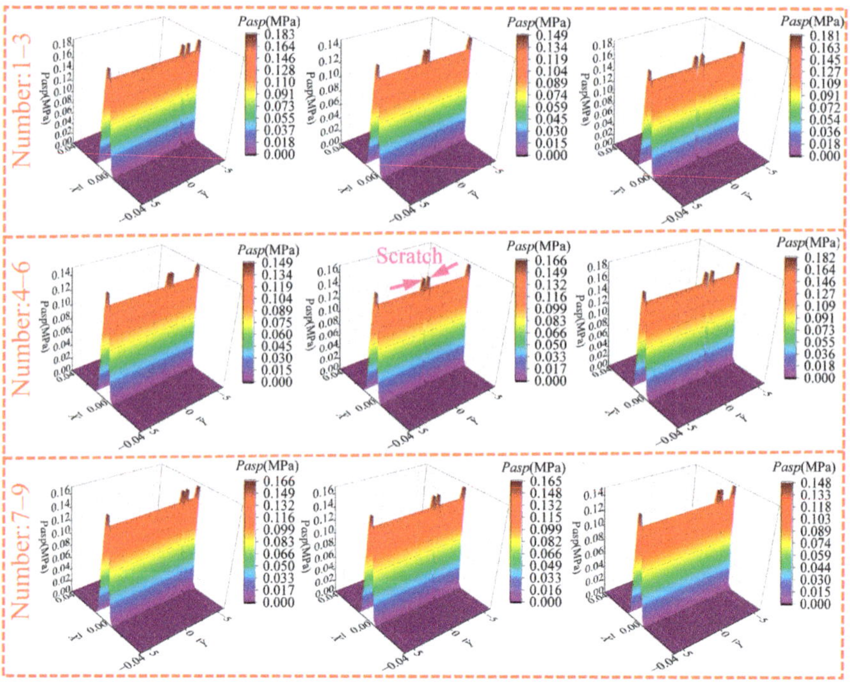

(**b**)

Figure 12. Simulation results of water film pressure and asperity contact pressure. (**a**) Water film pressure. (**b**) Asperity contact pressure.

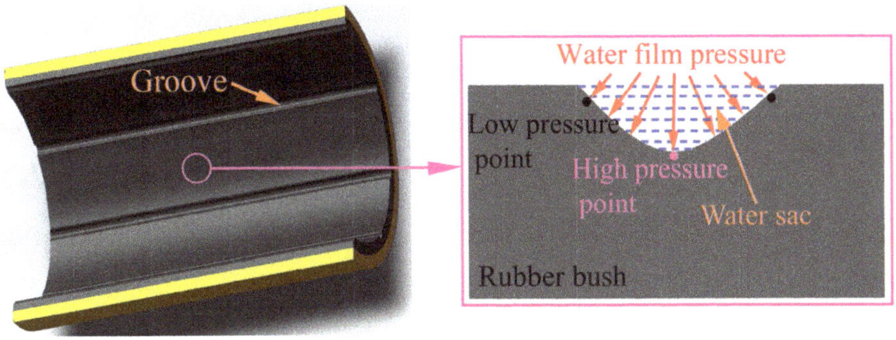

Figure 13. Water sac structure in the water-lubricated bearing.

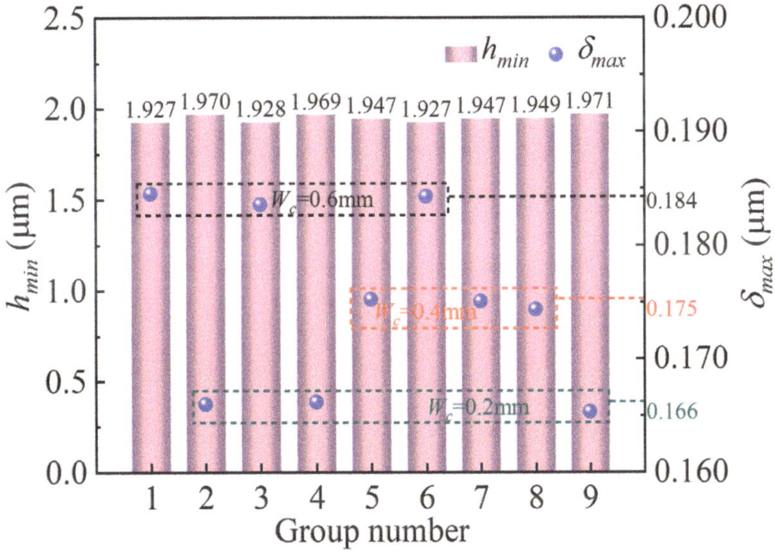

Figure 14. Simulation results of minimum film thickness and maximum deformation.

4.3. The Effect of Scratch Width on the Tribological Performance and Wear Performance of Block-on-Ring System

As described in Section 4.2, the scratch width has the greatest effect on the tribological performance (especially on friction coefficient and critical speed) of the block-on-ring system among the three scratch parameters. Therefore, to further study the influence of scratch width on the above two parameters, this part conducts block-on-ring friction experiments under different scratch widths (0.2~1.0 mm). In the experiment, the scratch depth h_c is 0.1 mm, and the scratch location L_c is $1/6L$. Figure 15 presents the experimental results of the friction coefficient μ and the simulation results of the proportion γ_0 of contact load capacity. When the rotation speed is 100 r/min, friction coefficients among different scratch widths are slightly different, which is due to the weak hydrodynamic pressure effect at a low shear rate. The scratch width has little influence on the friction coefficient at this low speed, but when the speed range is (300~900 r/min), the friction coefficient varies greatly with different scratch widths. The total load capacity consists of two parts, i.e., contact load capacity and water film load capacity. The variable γ_0 is defined as the proportion of contact load capacity in the total load capacity. From the column chart of the proportion of contact load capacity, it can be found that the proportion γ_0 is greater than 80% when v = 100 r/min, which indicates that the main support body is asperity contact rather than water film. The proportion γ_0 is inversely proportional to the rotation speed. However, it is directly proportional to the scratch width, i.e., widening scratch results in more severe asperity contact and higher friction coefficient of the block-on-ring system. When the rotation speed is raised further to 1300 r/min or 1500 r/min, the proportion of contact load capacity is all 0, illustrating that all five groups of block-on-ring systems completely rely on water film to support the load.

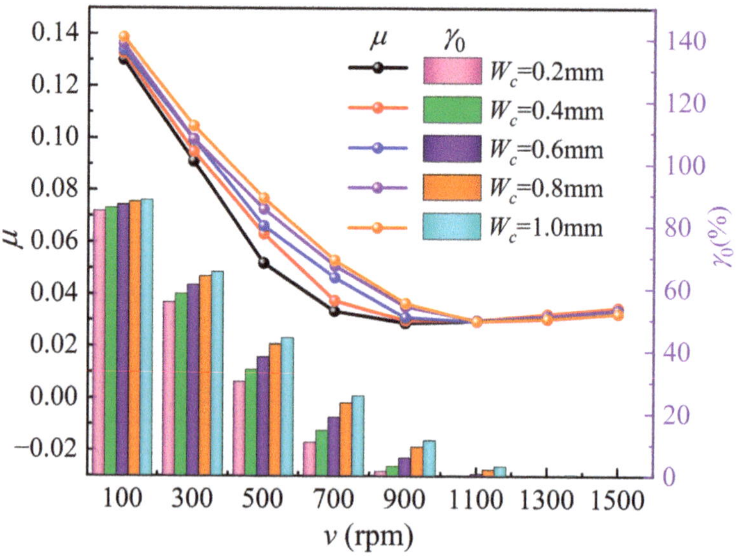

Figure 15. Friction coefficient and proportion of contact load capacity under different scratch widths.

The film thickness ratio λ is provided in Figure 16. It can be seen that the increase in film thickness ratio shows up with an increase in rotation speed, and the differences between neighboring film thickness ratio curves also increase with a rise in rotation speed. However, the film thickness ratio is inversely proportional to the scratch width. When the scratch width is 0.2 mm, the block-on-ring system enters the elastohydrodynamic lubrication state at a rotation speed of 700 r/min. When the scratch width grows to 1.0 mm, it is not until the rotation speed reaches 1000 r/min that the system can enter the EHL state. The pressure distribution at the middle section of the test ring under different rotation speeds is also provided in Figure 16. When the rotation speed is 100 r/min, the block-on-ring system mainly relies on the asperity contact to support the load, and the scratch width has little influence on the film thickness and film pressure. Therefore, the distribution of water film pressure is almost the same under different scratch widths (see pressure distribution image in the lower left corner). However, with the increased speed and enhanced hydrodynamic pressure effect, the frictional pairs enter the elastohydrodynamic lubrication state, and the scratch width gradually has a significant effect on the continuous distribution and thickness of the water film. Therefore, there are significant differences (see pressure distribution image in the lower right corner) in the pressure distribution curves of the water film under different scratch widths when v = 1500 rpm.

Figure 17 shows the variation of the maximum contact pressure p_{asp}^{max} of the test block with rotation speed. The following conclusions can be gained from the figure: (1) When the rotation speed is less than 1100 r/min, the maximum contact pressure falls with the growth of speed. This is because the block-on-ring system operates in the mixed-lubrication state at low speeds, and increasing speed enhances the fluid hydrodynamic pressure effect and water film load capacity, which reduces the contact load capacity and the maximum contact pressure. Under the same rotation speed, the maximum contact pressure rises with increasing scratch width. For example, at a speed of 500 r/min, the maximum contact pressure (169.8 KPa) under W_c = 0.4 mm is 1.33 times larger than that (127.2 KPa) under W_c = 0.2 mm. This is because widening the scratch significantly destroys the continuity of the fluid film, which reduces the proportion of water film load capacity and increases the proportion of contact load capacity. Accordingly, the maximum contact pressure also has a great increase. (2) When the rotation speed exceeds 1100 r/min, the maximum contact pressure of all groups drops to zero. All block-on-ring friction pairs enter the Elastohydrodynamic lubrication state, and the load is entirely supported by the water film rather than asperity contact. This is consistent with the variation trend of the film thickness ratio curve in Figure 16.

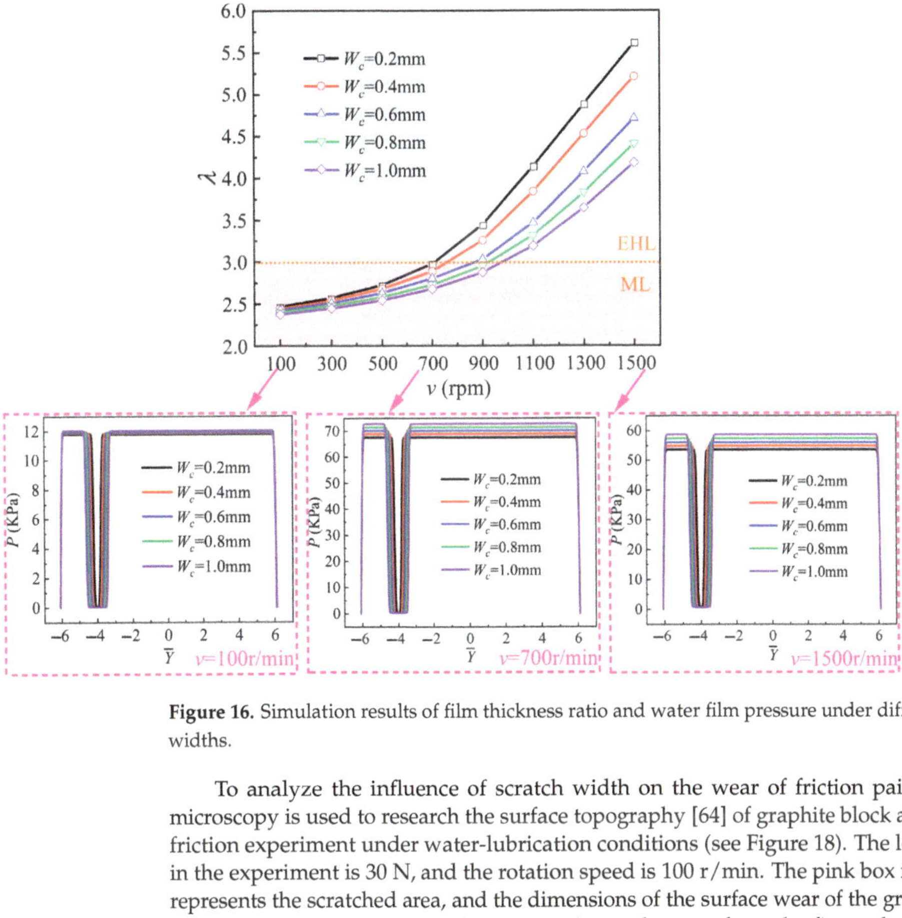

Figure 16. Simulation results of film thickness ratio and water film pressure under different scratch widths.

To analyze the influence of scratch width on the wear of friction pairs, confocal microscopy is used to research the surface topography [64] of graphite block after 600 s of friction experiment under water-lubrication conditions (see Figure 18). The load applied in the experiment is 30 N, and the rotation speed is 100 r/min. The pink box in the figure represents the scratched area, and the dimensions of the surface wear of the graphite block are given below the topography image. As can be seen from the figure, both the wear width and the wear depth of the graphite block increase with an increase in the scratch width on the test ring. When the scratch width increases from 0.2 mm to 1.0 mm, there is an increase of 15.23% in wear width (from 5290.9 μm to 6096.6 μm) of the graphite block and an increase of 54.28% in wear depth (from 108.7 μm to 167.7 μm). This is consistent with the previous theoretical analysis conclusion that the wider the scratch on the test ring surface, the more severe the surface wear on the graphite block.

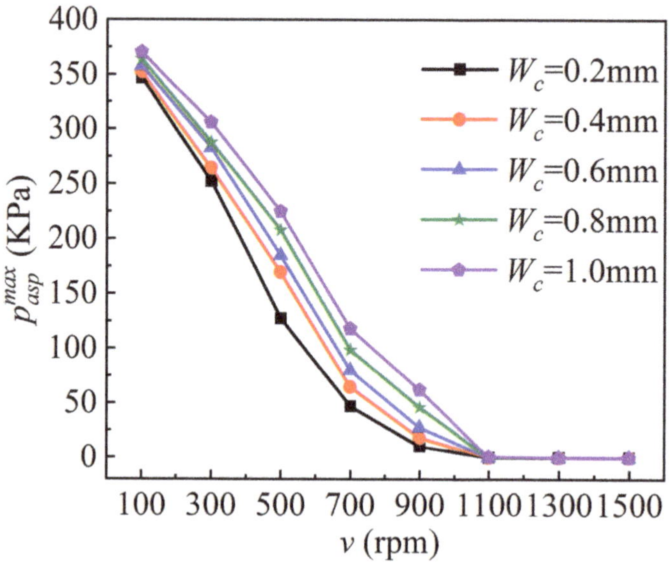

Figure 17. Simulation results of maximum contact pressure under different scratch widths.

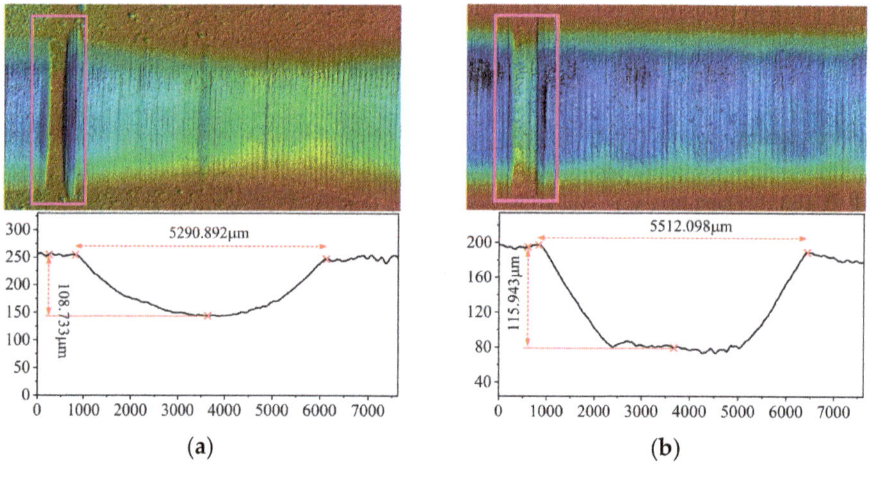

Figure 18. *Cont.*

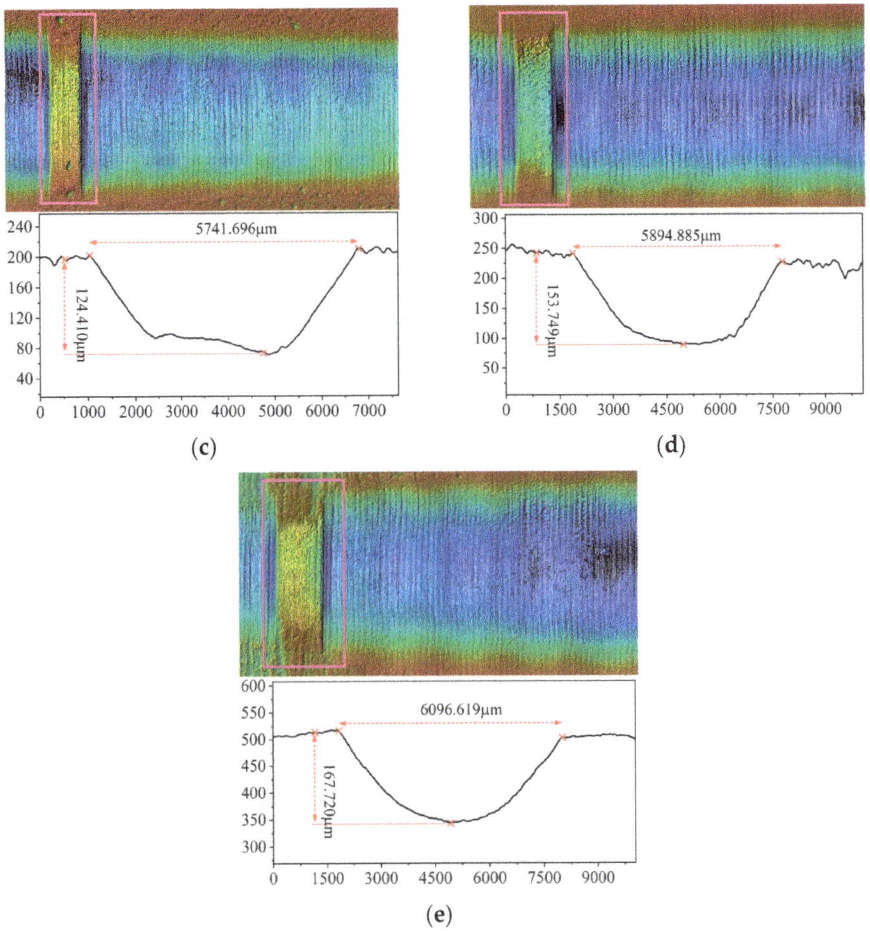

Figure 18. Surface wear of the graphite block. (**a**) W_c = 0.2 mm. (**b**) W_c = 0.4 mm. (**c**) W_c = 0.6 mm. (**d**) W_c = 0.8 mm. (**e**) W_c = 1.0 mm.

5. Conclusions

To study the effect of scratches on the tribological performance of friction pairs, this paper conducts several experiments and constructs a mixed-lubrication model of the block-on-ring system. Based on experimental results and theoretical models, the sensitivity of test blocks made of different materials to scratch parameters is studied. This paper also studies the relationship between the tribological performance of friction pairs and the structure parameters of scratches. The specific conclusions are as follows:

(1) Among the three types of test blocks, the tribological performance of the graphite block is the most sensitive to scratches. Scratches have a great effect on the friction coefficient of the graphite block, and the existence of scratches aggravates the microconvex contact and wear, which can severely reduce the service life of the friction pairs. Under the condition of one scratch (N = 1), the loading area of water film pressure is divided into two separate zones, i.e., a trapezoidal pressure zone and an extremely low-pressure zone. In addition, the variation of maximum water film pressure is determined by the positive effect (hydrodynamic pressure effect of fluid) and negative effect ("piercing effect" of the asperities).

(2) Compared with scratch depth and scratch location, the scratch width has the greatest influence on the tribological performance of the graphite block, especially on the friction coefficient. This is because a widening scratch can reduce the loading area of water film and solid contact. Supporting the load needs a smaller film thickness and a stronger asperity contact, and thus, the friction coefficient will significantly go up. The scratch depth and the scratch location have little influence on tribological performance (minimum film thickness, maximum deformation, maximum water film pressure, maximum contact pressure). The maximum contact pressure is located at both edges of the scratch due to the formation of a water sac structure.
(3) The scratch has a great influence on the transition of the lubrication state of the block-on-ring system. The existence of scratches increases the critical speed at which the block-on-ring system transits from the mixed-lubrication state to the elastohydrodynamic lubrication (EHL) state. For the stainless-steel ring-graphite block frictional pairs, the system enters the EHL state at the speed of 700 r/min when there are no scratches on the ring surface. However, when there is one scratch on the ring surface, the system does not enter the EHL state until the speed reaches 1100 r/min. The critical speed is directly proportional to the scratch width. A widening scratch will markedly weaken the hydrodynamic pressure effect.

Author Contributions: Q.L.: Numerical Computation, Methodology, Writing—original draft. P.L.: Supervision. F.G.: Conceptualization, Writing—review and editing. X.Z.: Formal analysis. S.L.: Validation. F.J.: Software. All authors have read and agreed to the published version of the manuscript.

Funding: This work was supported by National Natural Science Foundation of China (No. 52375190, No. 52175173, No. 52005278, No. 52205200); Youth Innovation Technology-support Program of Shandong Province Universities and Colleges (No. 2021KJ077); Shandong Provincial Natural Science Foundation (No. ZR2021ME198, No. ZR2022ME081); Shandong Province Science and Technology Small and Medium Enterprises Innovation Ability Enhancement Project (2023TSGC0612).

Data Availability Statement: The data that support the findings of this study are available from the corresponding author upon reasonable request.

Conflicts of Interest: The authors declare no conflict of interest.

Abbreviations

Nomenclature

O	center of the test ring
R	outer radius of the test ring, mm
ω	angular velocity, s^{-1}
Ω_1	unscratched area
Ω_2	scratched area
η	viscosity of water, Pa s
p	film pressure, MPa
h	nominal film thickness, mm
u_s	velocity of outer surface of shaft, m s^{-1}
σ	combined surface roughness, μm
σ_b	roughness of test block, μm
σ_r	roughness of test ring, μm
μ	friction coefficient
ρ	density of water, kg m^{-3}
h_{00}	rigid central film thickness, mm
(k_x, k_y)	turbulence factors
f	friction, N
(ϕ_x, ϕ_y)	pressure flow factors
ϕ_s	shear flow factor

ϕ_c	contact factor
h_c	scratch depth, mm
E'	composite elastic modulus, MPa
E_r	elastic modulus of ring, MPa
E_b	elastic modulus of block, MPa
ν_r	Poisson ratio of ring
ν_b	Poisson ratio of block
p_{asp}	contact pressure, MPa
β	curvature radius of asperities on either surface, μm
D	number of asperities per unit contact area
A_c	real contact area, mm^2
A_0	nominal contact area, mm^2
W	synthetic load capacity, N
F_h	load capacity generated by water film, N
F_c	load capacity generated by asperity contact, N
Subscripts and superscripts	
r	ring
b	block
asp	asperity contact

References

1. Ouyang, W.; Zhang, X.; Jin, Y.; Yuan, X. Experimental study on the dynamic performance of water-lubricated rubber bearings with local contact. *Shock Vib.* **2018**, *2018*, 6309727. [CrossRef]
2. Guo, Z.; Yuan, C.; Liu, A.; Jiang, S. Study on tribological properties of novel biomimetic material for water-lubricated stern tube bearing. *Wear* **2017**, *376*, 911–919. [CrossRef]
3. Shen, Y.; Zhang, Y.; Zhang, X.; Zheng, H.; Wei, G.; Wang, M. A Fluid-Structure Interaction Method for the Elastohydrodynamic Lubrication Characteristics of Rubber-Plastic Double-Layer Water-Lubricated Journal Bearings. *Lubricants* **2023**, *11*, 240. [CrossRef]
4. Litwin, W. Influence of local bush wear on water lubricated sliding bearing load carrying capacity. *Tribol. Int.* **2016**, *103*, 352–358. [CrossRef]
5. Zhou, G.; Wang, J.; Han, Y.; Wei, B.; Tang, B.; Zhong, P. An experimental study on film pressure circumferential distribution of water-lubricated rubber bearings with multiple grooves. *Tribol. Trans.* **2017**, *60*, 385–391. [CrossRef]
6. Li, X.; Guo, Z.; Huang, Q.; Yuan, C. Insight into the Lubrication Performance of Biomimetic Porous Structure Material for Water-lubricated Bearings. *J. Bionic Eng.* **2023**, *20*, 1905–1916. [CrossRef]
7. Yu, T.; Guo, F.; Zhang, X.; Ji, H.; Duan, W.; Liang, P. Water lubrication assisted by small-quantity silicone oil. *Tribol. Int.* **2022**, *173*, 107619. [CrossRef]
8. Wu, K.; Zhou, G.; Mi, X.; Zhong, P.; Wang, W.; Liao, D. Tribological and vibration properties of three different polymer materials for water-lubricated bearings. *Materials* **2020**, *13*, 3154. [CrossRef]
9. Zhang, X.; Yu, T.; Guo, F.; Liang, P. Analysis of the influence of small quantity secondary lubricant on water lubrication. *Tribol. Int.* **2021**, *159*, 106998. [CrossRef]
10. Wang, C.; Bai, X.; Guo, Z.; Dong, C.; Yuan, C. Friction and wear behaviours of polyacrylamide hydrogel microsphere/UHMWPE composite under water lubrication. *Wear* **2021**, *477*, 203841. [CrossRef]
11. Jia, J.; Zhou, H.; Gao, S.; Chen, J. A comparative investigation of the friction and wear behavior of polyimide composites under dry sliding and water-lubricated condition. *Mater. Sci. Eng. A* **2003**, *356*, 48–53. [CrossRef]
12. Chen, B.; Wang, J.; Yan, F. Comparative investigation on the tribological behaviors of CF/PEEK composites under sea water lubrication. *Tribol. Int.* **2012**, *52*, 170–177. [CrossRef]
13. Zhao, G.; Wang, T.; Wang, Q. Friction and wear behavior of the polyurethane composites reinforced with potassium titanate whiskers under dry sliding and water lubrication. *J. Mater. Sci.* **2011**, *46*, 6673–6681. [CrossRef]
14. Xiao, B.; Zheng, X.; Zhou, Y.; Yao, D.; Wan, Y. Tribological behaviors of the water-lubricated rubber bearings under different lubricated conditions. *Ind. Lubr. Tribol.* **2020**, *73*, 260–265. [CrossRef]
15. Huang, S.; Wang, Z.; Xu, L.; Huang, C. Friction and Wear Characteristics of Aqueous ZrO_2/GO Hybrid Nanolubricants. *Lubricants* **2022**, *10*, 109. [CrossRef]
16. Wang, L.; Guan, X.; Zhang, G. Friction and wear behaviors of carbon-based multilayer coatings sliding against different rubbers in water environment. *Tribol. Int.* **2013**, *64*, 69–77. [CrossRef]
17. Xiong, D. Friction and wear properties of UHMWPE composites reinforced with carbon fiber. *Mater. Lett.* **2005**, *59*, 175–179. [CrossRef]
18. Nobili, L.; Magagnin, L. DLC coatings for hydraulic applications. *Trans. Nonferrous Met. Soc. China* **2009**, *19*, 810–813. [CrossRef]
19. Xu, H.; Feng, Z.; Chen, J.; Zhou, H. Tribological behavior of the carbon fiber reinforced polyphenylene sulfide (PPS) composite coating under dry sliding and water lubrication. *Mater. Sci. Eng. A* **2006**, *416*, 66–73. [CrossRef]

20. Du, F.; Li, D.; Sa, X.; Li, C.; Yu, Y.; Li, C.; Wang, J.; Wang, W. Overview of friction and wear performance of sliding bearings. *Coatings* **2022**, *12*, 1303. [CrossRef]
21. Rezaei, A.; Paepegem, W.V.; Baets, P.D.; Ost, W.; Degrieck, J. Adaptive finite element simulation of wear evolution in radial sliding bearings. *Wear* **2012**, *296*, 660–671. [CrossRef]
22. Dobrica, M.B.; Fillon, M. Influence of scratches on the performance of a partial journal bearing. In Proceedings of the STLE/ASME, International Joint Tribology Conference, Miami, FL, USA, 20–22 October 2008; pp. 359–361. [CrossRef]
23. Dobrica, M.B.; Fillon, M. Performance degradation in scratched journal bearings. *Tribol. Int.* **2012**, *51*, 1–10. [CrossRef]
24. Chatterton, S.; Pennacchi, P.; Vania, A.; Hassini, M.A.; Kuczkowiak, A. Effect of Scratches on a Tilting-Pad Journal Bearing. In Proceedings of the ASME Turbo Expo 2020, Power for Land, Sea, and Air, Montreal, QC, Canada, 21–25 September 2020; p. V10BT29A006. [CrossRef]
25. Vo, A.T.; Fillon, M.; Bouyer, J. Numerical Study of a Journal Bearing with Scratches: Validation with Literature and Comparison with Experimental Data. *Lubricants* **2021**, *9*, 61. [CrossRef]
26. Chasalevris, A.C.; Nikolakopoulos, P.G.; Papadopoulos, C.A. Dynamic effect of bearing wear on rotor-bearing system response. *J. Vib. Acoust.* **2013**, *135*, 011008. [CrossRef]
27. Nicodemus, E.R.; Sharma, S.C. Influence of wear on the performance of multirecess hydrostatic journal bearing operating with micropolar lubricant. *J. Tribol.* **2010**, *132*, 021703. [CrossRef]
28. Cai, J.; Han, Y.; Xiang, G.; Wang, J.; Wang, L. Effects of wear and shaft-shape error defects on the tribo-dynamic response of water-lubricated bearings under propeller disturbance. *Phys. Fluids* **2022**, *34*, 077118. [CrossRef]
29. Lin, Y.; He, S.; Lai, D.; Wei, J.; Ji, Q.; Huang, J.; Pan, M. Wear mechanism and tool life prediction of high-strength vermicular graphite cast iron tools for high-efficiency cutting. *Wear* **2020**, *454*, 203319. [CrossRef]
30. Fan, B.; Yuan, K.; Chen, W.; Qi, S.; Liu, Y.; Liu, H. A numerical model with nonuniform grids for the analysis of dynamic characteristics of scratched tilting-pad bearings. *Ind. Lubr. Tribol.* **2023**, *75*, 817–829. [CrossRef]
31. Bouyer, J.; Fillon, M.; Helene, M.; Beaurain, J.; Giraudeau, C. Behavior of a two-lobe journal bearing with a scratched shaft: Comparison between theory and experiment. *J. Tribol.* **2019**, *141*, 021702. [CrossRef]
32. Bouyer, J.; Alexandre, Y.; Fillon, M. Experimental investigation on the influence of a multi-scratched shaft on hydrodynamic journal bearing performance. *Tribol. Int.* **2021**, *153*, 106543. [CrossRef]
33. Branagan, L.A. Influence of Deep, Continuous Circumferential Scratches on Radial Fluidfilm Bearings. In Proceedings of the 61st STLE Annual Meeting & Exhibition, Calgary, AB, Canada, 7–11 May 2006; pp. 15–21.
34. Branagan, L.A. Survey of damage investigation of babbitted industrial bearings. *Lubricants* **2015**, *3*, 91–112. [CrossRef]
35. Giraudeau, C.; Bouyer, J.; Fillon, M.; Hélène, M.; Beaurain, J. Experimental study of the influence of scratches on the performance of a two-lobe journal bearin. *Tribol. Trans.* **2017**, *60*, 942–955. [CrossRef]
36. Yan, X.; Fan, C.; Wang, W.; Liu, X.; Chen, B. Experimental and simulation study of the dynamic characteristics of friction force under third-body intrusion behaviour. *Mech. Syst. Signal Process.* **2022**, *168*, 108726. [CrossRef]
37. Dong, C.L.; Yuan, C.Q.; Bai, X.Q.; Yang, Y.; Yan, X. Study on wear behaviours for NBR/stainless steel under sand water-lubricated conditions. *Wear* **2015**, *332*, 1012–1020. [CrossRef]
38. Liang, P.; Li, X.; Guo, F.; Zhang, X.; Jiang, F. Influence of shaft scratch on static and transient behavior of water-lubricated bearing. *J. Tribol.* **2023**, *146*, 012201. [CrossRef]
39. Li, X.; Liang, P.; Guo, F.; Liang, Q.; Li, X.; Zhang, X.; Liu, X.; Jiang, F. Effect of shaft scratches on the dynamic performance of bearing systems resist to multiple load impacts. *Tribology*. **2022**, *accepted*. Available online: https://kns.cnki.net/kcms2/article/abstract?v=3uoqIhG8C45S0n9fL2suRadTyEVl2pW9UrhTDCdPD65olJTg3HQUF-xFhvAzkRg0r3za3dOD2XspSWeDR0zAdVi0fDNI1O2a&uniplatform=NZKPT (accessed on 27 July 2023).
40. Patir, N.; Cheng, H. An average flow model for determining effects of three-dimensional roughness on partial hydrodynamic lubrication. *J. Tribol.* **1978**, *100*, 12–17. [CrossRef]
41. Patir, N.; Cheng, H. Application of average flow model to lubrication between rough sliding surfaces. *J. Tribol.* **1979**, *101*, 220–229. [CrossRef]
42. Du, Y.; Lan, J.; Quan, H.; Sun, C.; Liu, X.; Yang, X. Effect of different turbulent lubrication models on the lubrication characteristics of water-lubricated rubber bearings at a high Reynolds number. *Phys. Fluids* **2021**, *33*, 065118. [CrossRef]
43. Liu, H.; Zhang, B.; Bader, N.; Guo, F.; Poll, G.; Yang, P. Crucial role of solid body temperature on elastohydrodynamic film thickness and traction. *Tribol. Int.* **2019**, *131*, 386–397. [CrossRef]
44. Taylor, G.I. VIII. Stability of a viscous liquid contained between two rotating cylinders. *Philos. Trans. R. Soc. Ser. A* **1923**, *223*, 289–343. [CrossRef]
45. Wang, Y.; Yin, Z.; Gao, G.; Zhang, X. Analysis of the performance of worn hydrodynamic water-lubricated plain journal bearings considering cavitation and elastic deformation. *Mech. Ind.* **2017**, *18*, 508. [CrossRef]
46. Ng, C.W.; Pan, C.H.T. A linearized turbulent lubrication theory. *J. Basic Eng.* **1965**, *87*, 675–682. [CrossRef]
47. Wu, C.; Zheng, L. An average Reynolds equation for partial film lubrication with a contact factor. *J. Tribol.* 1989. [CrossRef]
48. Greenwood, J.A.; Tripp, J.H. The contact of two nominally flat rough surfaces. *Proc. Inst. Mech. Eng.* **1970**, *185*, 625–633. [CrossRef]
49. Greenwood, J.A.; Williamson, J.B.P. Contact of nominally flat surfaces. *Proc. R. Soc. Ser. A* **1966**, *295*, 300–319. [CrossRef]
50. Liang, P.; Li, X.; Guo, F.; Cao, Y.; Zhang, X.; Jiang, F. Influence of sea wave shock on transient start-up performance of water-lubricated bearing. *Tribol. Int.* **2022**, *167*, 107332. [CrossRef]

51. Li, Y.; Wang, Q.; Wang, T.; Pan, G. Preparation and tribological properties of graphene oxide/nitrile rubber nanocomposites. *J. Mater. Sci.* **2012**, *47*, 730–738. [CrossRef]
52. Orndorff Jr, R.L. WATER-LUBRICATED rubber bearings, history and new developments. *Nav. Eng. J.* **1985**, *97*, 39–52. [CrossRef]
53. Wang, Y.; Zhang, L. Characteristics and outline of water-lubricated thordon bearing. *Adv. Mater. Res.* **2012**, *496*, 355–358. [CrossRef]
54. Wang, J.; Zhang, C.; Wang, Y.; Wang, D.; Jia, Q.; Cui, Y.; Yuan, X. Preparation and Experimental Study of Graphite Material for Water Lubricated Thrust Bearing of Nuclear Main Pump. *Mater. Sci. Forum* **2018**, *932*, 102–106. [CrossRef]
55. Clarke, C.G.; Allen, C. The water lubricated, sliding wear behaviour of polymeric materials against steel. *Tribol. Int.* **1991**, *24*, 109–118. [CrossRef]
56. Lu, X.; Khonsari, M.M.; Gelinck, E.R.M. The Stribeck curve: Experimental results and theoretical prediction. *J. Tribol.* **2006**, *128*, 789–794. [CrossRef]
57. Wang, Z.; Liu, Y.; Wang, Y.; Liu, X.; Wang, Y. Influence of squeezing and interface slippage on the performance of water-lubricated tilting-pad thrust bearing during start-up and shutdown. *Lubr. Sci.* **2018**, *30*, 137–148. [CrossRef]
58. Zhang, H.; Yuan, C.; Tan, Z. A novel approach to investigate temperature field evolution of water lubricated stern bearings (WLSBs) under hydrodynamic lubrication. *Adv. Mech. Eng.* **2021**, *13*, 1514–1519. [CrossRef]
59. Xing, H.; Duan, S.; Wu, Z. Bearing lubrication optimization for diesel engine based on orthogonal design method. In Proceedings of the 2011 Third International Conference on Measuring Technology and Mechatronics Automation, Shanghai, China, 6–7 January 2011; Volume 1, pp. 633–636. [CrossRef]
60. Xie, Z.; Rao, Z.; Ta, N.; Liu, L. Investigations on transitions of lubrication states for water lubricated bearing. Part II: Further insight into the film thickness ratio lambda. *Ind. Lubr. Tribol.* **2016**, *68*, 416–429. [CrossRef]
61. Galda, L.; Sep, J.; Olszewski, A.; Zochowski, T. Experimental investigation into surface texture effect on journal bearings performance. *Tribol. Int.* **2019**, *136*, 372–384. [CrossRef]
62. Peng, E.; Liu, Z.; Lan, F.; Zhang, S.; Dai, M. Research on noise generation mechanism of rubber material for water-lubricated bearings. *Appl. Mech. Mater.* **2011**, *84–85*, 539–543. [CrossRef]
63. Xiang, G.; Yang, T.; Guo, J.; Wang, J.; Liu, B.; Chen, S. Optimization transient wear and contact performances of water-lubricated bearings under fluid-solid-thermal coupling condition using profile modification. *Wear* **2022**, *502*, 204379. [CrossRef]
64. Nwaneshiudu, A.; Kuschal, C.; Sakamoto, F.H.; Anderson, R.R.; Schwarzenberger, K.; Young, R.C. Introduction to confocal microscopy. *J. Investig. Dermatol.* **2012**, *132*, 1–5. [CrossRef]

lubricants

MDPI

Article

Dual Network Co-Crosslinked HNBR Composites with Enhanced Tribological Properties under Water Lubrication

Hao Yu [1,2,3], Wuxuan Zheng [1,2,3], Caixia Zhang [2,3], Shoubing Chen [2,3], Guangke Tian [1,*] and Tingmei Wang [2,3,*]

1 National Engineering Research Center for Technology and Equipment of Green Coating, Lanzhou Jiaotong University, Lanzhou 730070, China; 15215425480@163.com (H.Y.); zwx13281896153@163.com (W.Z.)
2 Key Laboratory of Science and Technology on Wear and Protection of Materials, Lanzhou Institute of Chemical Physics, Chinese Academy of Sciences, Lanzhou 730000, China; zhangcaixia@licp.cas.cn (C.Z.); chenshoubing@licp.cas.cn (S.C.)
3 Center of Materials Science and Optoelectronics Engineering, University of Chinese Academy of Sciences, Beijing 100049, China
* Correspondence: tiangke@mail.lzjtu.cn (G.T.); tmwang@licp.cas.cn (T.W.); Tel.: +86-189-9305-3958 (G.T.)

Abstract: Water-lubricated bearings play a critical role in underwater propulsion systems but are often prone to failure due to mechanical wear and vibration, especially under high loads and prolonged friction. In response to this issue, our study introduces a novel approach: a dual network co-crosslinking strategy utilizing hydrogenated nitrile butadiene rubber (HNBR). This strategy connects the rubber network with the epoxy network through epoxidized Eucommia ulmoides gum. A comprehensive analysis was conducted to assess the resulting composite's damping, tribological, and mechanical properties. The results show that the material has excellent mechanical, damping, and tribological properties relative to pure HNBR, with a 65.9% increase in the damping temperature domain, a 78.5% increase in tensile strength, a low coefficient of friction of 0.022, and a high resistance to abrasion of 3.87×10^{-6} mm^3/Nm. The successful synthesis of HNBR-based composites via the dual network co-crosslinking strategy underscores their potential as a practical solution for improving the reliability and prolonging the service life of water-lubricated bearings.

Keywords: co-crosslinking; friction behaviors; water lubrication; rubber

Citation: Yu, H.; Zheng, W.; Zhang, C.; Chen, S.; Tian, G.; Wang, T. Dual Network Co-Crosslinked HNBR Composites with Enhanced Tribological Properties under Water Lubrication. *Lubricants* **2023**, *11*, 534. https://doi.org/10.3390/lubricants11120534

Received: 6 October 2023
Revised: 14 November 2023
Accepted: 18 November 2023
Published: 18 December 2023

1. Introduction

As an important component of underwater propulsion systems, water-lubricated bearings have attracted much attention in the mechanical field [1–3]. Compared with the traditional oil lubrication method, water-lubricated bearings have many advantages, such as safety, environmental protection, economy, and greenness, thus becoming an important direction of underwater bearing research [4–7]. However, compared with oil-lubricated bearings, water-lubricated bearings also have some disadvantages. For example, the poor load carrying capacity and low adsorption capacity of water lead to poor boundary lubrication performance [8–10]. Therefore, it is necessary to further improve the performance of water-lubricated bearings through technical improvement and design optimization, so as to enhance the reliability and service life of water-lubricated bearings and promote their application in non-underwater propulsion systems.

Rubber elastomers, as polymeric water lubrication materials, are effective in reducing bearing friction and wear under water lubrication conditions and have a low coefficient of friction [11–13]. In addition, the rubber material can absorb vibration and shock and reduce the generation of noise and vibration [14]. Among them, hydrogenated nitrile butadiene rubber (HNBR) is a product obtained by catalytic hydrogenation technology of nitrile rubber, which has higher oxidative stability, abrasion resistance, and smaller permanent compression deformation compared with traditional nitrile rubber [15,16].

When HNBR is used alone as a water-lubricated bearing, it can experience elevated temperatures under higher loads and conditions with reduced water availability or limited water film formation, which may lead to scorching and frictional failure [17]. For this reason, researchers have conducted numerous studies on HNBR water lubrication materials. Common ways to improve the friction performance include the addition of solid lubricants [18,19], microcapsules [20,21], blending [22,23], and surface modification [24,25]. Compared with pure HNBR, polymer blending is also a way to improve the performance of rubber materials. Chudzik et al. added modified epoxy diene resin to NBR and the addition of 10% unmodified resin (ED-20) resulted in the most significant reduction in friction of vulcanized NBR, with a reduction of 25% in friction [26]. Sang et al. grafted a silane coupling agent functional layer on the surface of plasma-functionalized polyamide (PA 6), the joining of PA 6 with hydrogenated nitrile butadiene rubber (HNBR) was achieved, and the heat resistance of the material was improved [27]. Zhou et al. investigated the effect of blending different fillers on the friction and wear properties, mechanical properties, and vulcanization properties of HNBR, and the UHMWPE/HNBR composites had the best friction properties with low-speed water lubrication [3]. People are committed to improving the performance of HNBR. Therefore, we believe that the introduction of high-performance polymers with good physicochemical properties can prepare composites with excellent performance.

Eucommia ulmoides gum (EUG) is a renewable natural rubber (NR) with excellent dynamic mechanical properties in a wide range of high-performance rubber materials, including tires, vibration-damping devices, and acoustic materials [28,29]. Epoxidized Eucommia ulmoides gum (EEUG) prepared by simple epoxy functionalization is effective in improving compatibility with polar ingredients. Wang et al. succeeded in improving the adhesive properties at the styrene butadiene rubber (SBR)/silicon dioxide (SiO_2) interface with mechanical properties exceeding those of other compatibilizers using epoxy dutasteride rubber as a compatibilizer [30]. Chen et al. developed an EEUG/epoxy (EP) composite coating, where the introduction of EEUG increased the crosslinking density, giving the coating excellent tensile strength and corrosion resistance [31]. Wang et al. prepared a tough biobased composite with a shape memory effect using a dynamic vulcanization technique [32]. Therefore, EEUGs enriched with epoxy groups and double bonds can effectively improve the compatibility of rubbers and polymers such as epoxy resins.

In this study, we report the development of a novel dual network co-crosslinked system that combines two networks, rubber and epoxy, by introducing epoxidized Eucommia ulmoides gum as an intermediate. We investigated the frictional properties of the material using a ring block friction and wear tester and revealed the mechanism of water lubrication in the dual network co-crosslinked system. Our results demonstrate that the introduced epoxy Eucommia ulmoides gum effectively links the separate crosslinking systems of rubber and epoxy resin and significantly improves the damping and tribological properties of the material. These findings suggest that our approach has potential applications in tribology and other engineering fields.

2. Materials and Methods

2.1. Materials

Hydrogenated nitrile butadiene rubber, LANXESS 4369, Germany, has an acrylonitrile content of 43% and a Menni viscosity (ML 1 + 4) of 97 at 100 °C. It was purchased from Meifu New Material Co., Ltd., (Suzhou, China). Eucommia ulmoides gum (EUG) was obtained from Beijing Huayan Shijia Quality Control Technology Co. (Beijing, China). Haine epoxy resin was obtained from Wuhan Lanabai Pharmaceutical & Chemical Co. (Wuhan, China) 4,4′-Methylene-bis(2-chloroaniline) (MOCA) was obtained from Anhui Xianglong Chemical Co. (Anqing, China) Stearic acid, sulfur, zinc oxide, carbon black (N220), tetramethylthiuram disulfide (TMTD), and 2-thiobenzothiazole (MBT) are all industrial grade. All chemicals, unless otherwise stated, were used without further purification.

2.2. *Preparation*

2.2.1. Preparation of Epoxidized Eucommia Ulmoides Gum

In the reaction vessel, 30 g of EUG was added to 600 mL of xylene with high-speed stirring at 40 °C to completely dissolve the EUG. A mixture of H_2O_2 (13.6 g) and formic acid (5.6 g) was added dropwise to the reaction vessel over a period of 2 h. The reaction was continued for 3 h. At the end of the reaction, the product was precipitated by addition of excess ethanol, then filtered, washed, and dried in a vacuum oven at 50 °C for 36 h to give EEUG.

2.2.2. Preparation of HNBR/EEUG Composites

The preparation process of the material is shown in Figure 1. HNBR and rubber additives such as zinc oxide, tetramethylthiuram disulfide (TMTD) 2-thiobenzothiazole (MBT), etc. were added to the mixer and mixed well to obtain the rubber master batch. The HNBR rubber master batch was chopped and added to a three-necked flask, and tetrahydrofuran (THF) was added at 1:10, accompanied by stirring for 12 h to dissolve it completely. EEUG was dissolved in a small amount of THF and subsequently added to the HNBR gum solution along with sulfur and MOCA with stirring for 1 h. After sufficient stirring, the solvent was placed on a far infrared graphite heating plate at 55 °C. Table 1 lists information on the components of the material. The obtained film was alternately triangularly wrapped and rolled 6 times on an open mill. The film was vulcanized in a plate vulcanizer at 150 °C and 10 MPa for 30 min. Finally, the samples were cured at 80 °C for 3 h; 120 °C for 3 h; 150 °C for 5 h under gradient temperature increase. The mass ratios of the prepared HNBR/EEUG composites, HNBR to EEUG, were 100:3, 100:5, 100:7, 100:9, and 100:11.

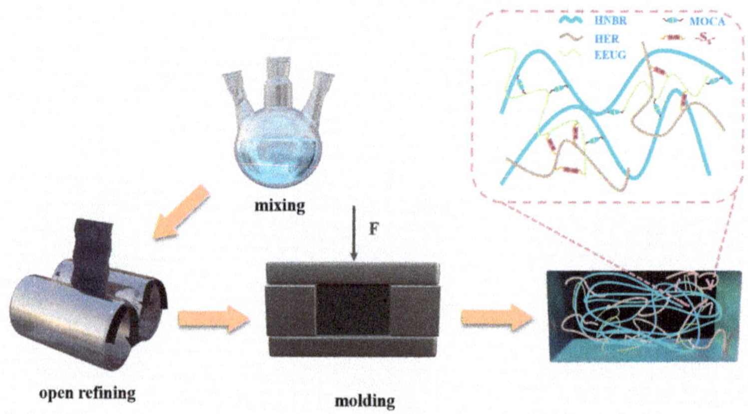

Figure 1. Schematic diagram of the preparation process of Dual Network Co-Crosslinked HNBR Composites.

Table 1. Materials and formulation.

Samples (phr)	HNBR	EEUG	HER	S	MOCA	SA	ZnO	TMTD	CB220	MBT
100:0:0	100	0	0	2	0	0.8	2.5	0.65	60	0.45
100:7:0	100	7	0	2.14	2.8	0.8	2.5	0.65	60	0.45
100:7:5	100	7	5	2.14	3.6	0.8	2.5	0.65	60	0.45

2.2.3. Preparation of Dual Network Co-Crosslinked Composites

The dual network co-crosslinked composites were prepared in a similar way to the former preparation method, in which HER and HNBR were dissolved together in THF, and based on the results of the later experiments, the composite of HNBR/EEUG 100:7 was selected as the matrix, and the ratios of HNBR/HER/EEUG were 100:7:3, 100:7:5, 100:7:7, and 100:7:9.

2.3. Tribological Tests

Friction and wear tests were carried out under water lubrication conditions using the MRH-3 ring wear tester produced by Era Test Metals Inc. and schematically shown in Figure 2. The friction partner was a tin-bronze ring (ZCuSn10Zn2) with the composition shown in Table 2. The surface of the specimen ring was ultrasonically cleaned in ethanol by grinding the specimen ring surface with 800 mesh metallographic sandpaper until smooth. At room temperature, the experimental speed was set at 140 revolutions per minute (rpm), the test force was 132 N, and the test drops were added with water on the surface of the friction subsurface. The acceleration rate of all the test drops was 9 mL/min, and the water was refilled at the end of each test. The stable friction value during the last 30 min of the experiment was taken as the average friction coefficient. The surface of the block material was wiped with anhydrous ethanol before and after friction and dried at 60 °C for 2 h. The specific wear rate (Ws, mm^3/Nm) was calculated by the following relation:

$$W_S = V_S / F * 2\pi R * n * nt * t \ (\text{mm}^3/\text{Nm}) \tag{1}$$

where Vs (mm^3) represents the wear volume of each friction rubber block and is given in the following equation. R denotes the outer diameter of the tin bronze ring, and F, n, and t represent the normal load (N), rotational speed (rpm), and test time (min), respectively.

$$V_S = \frac{D^2}{8} t \left[2\sin^{-1}\frac{b}{D} - \sin\left(2\sin^{-1}\frac{b}{D} \right) \right] \left(\text{mm}^3 \right) \tag{2}$$

where D (mm) is the outer diameter of the tin bronze ring and b (mm) and t (mm) are the average width of the abrasion mark and the width of the rubber block, respectively. To ensure the reliability of the experimental data, the average of three measurements is taken as the experimental data.

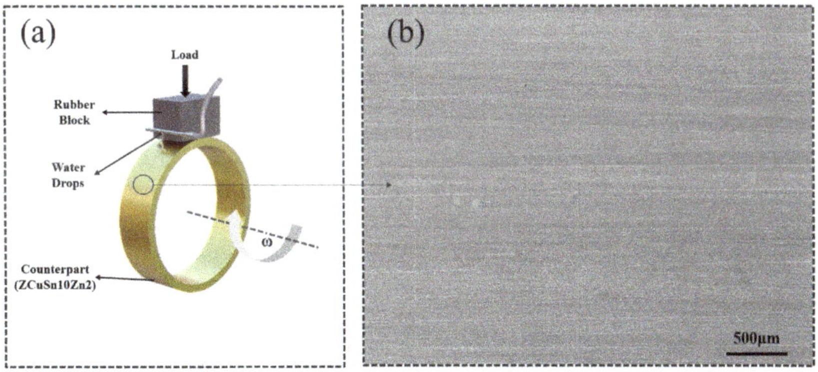

Figure 2. (**a**) Schematic diagram of the friction device; (**b**) SEM micrograph of the surface of the paired rings.

Table 2. Chemical composition of ZCuSn10Zn2.

Ingredient	Cu	Sn	Zn	Pb	Ni	Fe	Mn	Sb	S
Concentration (phr)	83.65	10	2	1.5	2	0.25	0.2	0.3	0.1

2.4. Characterizations

^{1}H NMR spectra were recorded at 25 °C on a nuclear magnetic resonance spectrometer (Bruker AVIII 400 MHz, USA). EUG and EEUG were dissolved in $CDCl_3$ and transferred to NMR tubes for subsequent determinations. Samples were analyzed using Fourier transform infrared spectroscopy (FT-IR, Bruker S V70). The mechanical properties of the materials were characterised by an AD-X (5000 N) universal testing machine (Shimadzu, Japan) and at least three samples were tested at a tensile speed of 500 mm/minute to obtain the average value. In order to clearly observe the micro-morphology of the material and to analyse its phase behaviour, we characterised the surface of the material using a Multimode 8 atomic force microscope (AFM) (Bruker, USA). The friction surface morphology of the material was observed by SEM, and the friction surface was sprayed with gold to increase the electrical conductivity of the material surface before testing. The contact angle (CA) of water on the surface of the material was measured on a DECCA-100 optical contact angle meter (DECCA Precision Instruments Co., Ltd., Shenzhen, China) using the sessile drop method. To obtain a more accurate value, the CA was measured randomly at three different positions on the friction surface of the sample, and then the arithmetic mean was obtained. The damping properties of the materials were characterized by DMA with the following parameters: the mode was tensile mode, the fixed frequency was 10 Hz, the temperature range was from −60 °C to 60 °C, and the temperature increase rate was 5 °C/min.

3. Results and Discussions

3.1. EEUG Characterization

Eucommia ulmoides gum, which is mainly composed of trans-isoprene, is poorly compatible with polar hydrogenated nitrile rubber because of its high degree of crystallinity and non-polar nature, and epoxy functionalization has become a simple and effective way to improve its compatibility. The successful epoxidation of dulcimer rubber is proved by the FT-IR spectra shown in Figure 3a, which shows that the peak of C=C telescopic vibration at 1665 cm^{-1} decreases after epoxide functionalization [33], the symmetric and asymmetric vibrational absorption peaks of C-O-C are enhanced at 1263 cm^{-1} and 870 cm^{-1}, 1665 cm^{-1} is C=C telescopic vibration, which decreases the peak after epoxidation, and the C-O-C symmetric and asymmetric telescopic vibrational peaks are observed at 1264 cm^{-1} and 870 cm^{-1} [34]. In Figure 3b, the signals at 1.61, 1.98–2.09, and 5.13 ppm are attributed to CH_3, CH_2, and alkene protons, respectively, in the trans-1,4 structure. The peak at 1.29 ppm is attributed to the methyl peak of the epoxide group. The peak at 2.71 ppm is due to the proton resonance of C-O-C, which suggests that some of the double bonds on the EUG chain were successfully epoxidized [35]. Epoxidized Eucommia ulmoides gum has both C=C and epoxy bonds, which can be crosslinked by both MOCA and S at the same time, providing conditions for HNBR to "connect" with HER.

3.2. Mechanical Properties of Composite Materials

Mechanical properties have an important impact on the application of materials, and the introduction of EEUG and HER into HNBR will inevitably cause changes in its mechanical properties. For this purpose, tensile measurements were performed, and the results obtained are shown in Figure 4. The tensile strength and elongation at break of the composites increased from 278.52% and 20.41 MPa to 598.4% and 29.4 MPa, respectively, with the addition of EEUG. For rubber composites with different additions of EEUG, the elongation at break and tensile strength increased with the increase in EEUG content. This indicates that the addition of EEUG improves the mechanical properties of pure HNBR. This is due to the good compatibility between EEUG and HNBR, and the incorporation of

EEUG as a physical crosslinking point increases the entanglement between molecular chain segments. With the addition of HER, a dual network co-crosslinking system was formed under the dual crosslinking of sulfur and MOCA. The maximum tensile strength of the material is 36.17 MPa, which is 78.5% higher than that of pure HNBR. This is attributed to the incorporation of HER to form a second network, and in the double crosslinking system, the chain segments of the polymers are chemically crosslinked and physically entangled, resulting in the formation of a denser structure. The hardness of the material increases at the macroscopic level.

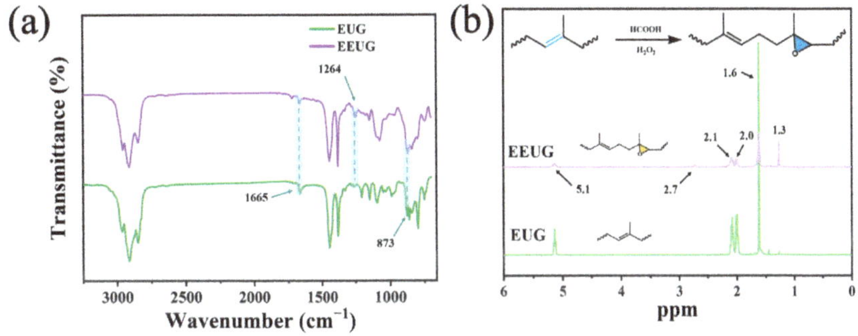

Figure 3. IR image (**a**) and ^{1}H NMR spectrum (**b**) of EEUG.

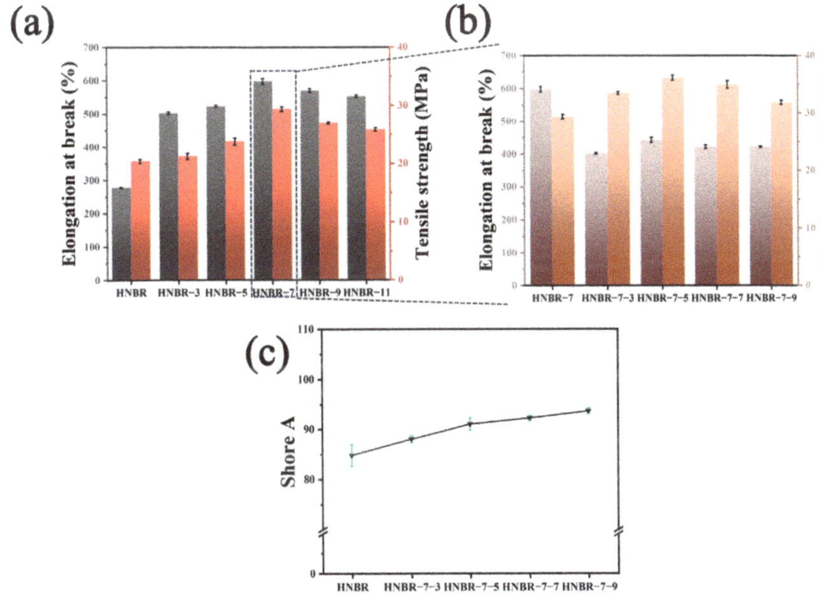

Figure 4. (**a**) Stress-strain diagram of the HNBR/EEUG composite. (**b**) Stress-strain diagram of the dual network co-crosslinked composite. (**c**) Shore hardness of the dual network co-crosslinked composite.

3.3. Damping Analysis of Composites

Bearings under water lubrication conditions experience significant vibration, which is not conducive to the actual service life of the material. Therefore, studying the damping properties of composite materials for bearings in vibration damping is of great significance. The damping properties of polymers are closely related to the internal friction generated by the relative slip of the polymer molecular chains [36]. Figure 5 shows the variation of energy storage modulus (E′) and loss factor (Tanδ) with temperature for different formulations of HNBR at 10 Hz, and Table 3 shows the material effect damping temperature domain. The polymer effect damping temperature range is defined as the temperature range where Tanδ exceeds 0.3. With the addition of soft EEUG, the energy storage modulus of the material decreases and is lower than that of the HNBR, but the effective damping range becomes wider. Having the largest energy storage modulus when HNBR/EEUG is 100:7, there is only a single peak in Tanδ for all the samples, which indicates that epoxidation gives good compatibility between dulcimer rubber and HNBR. With the addition of HER, the energy storage modulus gradually increases, while the loss factor is shifted to the high-temperature direction, and the maximum damping temperature domain is increased by 65.9% compared with that of pure HNBR. This phenomenon may be due to two reasons: one is that the EEUG acts as a "bridge" between HNBR and HER, and the two, through chemical crosslinking to connect the polymer after crosslinking, further reduce the polymer molecular chain of the internal rotation and lead to a further increase in the molecular chain of the internal friction, ultimately accelerating the internal energy dissipation of the polymer, which improves the damping performance of the material; second, it is caused by the greater rigidity brought by the HER five-membered nitrogen heterocycles, so the increase in the HER content leads to the enhancement of the interaction force between the two, which restricts the movement of the molecular chains (chain segments) in the free volume and leads to the increase in Tg. The high-HER component leads to the electrostatic repulsive force between the polar groups exceeding the attractive force, resulting in an increase in the intermolecular distance between the molecular chains and a decrease in the glass transition temperature.

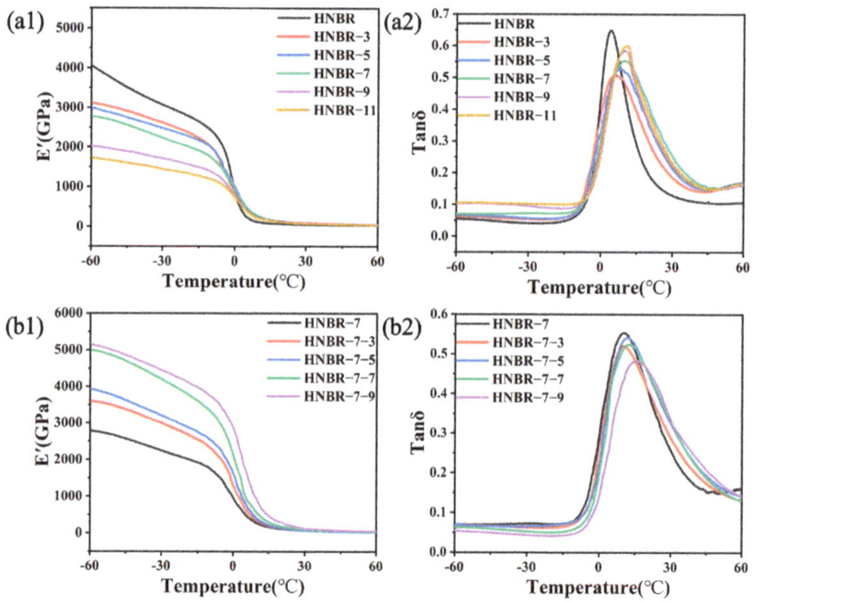

Figure 5. DMA curves of different composites: E′ (**a1**) and Tanδ (**a2**) for HNBR/EEUG; E′ (**b1**) and Tanδ (**b2**) for HNBR/EEUG/HER.

3.4. Tribological Properties of Composites

3.4.1. Tribological Properties of HNBR/EEUG

The tribological properties of composites have a decisive influence on their service conditions. By investigating the frictional properties of HNBR and EUG at different ratios, the frictional wear curves of HNBR/EEUG composites at different ratios under water lubrication conditions are shown in Figure 6. From Figure 6a, it can be seen that the friction curves showed very different trends after the addition of EEUG. The HNBR showed an increasing and then decreasing trend at the beginning of the friction, while the friction coefficients all showed a rapid decrease after the addition of EEUG, and then they all leveled off. In the early stages of friction, pure HNBR experiences a boundary lubrication state due to the growth of the transfer film and an increase in the friction vice contact area. This leads to an initial rise followed by a subsequent decrease in the friction coefficient. HNBR/EEUG composites show a different behavior during the initial stages of sliding, where the formation of a water film leads to a rapid decrease in the surface friction coefficient from boundary lubrication to mixed lubrication [37]. With water film stabilization, the friction coefficient of the decline gradually slows down and eventually stabilizes. With the increase in EEUG content, the friction coefficient appeared to decrease and then increase. At 7%, the material's coefficient of friction reached a minimum of 0.047, and it is exciting that the friction coefficient with the addition of EEUG is less than that of the pure HNBR. It can be concluded that the addition of EEUG is of great significance in improving the tribological properties of HNBR. The reduction in the friction coefficient is mainly attributed to the plasticizing and softening effect of EEUG on HNBR, which makes the contact area between the friction pair increase, and the larger friction surface makes the water film have a larger carrying capacity [38]. As shown in Figure 6b, the wear rate increases gradually with the addition of EEUG. The reduction in hardness makes the material more likely to adhere to the pair of surfaces under shear, which reduces the load-bearing capacity of the water film rupture, increases the coefficient of friction, and makes it more susceptible to wear [39].

Table 3. Damping properties of different composites.

HNBR/EEUG	$Tan\delta_{max}$	ΔT ($Tan\delta > 0.3$)	HNBR/EEUG /HER	$Tan\delta_{max}$	ΔT ($Tan\delta > 0.3$)
100:0	0.65	16.42	100:7	0.55	26.82
100:3	0.51	22.07	100:7:3	0.52	28.45
100:5	0.53	24.95	100:7:5	0.54	30.80
100:7	0.55	26.82	100:7:7	0.52	29.85
100:9	0.58	23.87	100:7:9	0.48	27.24
100:11	0.60	24.08	-	-	-

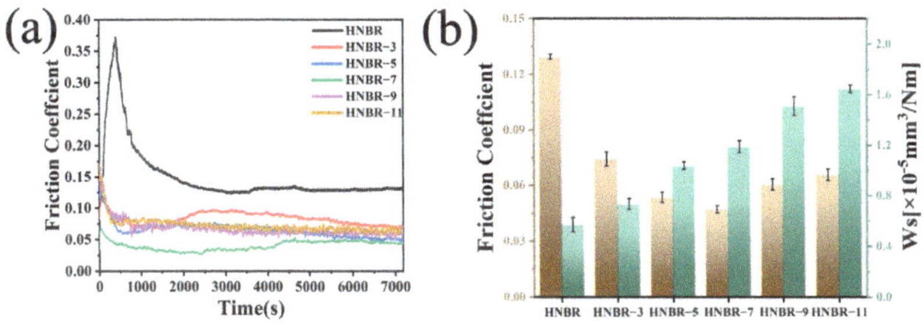

Figure 6. (**a**) Friction profile coefficients. (**b**) Average COF and Ws of HNBR/EEUG with different blending ratios.

To study the microscopic wear of the friction material, a scanning electron microscope was used to observe the friction surface. As shown in Figure 7a, the pure HNBR friction surface has stripping and microcracks along the friction direction, which is mainly dominated by adhesive wear. As shown in Figure 7b–f, after the addition of EEUG, the friction surface shows plow furrows, which are mainly dominated by abrasive wear, which is caused by the microcutting of the polymer by the hard tin bronze surface. With the addition of EEUG, the modulus of the material gradually decreases, which reduces the abrasion resistance of HNBR and aggravates the abrasion of the material under strong shear force.

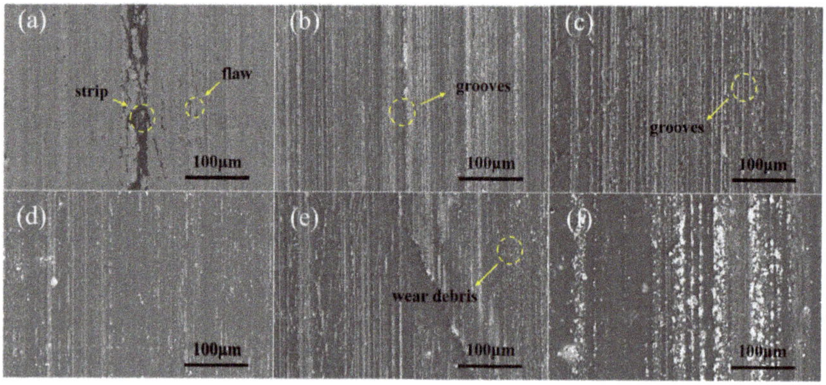

Figure 7. (**a**) SEM micrographs of the wear surfaces of pure HNBR and different HNBR/EEUG ratios: (**b**) 100:3, (**c**) 100:5, (**d**) 100:7, (**e**) 100:9, and (**f**) 100:11.

The results show that the incorporation of EEUG effectively improves the friction properties under water lubrication conditions, and the coefficient of friction of the composites decreases by nearly two thirds relative to that of pure HNBR at an EEUG content of 7%, but the incorporation of EEUG reduces the wear resistance of pure HNBR. Therefore, we propose a dual network co-crosslinking system based on the addition of HNBR/EEUG 100:7 with the addition of polar HER and expect these dual network co-crosslinked composites to have excellent tribological properties.

3.4.2. Frictional Properties of Dual Network Co-Crosslinked Composites

The curves of the friction coefficients of different dual network co-crosslinked systems as a function of test time and the effect of HER content on the average friction coefficients and wear rates are shown in Figure 8. Figure 8a shows images of the friction coefficient with time for dual network co-crosslinked systems with different HER contents. In the early stage of friction, with the addition of HER content, the friction time is gradually shortened, and the fluctuation of the friction coefficient decreases due to the better damping property of the material. The friction coefficient decreases rapidly and stabilizes, which is attributed to the uniform multiphase structure of the matrix material, and the friction coefficient decreases to 0.022. With the addition of HER to form a double network co-crosslinking system, the hardness and modulus of the material gradually increased, greatly improving the wear resistance of the material. The coefficient of friction is reduced, and at the same time, there is a better improvement in the wear resistance, and the wear rate is reduced to 3.87×10^{-6} mm^3/Nm. There is a significant reduction in the coefficient of friction and wear rate of the dual network co-crosslinked composites in Figure 8 relative to Figure 6. The enhanced friction performance is mainly attributed to the following two aspects. First, the synergistic effect of HER incorporation into the gradually formed dual network enables the water lubrication process from boundary lubrication to mixing lubrication to progress quickly, resulting in smaller fluctuations in the material friction coefficient. Second, the gradual formation of the second network leads to a gradual increase in the modulus and hardness of the composite material, smaller plastic deformation makes the material more resistant to shear, and the boundary lubrication performance of the material gradually increases, making the material more wear resistant.

Figure 9 shows SEM images of the friction surfaces of the dual network co-crosslinked composites at different mass ratios. As shown in Figure 9b–d, a small number of grooves gradually disappeared after the formation of the dual network co-crosslinking system, and the friction surface became flatter, mainly dominated by slight plastic deformation. The flat friction surface makes the water film stronger and more complete, which is more conducive to the separation of the water film from the friction subsurface, while the material friction surface produces only a small amount of wear debris [40]. As Figure 9c shows, the friction surface is flat with only a small amount of abrasive debris, which is mainly dominated by slight abrasive wear [41]. All the above results indicate that the dual network co-crosslinked composites possess a low coefficient of friction and wear rate.

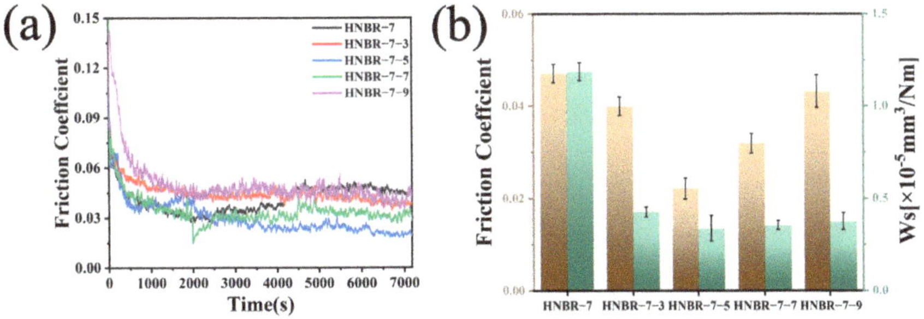

Figure 8. (**a**) Friction profile coefficients and (**b**) average COF and Ws for double network co-crosslinked composites at different ratios.

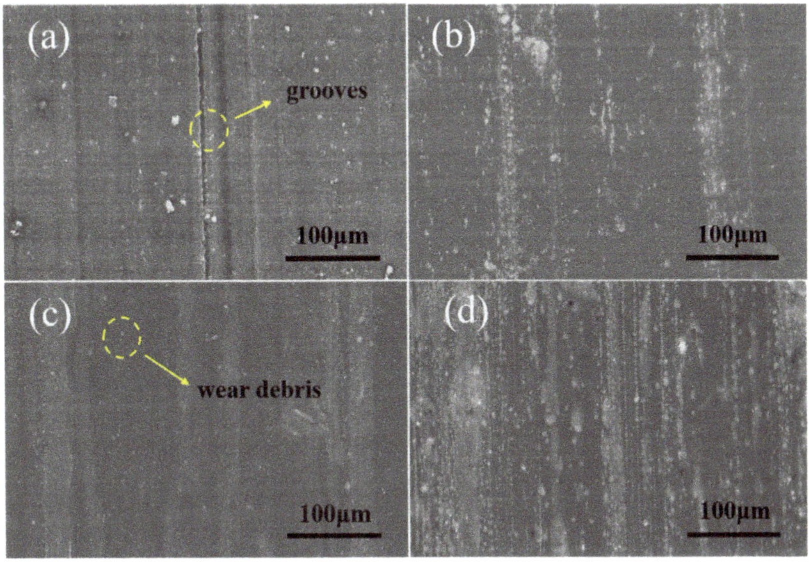

Figure 9. SEM micrographs of worn surfaces of different ratios of double network co-crosslinked composites: (**a**) 100:7:3, (**b**) 100:7:5, (**c**) 100:7:7, and (**d**) 100:7:9.

3.5. *Microscopic Morphology Analysis*

The phase morphology of the friction surface was examined using AFM in tap mode. The surface roughness and surface properties of the composite material are critical for the friction properties of the material. The AFM image is 256 × 256 pixels with a scanning range of 5 μm × 5 μm. The AFM image of the material is shown in Figure 10. From the comparison of the AFM morphology of the friction surface after friction with the phase image, Figure 10(a1–c1) show that in the material surface images, the material rms roughness Rq is 9.85 nm, 8.01 nm, and 7.92 nm in order, and the roughness decreases in order, which corresponds to the friction coefficient of the material [42]. The smoother surface may be an important factor in the formation of a stable water film during the friction process, avoiding large protrusions that can pierce the water film and weaken the lubrication effect. The low roughness reduces the mutual resistance between the friction partners and thus improves the boundary lubrication, resulting in a material with a lower coefficient of friction and wear rate [43]. The corresponding phase diagrams show that the latter has changed considerably from pure HNBR, as shown in Figure 10(a2–c2), where agglomeration of carbon black particles is unavoidable in pure HNBR, and the addition of EEUG improves the compatibility of carbon black. No obvious phase separation was observed in the phase diagrams of the latter two, which proves that the components of the prepared materials have good compatibility.

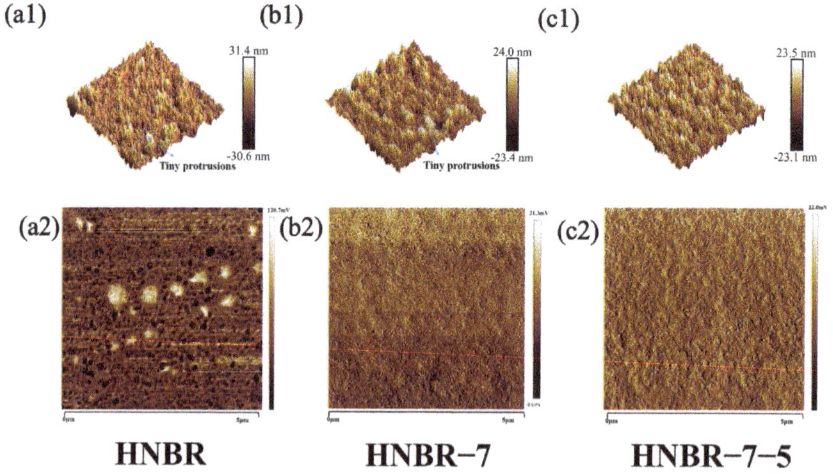

Figure 10. AFM topography of different samples (**a1–c3**), phase diagrams (**a2–c2**).

3.6. Water Contact Angle

The water contact angle was tested on the friction surface of the samples. As shown in Figure 11, the water contact angle decreases sequentially, and the water contact angle of the dual network co-crosslinked system decreases by 22.8° with respect to the pure HNBR. The hydrophilicity of the material improves due to the introduction of both EEUG and HER with hydrophilic groups [44]. The hydrophilic surface facilitates the formation of a water film during friction, which can better separate the friction partners, thus reducing the coefficient of friction and wear rate.

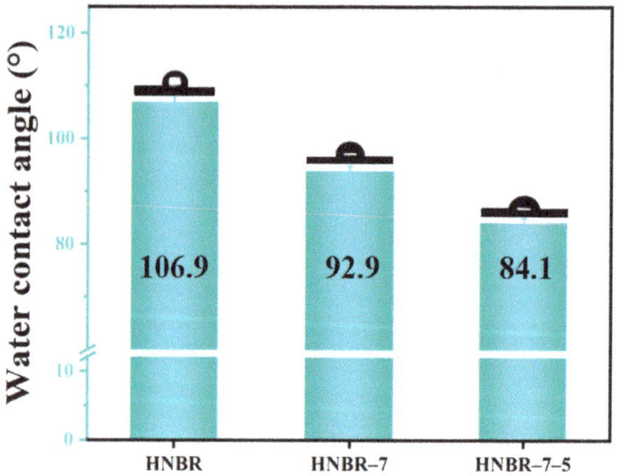

Figure 11. Water contact angle images of samples.

3.7. Long-Term Wear Test

To understand the wear resistance of rubber matrix composites co-crosslinked with dual networks, long-term friction and wear experiments were carried out, and the experimental conditions were set at a 132 N load and 140 rpm, and the experimental time was 12 h. As shown in Figure 12a, the double network co-crosslinked rubber matrix composites have smoother surfaces and lower friction coefficients under prolonged friction, and the SEM friction surface images are shown in Figure 12b–d. HNBR-7-5 has the smoothest and flattest friction surface, which is direct proof of the material's good abrasion resistance. Overall, HNBR-7-5 still has good tribological properties after the long-term wear test.

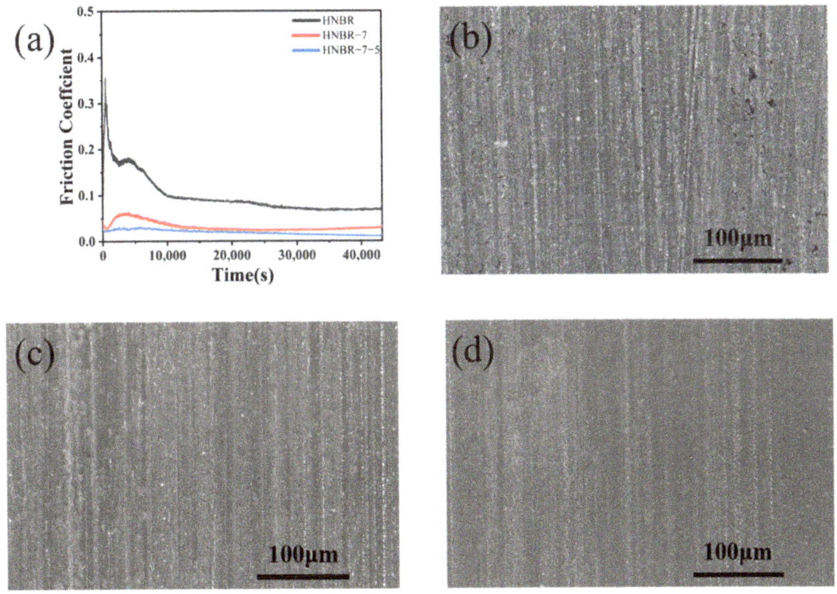

Figure 12. Comparison of friction curves for different samples over 12 h (**a**) and the after-wear surface micrograph for samples HNBR (**b**), HNBR-7, (**c**) and HNBR-7-5 (**d**).

4. Conclusions

In this paper, a dual network co-crosslinked rubber matrix composite is developed. By using epoxidized Eucommia ulmoides gum as a "bridge" between HNBR and HER, the two are chemically crosslinked. Then, the mechanical properties, damping properties, friction properties under water lubrication conditions, and microscopic morphology of the composites were systematically investigated. Atomic force microscopy results show that the dual network co-crosslinked composites have a uniform phase morphology, friction properties, and damping properties and a flatter friction surface. The frictional properties of the dual network co-crosslinked composites were significantly improved with the formation of dual networks, the friction coefficients of COF and Ws were reduced to 0.022, and the wear rate was reduced to 3.87×10^{-6} mm^3/Nm in HNBR/EEUG/HER 100:7:5. The effective damping temperature domain of the composites is increased by 87.6% compared with that of pure HNBR. New ideas and practices are provided for the design of composites with good water lubrication and vibration-damping properties in the field of underwater propulsion.

Author Contributions: Conceptualization, H.Y., G.T. and T.W.; methodology, S.C.; validation, C.Z. and W.Z.; investigation, S.C.; writing—original draft preparation, H.Y.; writing—review and editing, G.T.; visualization, W.Z.; supervision, T.W.; project administration, G.T.; funding acquisition, T.W. All authors have read and agreed to the published version of the manuscript.

Funding: This research was funded by the Strategic Priority Research Program of the Chinese Academy of Sciences (NO. XDB0470102).

Data Availability Statement: On account of legal or ethical reasons, all data involved in this paper are not available at present.

Conflicts of Interest: The authors declare no conflict of interest.

References

1. Wang, J.; Yan, F.; Xue, Q. Tribological behaviors of some polymeric materials in sea water. *Sci. Bull.* **2009**, *54*, 4541–4548. [CrossRef]
2. Litwin, W. Experimental research on water lubricated three layer sliding bearing with lubrication grooves in the upper part of the bush and its comparison with a rubber bearing. *Tribol. Int.* **2015**, *82*, 153–161. [CrossRef]
3. Zhou, G.; Wu, K.; Pu, W.; Li, P.; Han, Y. Tribological modification of hydrogenated nitrile rubber nanocomposites for water-lubricated bearing of ship stern shaft. *Wear* **2022**, *504–505*, 204432. [CrossRef]
4. Masuko, M.; Suzuki, A.; Sagae, Y.; Tokoro, M.; Yamamoto, K. Friction characteristics of inorganic or organic thin coatings on solid surfaces under water lubrication. *Tribol. Int.* **2006**, *39*, 1601–1608. [CrossRef]
5. Yamamoto, K.; Matsukado, K. Effect of hydrogenated DLC coating hardness on the tribological properties under water lubrication. *Tribol. Int.* **2006**, *39*, 1609–1614. [CrossRef]
6. Mou, Z.; Yan, R.; Peng, J.; Li, Y.; Huang, Z.; Wang, Z.; Zhao, B.; Xiao, D. Synthesis of polyzwitterionic carbon dots with superior friction and fatigue control behaviors under water lubrication. *Chem. Eng. J.* **2023**, *465*, 142986. [CrossRef]
7. Vadivel, H.S.; Golchin, A.; Emami, N. Tribological behaviour of carbon filled hybrid UHMWPE composites in water. *Tribol. Int.* **2018**, *124*, 169–177. [CrossRef]
8. Liu, S.; Luo, J.; Li, G.; Zhang, C.; Lu, X. Effect of surface physicochemical properties on the lubricating properties of water film. *Appl. Surf. Sci.* **2008**, *254*, 7137–7142. [CrossRef]
9. Wu, Z.; Guo, Z.; Yuan, C. Influence of polyethylene wax on wear resistance for polyurethane composite material under low speed water-lubricated conditions. *Wear* **2019**, *426–427*, 1008–1017. [CrossRef]
10. Wang, Q.; Zhou, F. Progress in Tribological Properties of Nano-Composite Hard Coatings under Water Lubrication. *Lubricants* **2017**, *5*, 5. [CrossRef]
11. Qu, C.; Zhang, N.; Wang, C.; Wang, T.; Wang, Q.; Li, S.; Chen, S. MoS2/CF synergistic reinforcement on tribological properties of NBR/PU/EP interpenetrating polymer networks. *Tribol. Int.* **2022**, *167*, 107384. [CrossRef]
12. Zang, L.; Chen, D.; Cai, Z.; Peng, J.; Zhu, M. Preparation and damping properties of an organic–inorganic hybrid material based on nitrile rubber. *Compos. Part B Eng.* **2018**, *137*, 217–224. [CrossRef]
13. Zhou, Y.-J.; Wang, D.-G.; Guo, Y.-B. The Reduction of Static Friction of Rubber Contact under Sea Water Droplet Lubrication. *Lubricants* **2017**, *5*, 12. [CrossRef]
14. Litwin, W. Properties comparison of rubber and three layer PTFE-NBR-bronze water lubricated bearings with lubricating grooves along entire bush circumference based on experimental tests. *Tribol. Int.* **2015**, *90*, 404–411. [CrossRef]
15. Narynbek Ulu, K.; Huneau, B.; Verron, E.; Béranger, A.S.; Heuillet, P. Effects of acrylonitrile content and hydrogenation on fatigue behaviour of HNBR. *Fatigue Fract. Eng. Mater. Struct.* **2019**, *42*, 1578–1594. [CrossRef]
16. Yeo, Y.-G.; Park, H.-H.; Lee, C.-S. A study on the characteristics of a rubber blend of fluorocarbon rubber and hydrogenated nitrile rubber. *J. Ind. Eng. Chem.* **2013**, *19*, 1540–1548. [CrossRef]
17. Dong, C.; Shi, L.; Li, L.; Bai, X.; Yuan, C.; Tian, Y. Stick-slip behaviours of water lubrication polymer materials under low speed conditions. *Tribol. Int.* **2017**, *106*, 55–61. [CrossRef]
18. Marian, M.; Berman, D.; Nečas, D.; Emami, N.; Ruggiero, A.; Rosenkranz, A. Roadmap for 2D materials in biotribological/biomedical applications–A review. *Adv. Colloid Interface Sci.* **2022**, *307*, 102747. [CrossRef]
19. Chen, H.; Cai, T.; Ruan, X.; Jiao, C.; Xia, J.; Wei, X.; Wang, Y.; Gong, P.; Li, H.; Atkin, R.; et al. Outstanding Bio-Tribological Performance Induced by the Synergistic Effect of 2D Diamond Nanosheet Coating and Silk Fibroin. *ACS Appl. Mater. Interfaces* **2022**, *14*, 48091–48105. [CrossRef]
20. Zhang, L.; Xie, G.; Wu, S.; Peng, S.; Zhang, X.; Guo, D.; Wen, S.; Luo, J. Ultralow friction polymer composites incorporated with monodispersed oil microcapsules. *Friction* **2019**, *9*, 29–40. [CrossRef]
21. Li, Z.; Li, K.; Li, X.; Feng, Y.; Li, H.; Wang, H. Preparation of linseed oil-loaded porous glass bubble/wax microcapsules for corrosion- and wear-resistant difunctional coatings. *Chem. Eng. J.* **2022**, *437*, 135403. [CrossRef]
22. Zhang, J.; Wang, L.; Zhao, Y. Understanding interpenetrating-polymer-network-like porous nitrile butadiene rubber hybrids by their long-period miscibility. *Mater. Des.* **2013**, *51*, 648–657. [CrossRef]

23. Emerson, J.A.; Garabedian, N.T.; Moore, A.C.; Burris, D.L.; Furst, E.M.; Epps, T.H. Unexpected Tribological Synergy in Polymer Blend Coatings: Leveraging Phase Separation to Isolate Domain Size Effects and Reduce Friction. *ACS Appl. Mater. Interfaces* **2017**, *9*, 34480–34488. [CrossRef]
24. Liu, Y.; Shuai, C.; Lu, G.; Yang, X.; Hu, X. Preparation of polyethylene glycol brush grafted from the surface of nitrile butadiene rubber with excellent tribological performance under aqueous lubrication. *Mater. Des.* **2022**, *224*, 111310. [CrossRef]
25. Xu, J.; Zhang, C.; Luo, J. Hydration Lubrication Applicable to Artificial Joints through Polyelectrolyte-Embedded Modification on UHMWPE. *ACS Appl. Polym. Mater.* **2022**, *4*, 7487–7497. [CrossRef]
26. Chudzik, J.; Bielinski, D.M.; Demchuk, Y.; Bratychak, M.; Astakhova, O. Influence of Modified Epoxy Dian Resin on Properties of Nitrile-Butadiene Rubber (NBR). *Materials* **2022**, *15*, 2766. [CrossRef]
27. Sang, J.; Sato, R.; Aisawa, S.; Hirahara, H.; Mori, K. Hybrid joining of polyamide and hydrogenated acrylonitrile butadiene rubber through heat-resistant functional layer of silane coupling agent. *Appl. Surf. Sci.* **2017**, *412*, 121–130. [CrossRef]
28. Yang, F.; Dai, L.; Liu, T.; Zhou, J.; Fang, Q. Preparation of high-damping soft elastomer based on Eucommia ulmoides gum. *Polym. Bull.* **2019**, *77*, 33–47. [CrossRef]
29. Dong, M.; Zhang, T.; Zhang, J.; Hou, G.; Yu, M.; Liu, L. Mechanism analysis of Eucommia ulmoides gum reducing the rolling resistance and the application study in green tires. *Polym. Test.* **2020**, *87*, 106539. [CrossRef]
30. Wang, Y.; Liu, J.; Xia, L.; Shen, M.; Xin, Z.; Kim, J. Role of epoxidized natural Eucommia ulmoides gum in modifying the interface of styrene-butadiene rubber/silica composites. *Polym. Adv. Technol.* **2019**, *30*, 2968–2976. [CrossRef]
31. Chen, B.; Wu, Q.; Li, J.; Lin, K.; Chen, D.; Zhou, C.; Wu, T.; Luo, X.; Liu, Y. A novel and green method to synthesize a epoxidized biomass eucommia gum as the nanofiller in the epoxy composite coating with excellent anticorrosive performance. *Chem. Eng. J.* **2020**, *379*, 122323. [CrossRef]
32. Wang, Y.; Liu, J.; Xia, L.; Shen, M.; Xin, Z. Super-tough poly(lactic acid) thermoplastic vulcanizates with heat triggered shape memory behaviors based on modified natural Eucommia ulmoides gum. *Polym. Test.* **2019**, *80*, 106077. [CrossRef]
33. Qi, X.; Pan, C.; Zhang, L.; Yue, D. Bio-Based, Self-Healing, Recyclable, Reconfigurable Multifunctional Polymers with Both One-Way and Two-Way Shape Memory Properties. *ACS Appl. Mater. Interfaces* **2023**, *15*, 3497–3506. [CrossRef]
34. Li, D.; Zhou, J.; Kang, H.; Li, L.; Han, W.; Fang, Q.; Wang, N.; Yang, F. Synthesis and properties of an efficient self-healing material based on Eucommia ulmoides gum. *Ind. Crops Prod.* **2022**, *187*, 115385. [CrossRef]
35. Yue, P.-P.; Rao, J.; Leng, Z.-J.; Chen, G.-G.; Hao, X.; Peng, P.; Peng, F. An electrospun composite of epoxidized Eucommia ulmoides gum and SiO2-GO with ultraviolet resistance. *J. Mater. Sci.* **2022**, *57*, 4862–4875. [CrossRef]
36. Qu, C.; Wang, T.; Wang, Q.; Chen, S. A novel ternary interpenetrating polymer networks based on NBR/PU/EP with outstanding damping and tribological properties for water-lubricated bearings. *Tribol. Int.* **2022**, *167*, 107249. [CrossRef]
37. de Kraker, A.; van Ostayen, R.A.J.; Rixen, D.J. Calculation of Stribeck curves for (water) lubricated journal bearings. *Tribol. Int.* **2007**, *40*, 459–469. [CrossRef]
38. Yu, P.; Li, G.; Zhang, L.; Zhao, F.; Guo, Y.; Pei, X.-Q.; Zhang, G. Role of SiC submicron-particles on tribofilm growth at water-lubricated interface of polyurethane/epoxy interpenetrating network (PU/EP IPN) composites and steel. *Tribol. Int.* **2021**, *153*, 106611. [CrossRef]
39. Fukahori, Y.; Gabriel, P.; Liang, H.; Busfield, J.J.C. A new generalized philosophy and theory for rubber friction and wear. *Wear* **2020**, *446–447*, 203166. [CrossRef]
40. Yin, T.; Wei, D.; Wang, T.; Xie, Z. Thermal compression and accumulation effect on lubrication regime transition mechanism of water seal. *Tribol. Int.* **2023**, *181*, 108285. [CrossRef]
41. Hale, J.; Lewis, R.; Carré, M.J. Rubber friction and the effect of shape. *Tribol. Int.* **2020**, *141*, 105911. [CrossRef]
42. Bai, C.; Qiang, L.; Zhang, B.; Gao, K.; Zhang, J. Optimizing the tribological performance of DLC-coated NBR rubber: The role of hydrogen in films. *Friction* **2021**, *10*, 866–877. [CrossRef]
43. Emami, A.; Khaleghian, S.; Taheri, S. Asperity-based modification on theory of contact mechanics and rubber friction for self-affine fractal surfaces. *Friction* **2021**, *9*, 1707–1725. [CrossRef]
44. Zhao, G.; Wang, T.; Wang, Q. Studies on wettability, mechanical and tribological properties of the polyurethane composites filled with talc. *Appl. Surf. Sci.* **2012**, *258*, 3557–3564. [CrossRef]

lubricants

Article

Analysis of Water-Lubricated Journal Bearings Assisted by a Small Quantity of Secondary Lubricating Medium with Navier–Stokes Equation and VOF Model

Xiaohan Zhang [1,2], Tao Yu [1], Hao Ji [1], Feng Guo [1,*], Wenbin Duan [1], Peng Liang [1] and Ling Ma [1]

1 School of Mechanical and Automotive Engineering, Qingdao University of Technology, Qingdao 266520, China; zhangxiaohan@qut.edu.cn (X.Z.); taoyu_official@126.com (T.Y.); 18852343390@163.com (H.J.); a15851736631@163.com (W.D.); liangpeng2009@126.com (P.L.); ml719210@163.com (L.M.)

2 Key Lab of Industrial Fluid Energy Conservation and Pollution Control (Qingdao University of Technology), Ministry of Education, Qingdao 266520, China

* Correspondence: mefguo@qut.edu.cn

Abstract: Due to the low viscosity of water, water-lubricated bearings are susceptible to significant wear and noise in demanding operating conditions. It has been demonstrated that a small quantity of secondary lubricating medium can improve the lubrication performance of water-lubricated contact surfaces and achieve the purpose of temporary risk aversion. As a further step, the feasibility of the proposed idea is experimentally validated on a water-lubricated bearing test bench. A numerical model that couples the N–S equation and the VOF model is then developed to investigate the behavior of the flow field lubricated by pure water and water with a small quantity of the secondary lubricating medium. This model provides the predictions of important quantities such as the load-carrying capacity, the secondary lubricating medium volume fraction and the contact pressure under different lubricated conditions. The results show that the secondary lubricating medium can enter into the contact region and improve the lubrication performance of water-lubricated bearings, especially at lower shaft rotational speeds. Therefore, the feasibility of our proposed idea is verified, which provides a promising approach to reduce the wear and friction of water-lubricated bearings when they encounter short-time severe working conditions.

Keywords: water-lubricated bearings; numerical model; secondary lubricating medium

Citation: Zhang, X.; Yu, T.; Ji, H.; Guo, F.; Duan, W.; Liang, P.; Ma, L. Analysis of Water-Lubricated Journal Bearings Assisted by a Small Quantity of Secondary Lubricating Medium with Navier–Stokes Equation and VOF Model. *Lubricants* **2024**, *12*, 16. https://doi.org/10.3390/lubricants12010016

Received: 15 December 2023
Revised: 4 January 2024
Accepted: 5 January 2024
Published: 7 January 2024

1. Introduction

Water-lubricated bearings are increasingly popular due to their environmentally friendly properties. However, the water film thickness can easily be broken when water-lubricated bearings encounter short-time severe working conditions (sudden impact of the external load, the start-up and shut down period) [1] owing to the low viscosity of water; thus, the bearing wear failure, as well as the vibration and the noise of water-lubricated bearings, can be induced.

To address these issues and enhance the performance of water-lubricated bearings, researchers have conducted extensive studies from various perspectives, including bearing materials, bearing theory, bearing structure and the usage of new lubricating fluids [2,3].

The bearing bushing is a crucial component in water-lubricated bearings, which is designed to enhance efficiency and minimize the wear and vibration of water-lubricated bearings. Various fixed structures with grooves have been developed for this purpose. The key aspect of structural design lies in the groove and lubrication analysis of the bearing bushings, including the effect of journal misalignment on lubrication properties [4–6], the calculation of static and dynamic characteristics of water-lubricated bearings [7–9] and the influence of groove number and size on the stability of water-lubricated bearings [10–12].

Among these designs, the design of the bushing structure has always been controversial. The straight-fluted bearing is the most commonly used [13], featuring load-carrying lands or staves that facilitate partial hydrodynamic lubrication and low load capacity. Spiral grooved bushings, studied by Zhou [14], displayed a vortex effect inside the grooves that facilitates impurity discharge such as sediment. Herringbone groove bearings, as found by Feng [15], offered maximum load capacity, followed by straight groove and spiral groove bearings. Dong's study [16] revealed that soft materials, like rubber, exhibited stick–slip friction and a fold morphology under water lubrication, but it remained unclear whether this phenomenon applies to water-lubricated bearings. All of these types of structures play important roles in improving the performance of water-lubricated bearings, but instability generated by the complex structure urges researchers to find simpler ways to reduce the abrasion of water-lubricated bearings.

The development of new materials for bushings is another commonly used approach by researchers to improve the lubrication performance of water-lubricated bearings. Common materials used for water-lubricated bearings include lignum vitae, rubber, modified polytetrafluoroethylene and composite polymer materials. In addition to these, high-performance materials for water-lubricated bearings are constantly being synthesized, such as the Thordon and the Feroform, which are widely used in water-lubricated bearings [17]. Yang [18] synthesized urea formaldehyde (UF) microcapsules using in situ emulsion polymerization and added them to a high-density polyethylene (HDPE) matrix to reduce the coefficient of friction (hereafter 'COF') and adhesion wear of water-lubricated bearings. Liang [19] introduced polyacrylonitrile (PAN) fiber as a reinforcement material to further enhance the tribological properties of water-lubricated bearings, thus improving their service life. Although these materials have been proven to have a low COF and better wear resistance, the potential for existing materials to improve the performance of water-lubricated bearings is limited, and new materials need to be developed.

Traditional theoretical analyses of water-lubricated bearings typically utilize lubrication theory based on the Reynolds equation [1,20,21]. While these methods can yield reasonable results, complex bearing models and potential parameter selection errors may lead to simulation result deviations. Additionally, traditional methods often focus on single-phase flow, rather than multiphase flow, even though water-lubricated bearings can easily wear out when working in water with high sediment content or a cavitation environment.

In contrast to traditional methods, our previous work [22,23] proposed a new approach: adding a small quantity of secondary lubricating medium in real time to the water when the water-lubricated contact surfaces encounter severe working conditions; thus, the film-forming ability of the water can be enhanced and the COF between the shaft surface and bearing surface can be decreased.

We have demonstrated the feasibility of this approach in a block-on-ring test rig. In this work, we conduct experiments on a water-lubricated bearing test bench to investigate the applicability of this approach in water-lubricated bearings.

According to the experiments, the widely used VOF method [24] is chosen to track the interface between the water and the secondary lubricating medium. By combining it with the Navier–Stokes (hereafter 'N–S') equation, a numerical model is established with Matlab R2022b to simulate the flow field in water-lubricated bearings after a small amount of secondary lubricating medium is injected, thus theoretically verifying the feasibility of the proposed approach for water-lubricated bearings. Additionally, the influence of the proposed idea on various parameters (the pressure distribution, the load-carrying capacity, etc.) is also explored.

2. Geometry Description

Figure 1 depicts two different types of water-lubricated bearings. Figure 1a displays the typical geometry of commonly used water-lubricated bearings, while Figure 1b shows the new-type water-lubricated bearing geometry studied in our work, which includes a small quantity of secondary lubricating medium supply structure added to the bearing.

Pure water is supplied from one side for lubrication (as shown in Figure 1), and the oil is supplied from the oil inlet to mix with pure water for lubrication (as shown in Figure 1b).

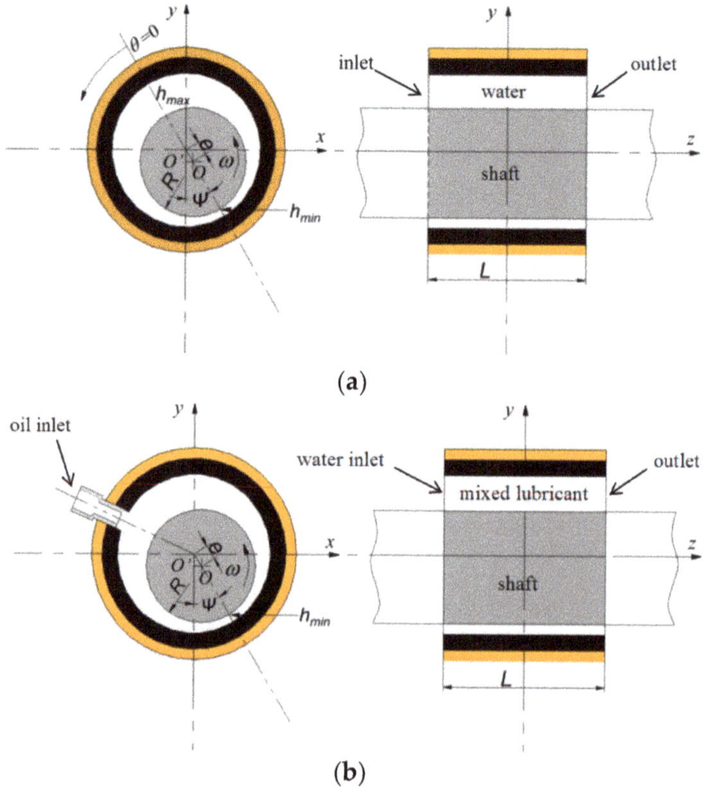

Figure 1. Schematic structure of water-lubricated bearings studied in this work. (**a**) Typical geometry of commonly used water-lubricated bearing; (**b**) new-type water-lubricated bearing geometry studied in present work.

Table 1 presents the relative parameters of these two types of water-lubricated bearings, which are the same as those used in industrial applications.

Table 1. Parameters of the two types of water-lubricated bearings.

Parameters	Values/Comments
Shaft material	316 stainless steels
Bushing material	3606 NBR
Bushing length	80 mm
Bearing inner diameter	100 mm
Radial clearance	0.07 mm
Oil inlet diameter	10 mm

3. Experimental Verification

The feasibility of the proposed idea on the water-lubricated bearing is carried out. The schematic diagram of the water-lubricated bearing test bench used in the experiment is shown in Figure 2.

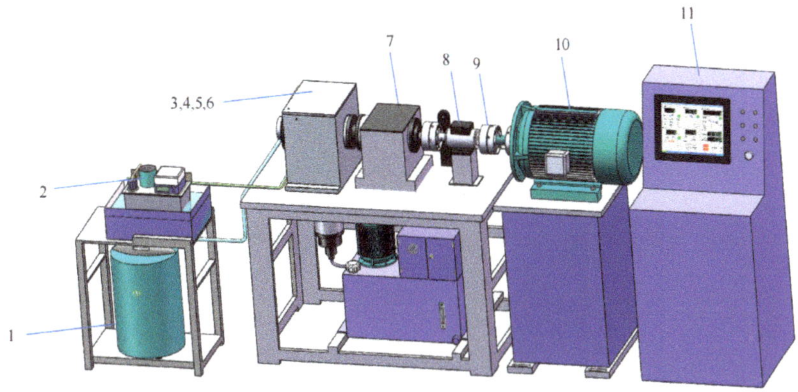

Figure 2. Schematic diagram of water-lubricated bearing test bench used in the experiment. 1—water supply system, 2—small amount of secondary lubricating medium supply device, 3—bearing, 4—hydraulic cylinder control system, 5—sealing chamber, 6—bearing external fixing shell, 7—bearing housing, 8—torque–velocity meter, 9—flexible pin coupling, 10—variable frequency motor, 11—control cabinet.

It has been observed that the lubrication performance of the micro-emulsified oil droplets is effectively utilized by using O/W (oil-in-water) emulsion as a lubricant, and the application of O/W emulsion is already found in many fields such as metal processing. The film-forming theories about the O/W (oil-in-water) emulsion have been proposed in [25,26], such as plate-out, dynamic concentration and secondary emulsification. It should be noted that within the framework of our proposed concept, the aforementioned O/W film-forming theories undeniably contribute to our research by employing emulsifying oil as a small quantity of secondary lubrication medium. Meanwhile, our previous work [27] has also proved the effectiveness of emulsifying oil in decreasing the COF between two surfaces on the block-on-ring test rig. Therefore, the secondary lubricating medium used in the experiment is an emulsifying oil (provided by Zhongkerunmei Lubrication Materials Technology Co., Ltd., Qingdao, China).

As we mentioned in our previous work [22], contact between the bearing surface and the shaft surface occurs under the mixed lubrication regime at some severe working conditions, such as heavy impact load and low rotational speeds, and the ring rotational speed for the mixed lubrication regime lies in the interval of 50 r/min to 1500 r/min under two different applied loads. In an effort to determine whether a modest injection of secondary lubricating medium within this contact can alter the COF between the shaft and the bearing surfaces, shaft rotational speeds of 100 r/min and 200 r/min are selected with the applied load of 500 N, while variations in the COF are subsequently monitored.

Before the test bench starts up, water from the water supply system (No.1 in Figure 2) is pumped into the sealing chamber (No.5 in Figure 2) through the inlet at the sealing chamber, and another water pump in the water supply system is used to draw the water from the chamber into the waste tank through the outlet. Thus, the open-water environment can be simulated.

Once the test bench has started up (ω = 100 r/min, P = 500 N) and the COF has become steady (after a running-in period of approximately 300 s), 8.8 mL of emulsifying oil is injected at a constant rate (2.2 mL/s) over a period of 4 s, and the oil feeding pressure is 2 MPa. The experiment is conducted three times while paying close attention to the careful cleaning of both the bearing and shaft surfaces using petroleum ether and ethyl alcohol, with the COF being recorded on each occasion. In order to depict detailed information about the test curve, data from the second repetition test of the experiment are selected as the typical outcome for plotting the COF curve, which is consistent with our previous work [22].

In Figure 3a, we can observe the variation in COF over a period of 600 s when a small amount of emulsifying oil is injected into the water-lubricated bearing at the shaft rotational speed of 100 r/min. The injection of emulsifying oil instantaneously reduces the COF between the two surfaces in contact from 0.24 to roughly 0.19 and becomes stable, which is lower than the COF observed during pure water lubrication.

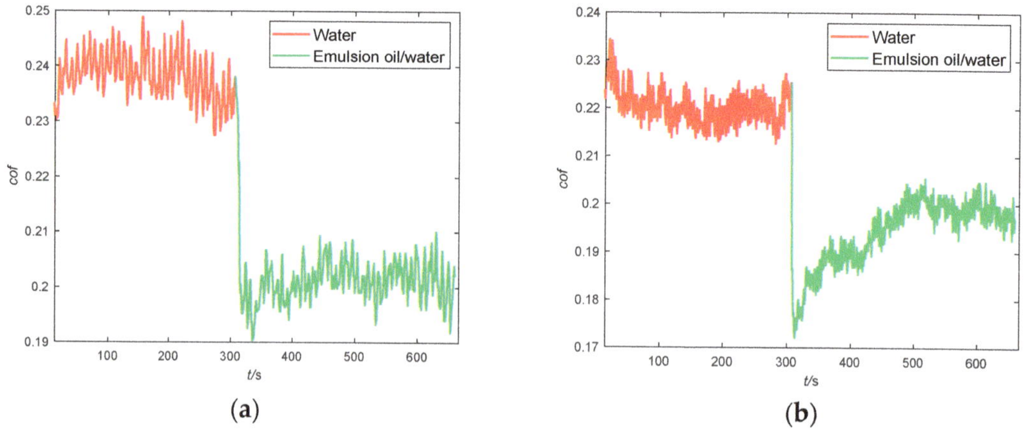

Figure 3. COF variations when a small quantity of emulsifying oil is injected into the water-lubricated bearing. (**a**) ω = 100 r/min; (**b**) ω = 200 r/min.

By increasing the shaft rotational speed to 200 r/min while keeping the other experimental conditions unchanged, the variation in the COF is recorded, as shown in Figure 3b. Similar to the observations in Figure 3a, the COF initially decreases from 0.23 to 0.175 upon injecting the emulsifying oil. Following some fluctuations, the COF subsequently increases before stabilizing around 0.196, which is still lower than the COF in pure water lubrication.

Figure 4 shows the conditions of the journal and bearing after the experiment. It can be observed that some emulsified oil is stuck on the surfaces of the bearing and shaft, and this adherence continues to contribute to lubrication even after the injection of emulsified oil is ceased. This explains why, as depicted in Figure 3, the stable COF following the injection of emulsified oil remains lower than that of pure water lubrication.

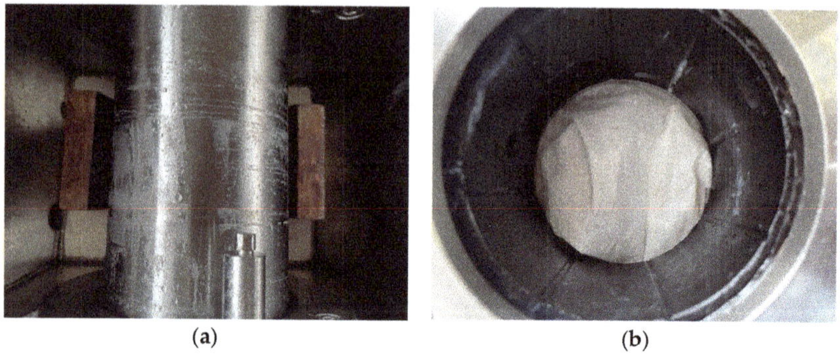

Figure 4. Stuck emulsified oil on the surfaces of shaft and bearing after the experiment. (**a**) Stuck emulsified oil on the shaft surface; (**b**) stuck emulsified oil on the bearing surface.

These observations suggest that the emulsifying oil can assist with the lubrication of the contact surfaces and increase the film-forming capacity of water. Therefore, our proposed idea has been validated regarding water-lubricated bearings.

For a better understanding of the synergistic lubrication of water and the secondary lubricant in water-lubricated bearings, after a small quantity of secondary lubricating medium has been injected, a numerical model is established. It should be noted that when a small quantity of secondary lubricating medium is introduced into water-lubricated bearings, the dispersion of the secondary lubricating medium is dependent on the velocity of the flow field. Furthermore, the combined effect between the water and the secondary lubricating medium can impact the density and viscosity of the mixed lubricant. To model these phenomena, the N–S equation and VOF model are integrated, with Matlab 2022Rb programming utilized.

4. Methodology

4.1. Governing Equations

4.1.1. Navier–Stokes (N–S) Equation

In this work, we mainly focused on the spatial distribution of the oil; the incompressible steady-state N–S equations are then used in this paper, which are

$$\nabla\cdot\left(\rho\vec{U}\vec{U}\right) = -\nabla p + \mu\nabla^2\vec{U} \tag{1}$$

$$\nabla\cdot\left(\rho\vec{U}\right) = 0, \tag{2}$$

where p is pressure; ρ is density; μ is viscosity and U is the velocity of the flow.

The integrated form of Equations (1) and (2) in any closed region are

$$\oint\left(\rho\vec{U}\left(\vec{U}\cdot d\vec{s}\right)\right) = -\oint\left(pd\vec{s}\right) + \oint\left(\mu\nabla\vec{U}\cdot d\vec{s}\right) \tag{3}$$

$$\oint\left(\rho\vec{U}\cdot d\vec{s}\right) = 0 \tag{4}$$

By using the finite volume method, Equations (3) and (4) can be discretized as

$$\sum_{k=0}^{nfaces}\vec{U}_{ik}\left(\rho\vec{U}_{ik}\cdot\vec{S}_{ik}\right) = -\sum_{k=0}^{nfaces}\left(\rho\vec{S}_{ik}\right) + \sum_{k=0}^{nfaces}\left(\mu\nabla\vec{U}\cdot\vec{S}_{ik}\right) \tag{5}$$

$$\sum_{k=0}^{nfaces}\left(\rho\vec{U}_{ik}\cdot\vec{S}_{ik}\right) = 0, \tag{6}$$

where i represents the number on the grid and k is the number on the edge of the grid.

In order to prevent oscillations in pressure during the discretization, the staggered grid method (Figure 5) is used, which stores velocities at locations offset from the pressure storage locations in their respective directions, and each of the pressure and the velocity components have separate control volumes.

Due to the influence of water velocity on the distribution of the secondary lubricating medium, not only the flow velocity change along the circumferential and the axial directions should be considered, but the flow velocity change along the film thickness direction needs to be considered in this situation. However, the dimension in the direction of film thickness is smaller than the dimensions in both circumferential and axial directions, and the computation cost of a 3D model is very high in the present work. Thus, a quasi-2D model is then established, and Equations (5) and (6) can then be written as

$$\sum_{k=0}^{nfaces}\vec{U}_{ik}\varphi S_{ik}h_{ik} = -\sum_{k=0}^{nfaces}\left(\rho\vec{S}_{ik}h_{ik}\right) + \sum_{k=0}^{nfaces}\left(\mu\nabla\vec{U}\cdot\vec{S}_{ik}h_{ik}\right) \tag{7}$$

$$\sum_{k=0}^{nfaces}\varphi_{ik}S_{ik} = 0, \tag{8}$$

where φ represents normal flux and can be calculated as follows:

$$\varphi = \frac{\rho\vec{U}_i + \rho\vec{U}_k}{2} \cdot \frac{\vec{S}_{ik}}{S_{ik}} \tag{9}$$

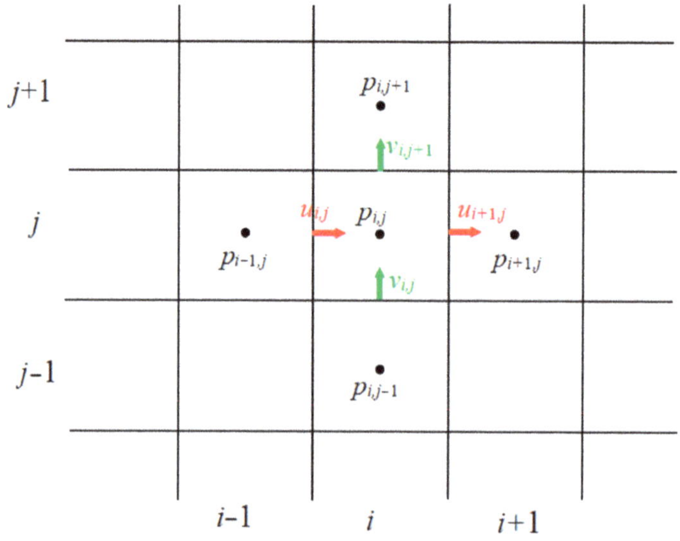

Figure 5. Scheme of the staggered grid method.

Due to the roughness and the nonparallelism of the shaft surface and the bearing surface, additional fluxes can be caused by the pressure and the viscous forces, which will also modify the momentum source term in Equation (7) to the form:

$$\sum \mu \nabla \vec{U} \cdot \vec{S} = \sum \mu A \frac{\partial \vec{U}}{\partial n} \tag{10}$$

In addition, the distribution of the flow velocity along the film thickness direction can be supposed to satisfy the Couette flow, then the flow velocity can be written as

$$\vec{U} = \vec{a} y^2 + \vec{b} y + \vec{c} \tag{11}$$

By considering the rotational speed of the bearing as zero, the rotational speed of the shaft is $\vec{r} \times \vec{\omega}$, and the average flow velocity between the shaft and the bearing is regarded as the flow velocity along the film thickness, so that

$$\vec{U} = \frac{3\vec{r} \times \vec{\omega} - 6\vec{\overline{U}}}{h^2} y^2 + \frac{6\vec{\overline{U}} - 2\vec{r} \times \vec{\omega}}{h} y \tag{12}$$

By combining Equations (10) and (12), the momentum source can be written as

$$\sum \mu A \frac{\partial \vec{U}}{\partial n} = \mu A \frac{6\vec{r} \times \vec{\omega} - 12\vec{\overline{U}}}{h} \tag{13}$$

Then, Equation (7) is revised as

$$\sum_{k=0}^{nfaces} \vec{U}_{ik} \varphi S_{ik} h_{ik} = -\sum_{k=0}^{nfaces} \left(\rho \vec{S}_{ik} h_{ik} \right) + \mu A \frac{6\vec{r} \times \vec{\omega} - 12\vec{\overline{U}}}{h_{ia}} h_{ik}, \quad (14)$$

where a represents the average value.

The film thickness equation used in the calculation is:

$$h = c + e\cos(\theta - \psi), \quad (15)$$

where c is the radial clearance, e is the eccentric distance and ψ is the attitude angle.

4.1.2. Volume of Fluid (VOF) Method

As we mentioned in our previous work [22], the volume fraction of a phase is indicated by α in the VOF method, which is defined as

$$\alpha = \begin{cases} 1 \text{ computational cell is fully filled with the water} \\ 0 \text{ computational cells are fully filled with the secondary lubricating medium} \\ 0 < \alpha < 1 \text{ interface cells} \end{cases}$$

It can be calculated by using the steady-state advection equation:

$$\nabla \cdot \left(\alpha \vec{u} \right) = 0 \quad (16)$$

By discretizing the advection equation in the quasi-2D model, Equation (16) becomes

$$\sum_{k=0}^{nfaces} \alpha_{ik} \varphi_{ik} S_{ik} \frac{h_{ik}}{h_{ia}} = 0 \quad (17)$$

The density and the viscosity of the fluid then can be calculated using the following equations:

$$\rho = \alpha_1 \rho_1 + \alpha_2 \rho_2 \quad (18)$$

$$\mu = \alpha_1 \mu_1 + \alpha_2 \mu_2, \quad (19)$$

where the subscripts 1 and 2 represent two different fluids. For each computational cell, $\alpha_o + \alpha_w = 1$, where o means oil and w means water.

Values of the density and the viscosity of the mixed lubricant, as solved by Equations (16)–(19), are then used in Equations (8), (9) and (14) to establish the quasi-2D two-phase medium lubrication model. Equations (8) and (14) are then used to calculate the distribution of the velocity and the pressure for the flow field in the bearing by implementing the SIMPLE method.

4.1.3. Load-Carrying Capacity

The load-carrying capacity generated by the fluid acting on the bearing can be expressed as [28,29]

$$\vec{F} = \int_{z_1}^{z_2} \int_{\theta_1}^{\theta_2} p d\theta dz, \quad (20)$$

where θ is circumference direction and z is the axial direction.

The flow chart of the entire calculation process is shown in Figure 6.

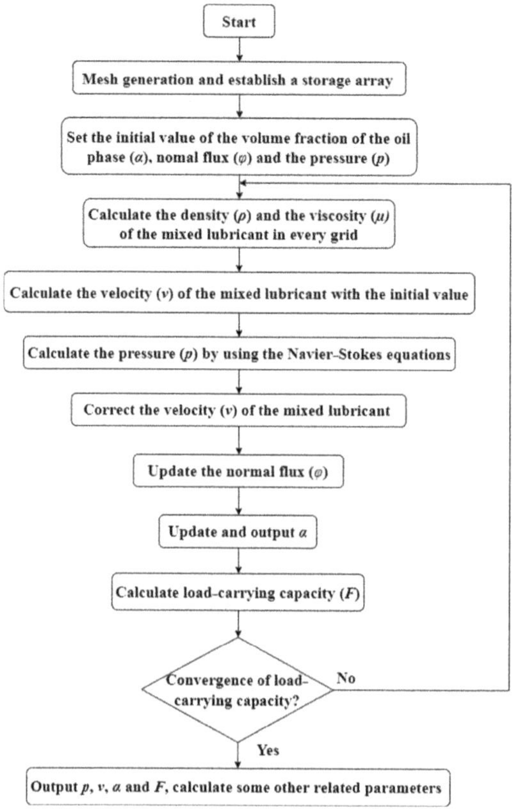

Figure 6. Flow chart of the whole calculation process.

4.2. Boundary Conditions

In this work, the Reynolds boundary conditions are used in the analysis:

$$p(\theta, 0) = p(\theta, L) = 0 \tag{21}$$

$$p(\theta_0, z) = 0, \frac{\partial p(\theta_0, z)}{\partial \theta} = 0, \tag{22}$$

where L is the length of the bearing and θ_0 is the film rupture position in the circumferential direction.

Using this boundary condition ensures that the pressure and its gradient are zero at the circumferential initial point of the journal bearing and that the rupture boundaries are satisfied.

4.3. Mesh Generation

Figure 7 illustrates the meshing of the flow field used in the present work. Due to the large number of configurations requiring calculation under different parameters, a two-dimensional grid generation program is developed specifically for this purpose. Due to the large amount of calculations near the oil inlet and boundary positions, denser mesh grids are employed to guarantee the simulation accuracy.

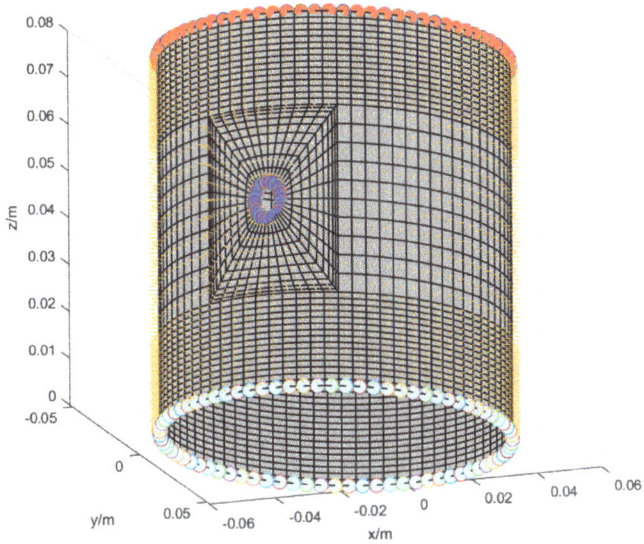

Figure 7. Mesh situation of the flow field in the present work.

5. Model Verification

FLUENT 2020 R2 is used to verify the reliability of the established quasi-2D model. The quasi-2D model and the FLUENT are both run under the same working conditions (ω = 300 rpm, v_w = 0.03 m/s, v_o = 0 m/s), and the comparison of the water flow velocity along the direction of the film thickness (red arrow in Figure 8) is depicted in Figure 9. It is clear that the water flow velocity calculated by the quasi-2D model is consistent with the values from the FLUENT, thereby verifying the accuracy of the established quasi-2D model.

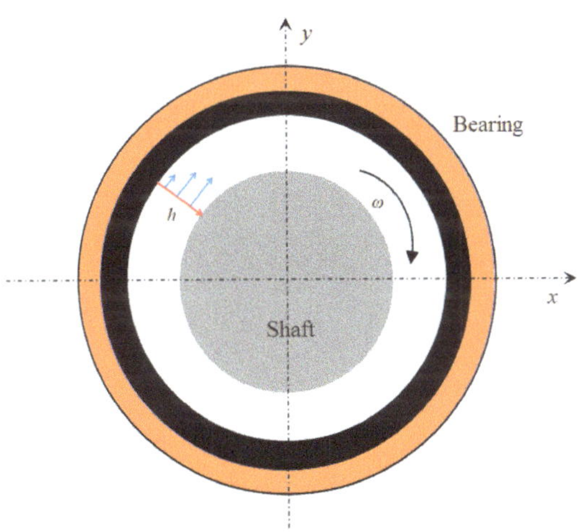

Figure 8. Diagram of chosen direction in comparing water flow velocity in the water-lubricated bearing system.

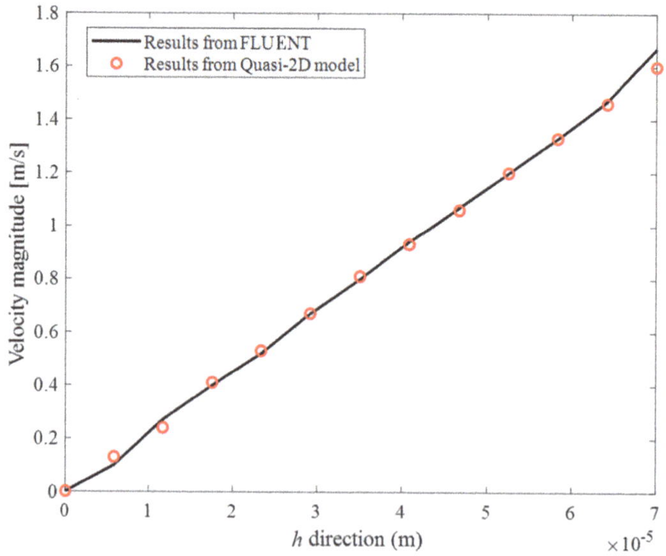

Figure 9. Comparison of the velocity of the water flow along the direction of the film thickness calculated by the quasi-2D model and the FLUENT.

6. Results and Discussion

Table 2 provides comprehensive information on the simulation parameters, wherein two phases considered in the simulation are water phase and emulsifying oil phase, the radial clearance is preselected as 70 μm and the eccentricity ratio is set to 0.9 during the simulation process. It should be recognized that although this gap size varies from the actual experimental ones in the CFD calculation, it can still provide valuable positions about flow field phenomena and the volume fraction of the oil between the bearing and the shaft for subsequent analysis.

Table 2. Simulation parameters.

Parameters	Values/Comments
Shaft rotational speed	100 r/min
Water density	1000 kg/m^3
Water dynamic viscosity	1 mPa·s
Emulsifying oil density	698 kg/m^3
Emulsifying oil viscosity(@ 21 °C)	63.79 mPa·s
Eccentricity ratio	0.9
Radial clearance	0.07 mm

6.1. Different Lubrication Conditions (ω = 100 r/min, v_w = 0.03 m/s, Oil Flow Rate = 125 μL/s)

Figure 10 presents the pressure distribution when the journal bearing is lubricated by pure water and water with a small quantity of emulsifying oil (water is set to enter from the bottom of the plot and exit from the top of the plot, as in other cases). Even if the simulation is under the hydrodynamic lubrication regime, the maximum value of pressure significantly increases as the secondary lubricating medium is injected into the water-lubricated bearing.

Meanwhile, the pressure distribution becomes more concentrated towards the center after injecting the emulsifying oil, creating a distinctive diamond-like pattern compared to the "cashew nut" shape shown in [30]. This difference could be attributed to the increased comprehensive viscosity of the mixed lubricant caused by the added emulsifying oil. Moreover, due to the use of Reynolds boundary conditions during the simulation, the pressure gradually decreases while spreading outwards as a result of the viscosity, leading to a distribution that is centered towards the bearing's center.

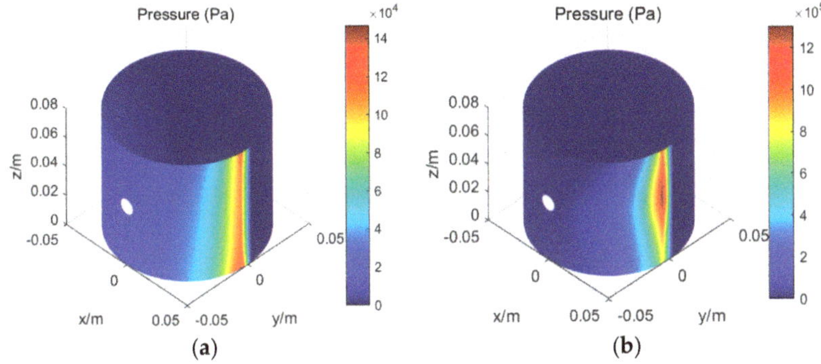

Figure 10. Pressure distribution when the journal bearing lubricated by pure water and water with a small quantity of emulsifying oil. (**a**) Journal bearing lubricated by pure water; (**b**) journal bearing lubricated by water with a small quantity of emulsifying oil.

According to the value of the pressure, the calculated load-carrying capacity in the journal bearing lubricated by pure water and water with a small quantity of emulsifying oil is 256 N and 1621 N, respectively, which illustrates that the load-carrying capacity becomes larger with the help of the injected secondary lubricating medium. Therefore, the load-carrying capacity will also be augmented along with the injection of the secondary lubricating medium when the water-lubricated bearing is under the mixed lubrication regime. The increase in load-carrying capacity usually accompanies an increase in the thickness of the lubrication film, thereby reducing the actual contact area between the contacting surfaces. Perhaps that is why the COF decreases immediately with the actual contact area reduction after injecting the emulsifying oil in the experiment.

6.2. Different Water Flow Velocity (ω = 100 r/min, Oil Flow Rate = 125 µL/s)

The emulsifying oil distribution under different water flow velocities is illustrated in Figure 11. It can be observed that there are side leakages and a decreasing distribution area in the secondary lubricating medium, which could be caused by the force that water exerts on the secondary lubrication medium becoming stronger as the water flow velocity increases, consequently preventing the entry of the secondary lubricating medium into the contact region. This indicates that the water flow velocity is a critical factor for the successful delivery of the secondary lubricating medium to the contact region.

Figure 12 displays the pressure distributions for different water flow velocities, where the diamond-shaped pressure distribution exists regardless of water flow velocities. The calculated load-carrying capacities with different water flow velocities for a water-lubricated bearing are 1867 N, 1621 N, 1664 N and 1601 N, respectively, displaying an overall decreasing trend, which is caused by the side leakage and decreasing distribution area of the secondary lubricating medium.

The velocity field of the fluid field for different water flow velocities is shown in Figure 13. It can be seen that the velocity direction of the flow field points towards the water outlet with an increase in water flow velocity, which indicates that the influence of the water flow on the overall flow field strengthens accordingly. At the same time, the velocity magnitude of the flow field also changes; the location of the maximum value of velocity in the overall flow field begins to shift from the contact region toward the water outlet as the flow progresses. However, there is always a region with the highest water flow velocity at the inlet of the water, which is because the water inlet at the water-lubricated bearing system is a necking area in the open-water environment, leading to the increase in the water flow velocity.

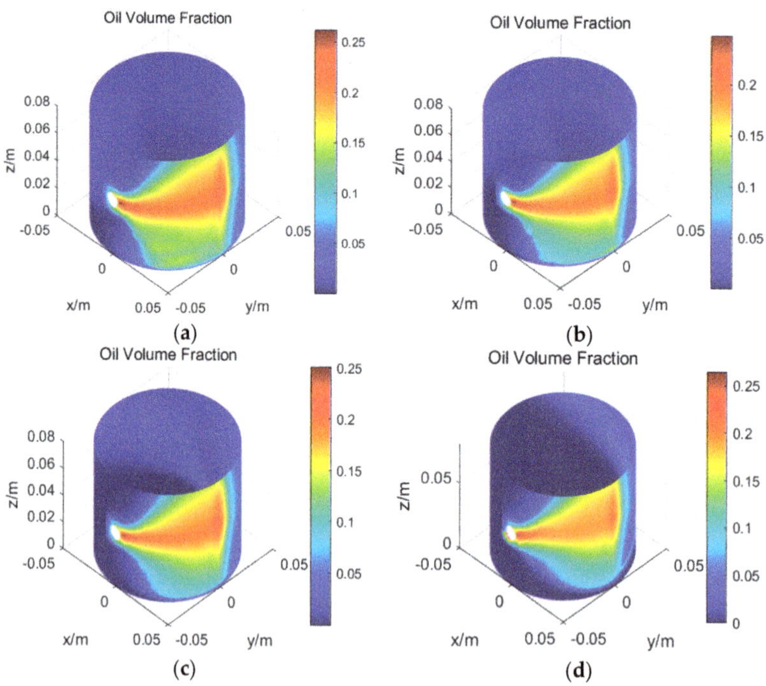

Figure 11. Emulsifying oil distribution under different water flow velocities. (**a**) v_w = 0.01 m/s; (**b**) v_w = 0.03 m/s; (**c**) v_w = 0.08 m/s; (**d**) v_w = 0.3 m/s.

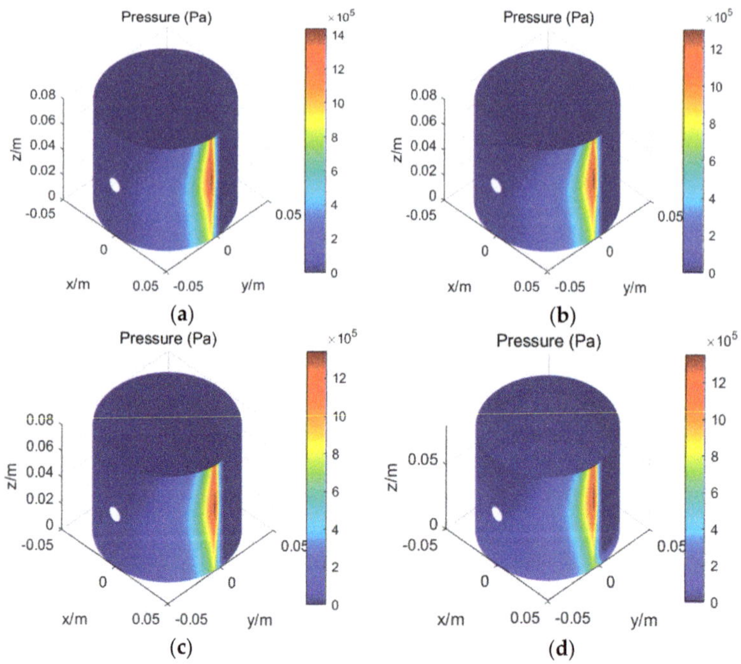

Figure 12. Pressure distribution under different water flow velocities. (**a**) v_w = 0.01 m/s; (**b**) v_w = 0.03 m/s; (**c**) v_w = 0.08 m/s; (**d**) v_w = 0.3 m/s.

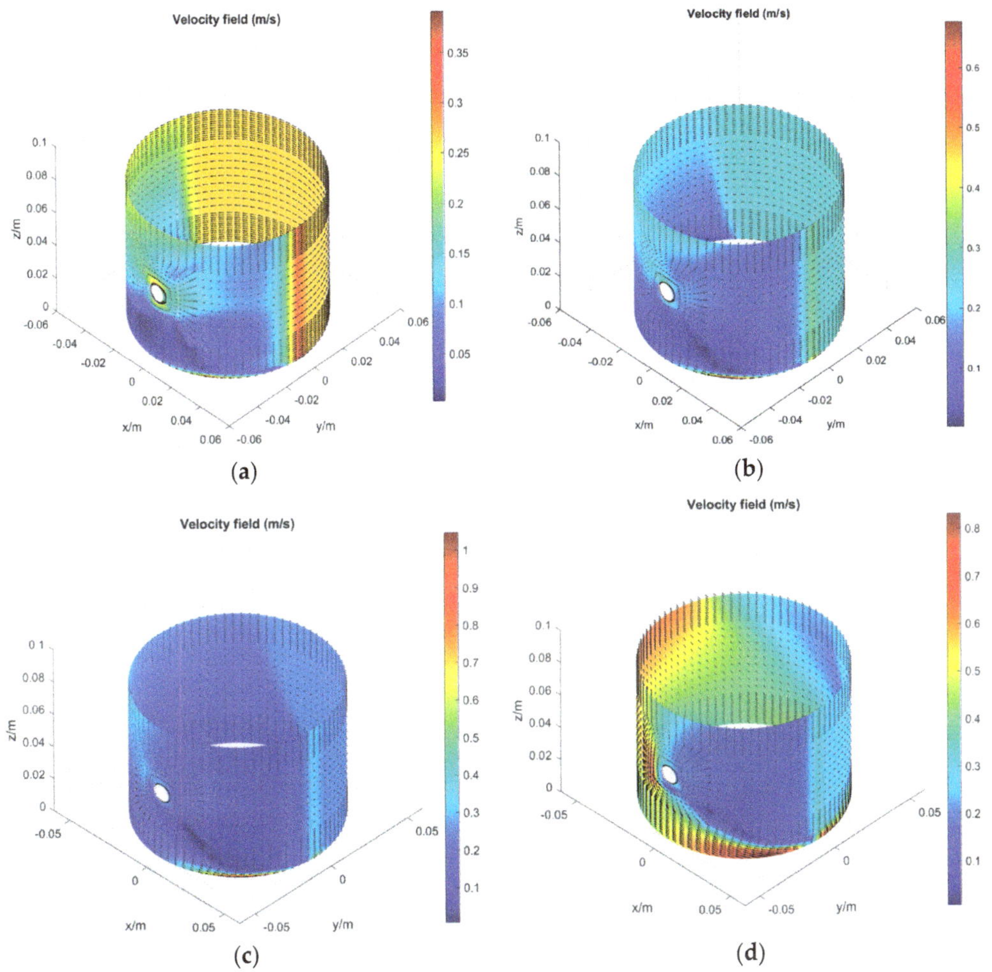

Figure 13. Distribution of velocity field under different water flow velocities. (**a**) v_w = 0.01 m/s; (**b**) v_w = 0.03 m/s; (**c**) v_w = 0.08 m/s; (**d**) v_w = 0.3 m/s.

Figure 14 shows the magnified velocity field around the secondary lubricating medium injection zone under different water flow velocities. It can be seen that reverse flow is mostly concentrated at the water inlet and around the injection zone regardless of the water flow velocities, and the reverse flow zone is increased and the reverse flow velocity is enhanced as the water flow velocity increases. This may be because when the secondary lubricating medium is injected into a stable flow field, the injected secondary lubricating medium acts as an obstacle in the stable flow field. This can lead to a sudden decrease in water flow velocity near the secondary lubricating medium injection zone and the formation of vortex structures, which can affect the surrounding fluid and cause reverse flow. As the water flow velocity increases, the intensity of vortices becomes stronger, resulting in the formation of more vortices in the fluid field. When these vortices continuously expand and merge, the reverse flow zone also grows. Additionally, the surrounding fluid is easily drawn into the vortices due to the inertia effect of high-speed water flow, further enhancing the reverse flow phenomenon.

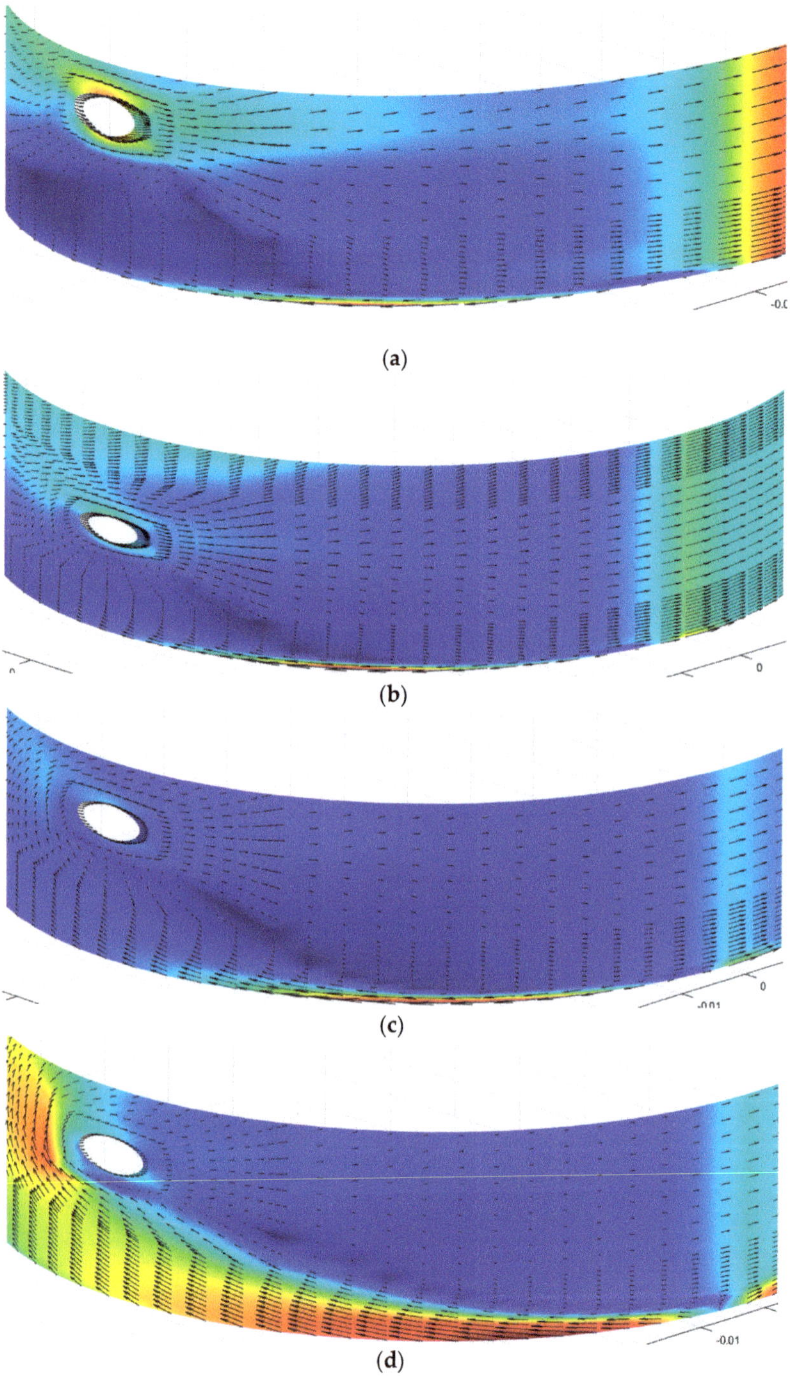

Figure 14. Magnified velocity field around the secondary lubricating medium injection zone under different water flow velocities. (**a**) v_w = 0.01 m/s; (**b**) v_w = 0.03 m/s; (**c**) v_w = 0.08 m/s; (**d**) v_w = 0.3 m/s.

Therefore, reverse flow should be avoided as much as possible when injecting the secondary lubricating medium, especially when the water flow velocity is high. The oil inlet where the secondary lubricating medium is injected in Figure 1 should be located in the middle of the bearing or near the outlet of the flow field.

It can also be concluded that there is still some mixed lubricant breaking through the reverse flow and entering the contact region to participate in the lubrication, which can reduce the friction and wear between contact surfaces.

6.3. Different Oil Flow Rate (ω = 100 rpm, v_w = 0.03 m/s)

The influence of the oil flow rate on the flow field is discussed in this part. Figure 15 shows the oil distribution for different oil flow rates. The distribution area of the secondary lubricating medium remains relatively stable across different oil flow rates, as the water flow velocity remains unchanged. However, the volume fraction of the secondary lubricating medium in the contact area increases with the increased flow rate of the secondary lubricating medium.

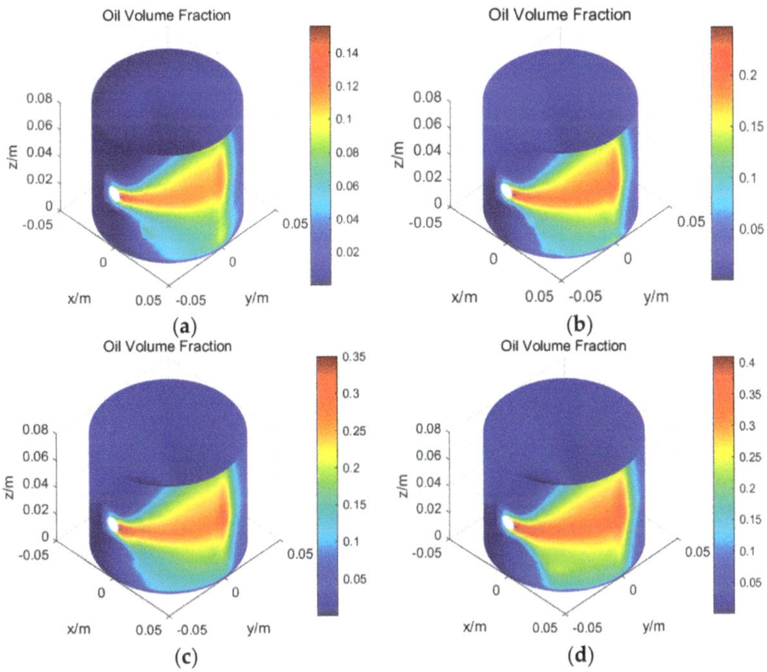

Figure 15. Emulsifying oil distribution under different oil flow rates. (**a**) Oil flow rate = 75 μL/s; (**b**) oil flow rate = 125 μL/s; (**c**) oil flow rate = 175 μL/s; (**d**) oil flow rate = 225 μL/s.

Figure 16 is the pressure distribution under different oil flow rates, and the corresponding load-carrying capacities are 989 N, 1621 N, 2247 N and 2778 N, which is proportional to the oil flow rate.

Figures 17 and 18 show the distribution of the velocity field in the flow field under different oil flow rates. It can be found that the distribution of the velocity vector and the reverse flow are basically the same under these four oil flow rates, indicating that the oil flow rate has little effect on the flow field. Therefore, if environmental friendliness is ensured, more secondary lubricating medium can be injected into the contact surfaces to improve the lubrication status between the bearing surface and the shaft surface.

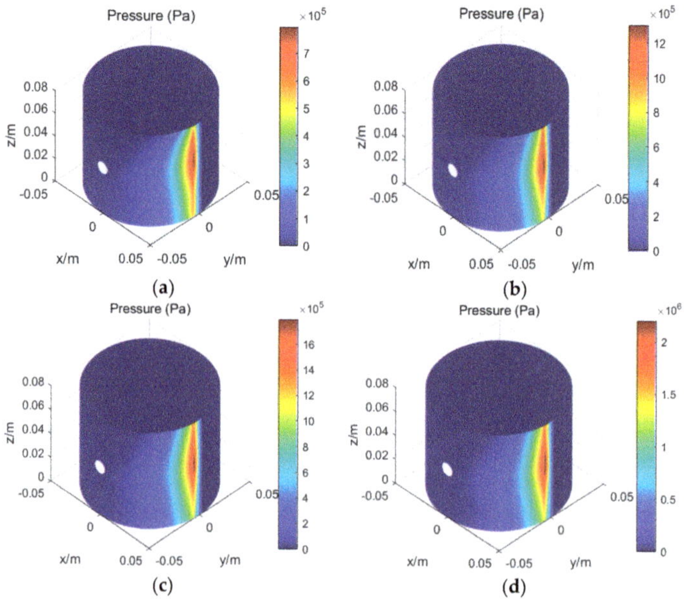

Figure 16. Pressure distribution under different oil flow rates. (**a**) Oil flow rate = 75 μL/s; (**b**) oil flow rate = 125 μL/s; (**c**) oil flow rate = 175 μL/s; (**d**) oil flow rate = 225 μL/s.

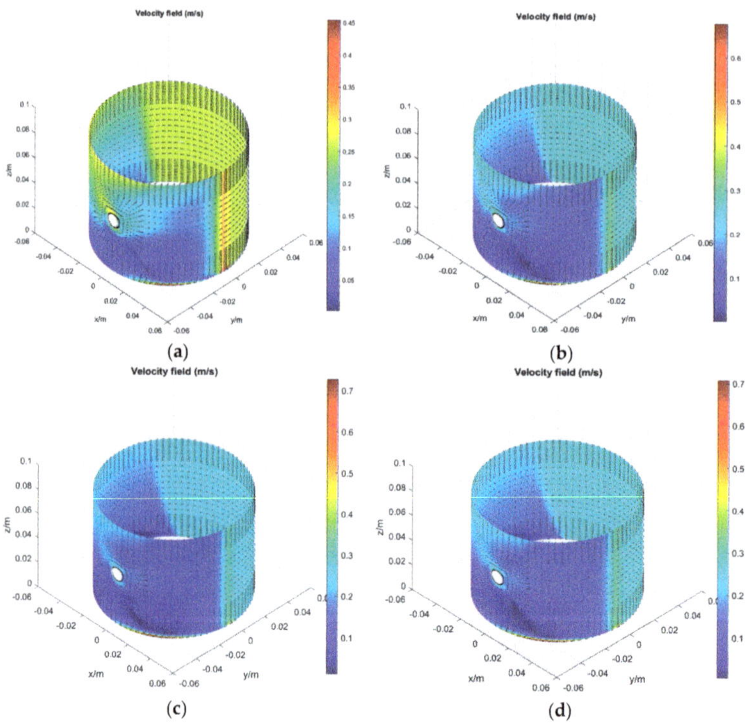

Figure 17. Distribution of velocity field under different oil flow rates. (**a**) Oil flow rate = 75 μL/s; (**b**) oil flow rate = 125 μL/s; (**c**) oil flow rate = 175 μL/s; (**d**) oil flow rate = 225 μL/s.

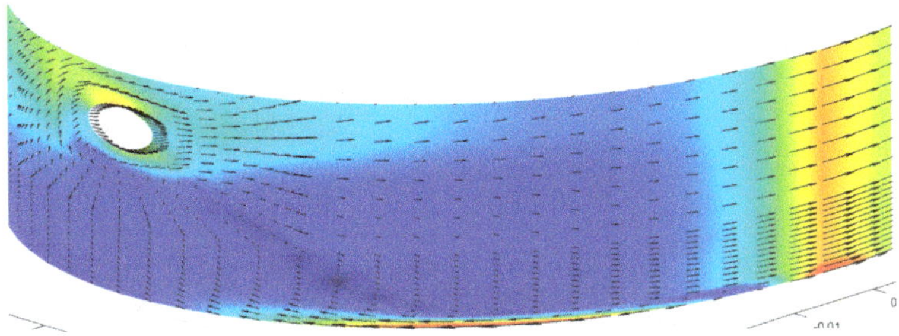

(**a**)

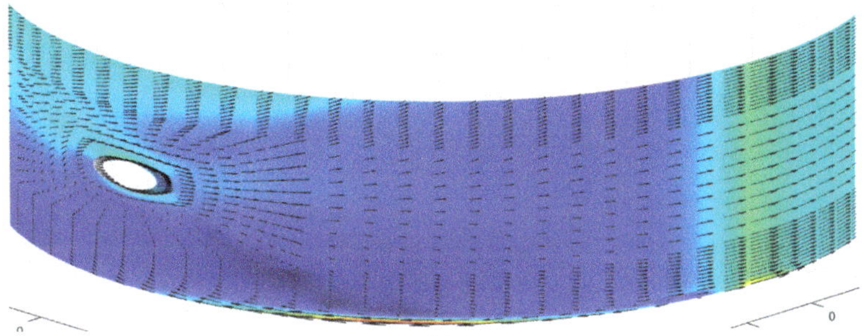

(**b**)

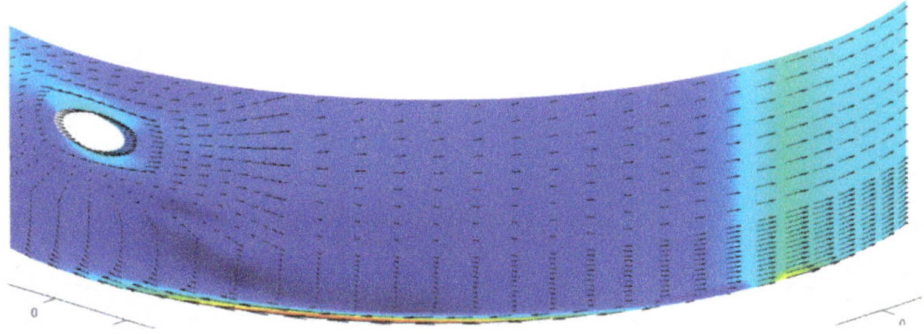

(**c**)

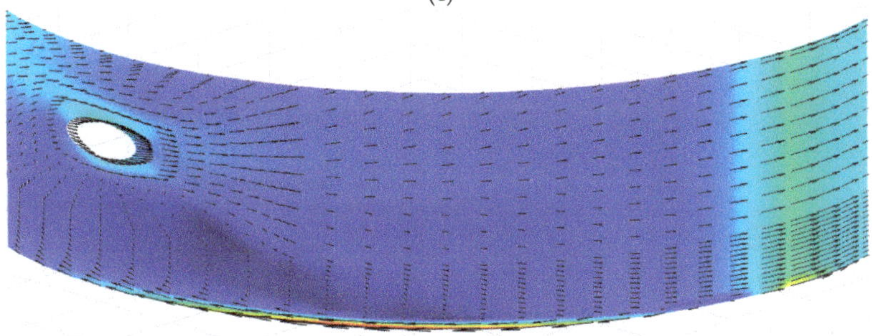

(**d**)

Figure 18. Magnified velocity field around the secondary lubricating medium injection zone under different oil flow rates. (**a**) Oil flow rate = 75 μL/s; (**b**) oil flow rate = 125 μL/s; (**c**) oil flow rate = 175 μL/s; (**d**) oil flow rate = 225 μL/s.

7. Conclusions

(1) Our proposed idea has been practically examined through experiments conducted on the water-lubricated bearing test bench. The results reveal that a small amount of emulsifying oil can significantly reduce the COF between the bearing surface and the shaft

surface under the high load and low shaft rotational speed conditions, demonstrating the effectiveness of the proposed idea on water-lubricated bearings.

(2) A numerical quasi-2D model using MATLAB R2022b has been established, which couples the N–S equation and VOF method for simulation purposes.Based on the simulation results, reverse flow exists around the injection zone regardless of the water flow velocities; the reverse flow is mostly concentrated at the water inlet and secondary lubricating medium injection zone, which gives guidance when choosing the secondary lubricating medium injection location.

(3) Despite the reverse flow, the secondary lubricating medium can still penetrate into the contact region and participate in lubrication under water environments, which can lead to a noticeable decrease in COF between the bearing surface and shaft surface.

(4) The flow field condition remains unchanged as the oil flow rate increases, indicating that the oil flow rate has little effect on the flow field.Therefore, if environmental friendliness is ensured, more secondary lubricating medium can be injected into the contact surfaces to improve the lubrication status between the bearing surface and the shaft surface.

(5) These simulation results can provide theoretical guidance for the further development of bench experiments.Since water-lubricated bearings are under a mixed lubrication regime when the encounter severe working conditions, the mixed lubrication model for water-lubricated bearings with secondary lubricating medium will be programmed in the future.

Author Contributions: Conceptualization, F.G. and X.Z.; methodology, X.Z.; software, X.Z.; validation, T.Y., H.J. and W.D.; formal analysis, X.Z. and P.L.; investigation, X.Z. and L.M.; writing—original draft preparation, X.Z.; writing—review and editing, X.Z., F.G. and P.L. All authors have read and agreed to the published version of the manuscript.

Funding: This research is supported by the National Natural Science Foundation of China (No. 52205200, No. 52375190 and No. 52375190); Taishan Scholar Talents Project from Shandong Province (No. TS20190943); Qingdao Postdoctoral Innovation Project (No. QDBSH20220202008) and Shandong Provincial Natural Science Foundation (No. ZR2021QE214).

Data Availability Statement: The data that support the findings of this study are available from the corresponding author upon reasonable request.

Conflicts of Interest: The authors declare that they have no known competing financial interests or personal relationships that could have appeared to influence the work reported in this paper.

Nomenclature

C	radial clearance
E	eccentric distance
F	load-carrying capacity
K	number on the edge of the grid
L	bearing length
P	pressure
I	number on the grid
U	velocity of the flow
v_o	velocity of the oil injection
v_w	velocity of the water flow
z	axial direction
ω	shaft rotational speed
μ	viscosity
θ	circumferential direction
θ_0	film rupture position in the circumferential direction
ρ	density
φ	normal flux
α	volume fraction of the phase
ψ	attitude angle

References

1. Gong, J.; Jin, Y.; Liu, Z.; Jiang, H.; Xiao, M. Study on influencing factors of lubrication performance of water-lubricated micro-groove bearing. *Tribol. Int.* **2019**, *129*, 390–397. [CrossRef]
2. Urreta, H.; Aguirre, G.; Kuzhir, P.; de Lacalle, L.N.L. Seals based on magnetic fluids for high precision spindles of machine tools. *Int. J. Precis. Eng. Manuf.* **2018**, *19*, 495–503. [CrossRef]
3. Urreta, H.; Aguirre, G.; Kuzhir, P.; de Lacalle, L.N.L. Actively lubricated hybrid journal bearings based on magnetic fluids for high-precision spindles of machine tools. *J. Intell. Mater. Syst. Struct.* **2019**, *30*, 2257–2271. [CrossRef]
4. Xie, Z.; Shen, N.; Zhu, W.; Tian, W.; Hao, L. Theoretical and experimental investigation on the influences of misalignment on the lubrication performances and lubrication regimes transition of water-lubricated bearing. *Mech. Syst. Signal. Process.* **2020**, *149*, 107211. [CrossRef]
5. Mallya, R.; Shenoy, S.B.; Pai, R. Static characteristics of misaligned multiple axial groove water-lubricated bearing in the turbulent regime. *Proc. Inst. Mech. Eng. Part J J. Eng. Tribol.* **2017**, *231*, 385–398. [CrossRef]
6. Wang, J.; Han, Y.; Geng, Z.; Xiang, G.; Wang, J.; Zhao, R. A profile design method to improve the wear performance of misaligned water-lubricated bearing. *Lubr. Sci.* **2021**, *33*, 215–228. [CrossRef]
7. Xu, B.; Guo, H.; Wu, X.; He, Y.; Wang, X.; Bai, J. Static and dynamic characteristics and stability analysis of high-velocity water-lubricated hydrodynamic journal bearings. *Proc. Inst. Mech. Eng. Part J J. Eng. Tribol.* **2021**, *236*, 1994–1996.
8. Fedorynenko, D.; Kirigaya, R.; Nakao, Y. Dynamic characteristics of spindle with water-lubricated hydrostatic bearings for ultra-precision machine tools. *Precis. Eng.* **2020**, *63*, 187–196. [CrossRef]
9. Sayed, H.; El-Sayed, T.A. Nonlinear dynamics and bifurcation analysis of journal bearings based on second order stiffness and damping coefficients. *Int. J. Non-Linear Mech.* **2022**, *142*, 103972. [CrossRef]
10. Brunetière, N.; Rouillon, M. Fluid flow regime transition in water lubricated spiral grooved face seals. *Tribol. Int.* **2021**, *153*, 106605. [CrossRef]
11. Wu, Z.; Yuan, C.; Guo, Z.; Huang, Q. Effect of the groove parameters on the lubricating performance of the water-lubricated bearing under low speed. *Wear* **2023**, *522*, 204708. [CrossRef]
12. Shinde, A.B.; Chavan, S.P. Parametric investigation of surface texturing on performance characteristics of water-lubricated journal bearing using, FSI approach. *SN Appl. Sci.* **2019**, *2*, 36. [CrossRef]
13. Pai, R.; Hargreaves, D.J. Water-lubricated, Bearings. In *Green Tribology*; Green Energy and Technology; Nosonovsky, M., Bhushan, B., Eds.; Springer: Berlin/Heidelberg, Germany, 2012.
14. Zhou, G.; Mi, X.; Wang, J.; Hu, R. Experimental comparison between the Stribeck curves of water-lubricated rubber bearing with straight and spiral grooves. *Ind. Lubric. Tribol.* **2018**, *70*, 1326–1330. [CrossRef]
15. Feng, W.; Han, Y.; Xiang, G.; Wang, J. Hydrodynamic lubrication analysis of water-lubricated bearings with partial microgroove considering wall slip. *Surf. Topogr. Metrol. Prop.* **2021**, *9*, 015019. [CrossRef]
16. Dong, C.; Yuan, C.; Xu, A.; Bai, X.; Tian, Y. Rippled polymer surface generated by stick-slip friction. *Langmuir* **2019**, *35*, 2878–2884. [CrossRef] [PubMed]
17. Ibrahim, S.A. *The Development of Computer Programme to Calculate the Hydrodynamic Lubrication Performance of a Ship Journal Bearing*; UniKL MIMET: Lumut, Malaysia, 2020.
18. Yang, Z.; Guo, Z.; Yuan, C. Effects of, MoS2 microencapsulation on the tribological properties of a composite material in a water-lubricated condition. *Wear* **2019**, *432–433*, 102919. [CrossRef]
19. Liang, X.; Guo, Z.; Tian, J.; Yuan, C. Development of modified polyacrylonitrile fibers for improving tribological performance characteristics of thermoplastic polyurethane material in water-lubricated sliding bearings. *Polym. Adv. Technol.* **2020**, *31*, 3258–3271. [CrossRef]
20. Feng, H.; Jiang, S.; Ji, A. Investigations of the static and dynamic characteristics of water-lubricated hydrodynamic journal bearing considering turbulent, thermohydrodynamic and misaligned effects. *Tribol. Int.* **2019**, *130*, 245–260. [CrossRef]
21. Narwat, K.; Kumar, V.; Singh, S.J.; Kumar, A.; Sharma, S.C. Performance of rough surface hydrodynamic circular and multi-lobe journal bearings in turbulent regimes. *Proc. Inst. Mech. Eng. Part J J. Eng. Tribol.* **2023**, *237*, 860–880. [CrossRef]
22. Zhang, X.; Yu, T.; Guo, F.; Liang, P. Analysis of the Influence of Small Quantity secondary lubricant on Water Lubrication. *Tribol. Int.* **2021**, *159*, 106998. [CrossRef]
23. Yu, T.; Guo, F.; Zhang, X.; Ji, H.; Duan, W.; Liang, P. Water Lubrication Assisted by Small-quantity Silicone Oil. *Tribol. Int.* **2022**, *173*, 107619. [CrossRef]
24. Katopodes, N. Volume of fluid method. In *Free-Surface Flow*; Katopodes, N., Ed.; Butterworth-Heinemann: Oxford, UK, 2019; pp. 766–802.
25. Wu, Y.; Sun, T.; He, Z.; Zeng, X.; Ren, T.; de Vries, E.; van der Heide, E. Study on the relationship between the tribological properties and oxidation degree of graphene derivatives in O/W emulsion. *Tribol. Int.* **2021**, *157*, 106875. [CrossRef]
26. Kumar, D.; Rajabi, H. Effect of lubrication on a surface parameter of strip in cold rolling with oil in water emulsion. *Int. J. Appl. Eng. Res.* **2019**, *14*, 3261–3267.
27. Yu, T.; Zhang, X.; Guo, F.; Jin, W.; Liang, P. Enhancing water lubrication by secondary assistant lubricant in small quantity. *Tribology* **2022**, *42*, 358–365. (In Chinese)
28. Xie, Z.; Jiao, J.; Wrona, S. The fluid-structure interaction lubrication performances of a novel bearing: Experimental and numerical study. *Tribol. Int.* **2023**, *179*, 108151. [CrossRef]

29. Xie, Z.; Jiao, J.; Zhao, B.; Zhang, J.; Xu, F. Theoretical and experimental research on the effect of bi-directional misalignment on the static and dynamic characteristics of a novel bearing. *Mech. Syst. Signal. Process.* **2024**, *208*, 111041. [CrossRef]
30. Xie, Z.; Jiao, J.; Yang, K. Experimental and numerical study on the mixed lubricant performances of a new bearing. *Tribol. Int.* **2023**, *182*, 108334. [CrossRef]

lubricants

Article

Influence of Non-Parallelism on the Micro-Interface Lubrication Mechanism of Water-Lubricated Bearings

Lin Sun [1], Jianchao Shi [1], Tao Jiang [1], Zhen Li [1], Yu Wang [2] and Zhaozeng Liu [2,*]

1 CNOOC EnerTech Equipment Technology Co., Ltd., Tianjin 300457, China
2 College of New Energy, China University of Petroleum (East China), Qingdao 266580, China
* Correspondence: liuzz@upc.edu.cn

Abstract: Water-lubricated bearings can effectively solve the pollution problem caused by lubricant leakage and are used in offshore engineering equipment for this reason. Aiming at the problems of unclear and undefined micro-interface lubrication mechanisms of water-lubricated bearings, this paper investigates the influence of non-parallel micro-cavities on the micro-interface lubrication mechanism of bearings. Based on a single micro-cavity model, the lubrication mechanism of micro-cavities is studied in this paper. Lubrication models of the non-parallel contact friction pairs model are built, and the effect of the non-parallelism on the lubrication performance of the micro-cavities is obtained using the computational fluid dynamics method. The results show that, under the same Reynolds number and cavitation pressure, the wedge effect caused by the non-parallelism causes the pressure at the inlet to rise, thus increasing the load-carrying capacity. The existence of non-parallelism limits the rise of the high pressure of the inertia effect on the micro-cavities and reduces the load-carrying capacity. The presence of non-parallelism decreases the area of the negative pressure proportion and increases the proportion of the positive pressure zone inside the micro-cavities, thus increasing the load-carrying capacity.

Keywords: lubrication mechanism; inertia effect; cavitation effect; micro interface; texture

Citation: Sun, L.; Shi, J.; Jiang, T.; Li, Z.; Wang, Y.; Liu, Z. Influence of Non-Parallelism on the Micro-Interface Lubrication Mechanism of Water-Lubricated Bearings. *Lubricants* **2024**, *12*, 49. https://doi.org/10.3390/lubricants12020049

Received: 6 December 2023
Revised: 23 January 2024
Accepted: 30 January 2024
Published: 8 February 2024

1. Introduction

Water-lubricated bearings can effectively solve the pollution problem of lubricant leakage from marine equipment. In particular, the use of seawater as the lubricating medium can eliminate the harsh sealing system, simplify the positioning and propulsion system, and improve working reliability. In most cases, sliding bearings work in the mode of hydrodynamic lubrication. At the same time, the wear of the working surfaces of the bed and sleeve is negligibly small. However, in the period of start-up and stoppage, the deposits that in normal exploitation work under conditions of hydrodynamic lubrication switch to work under conditions of boundary lubrication. In these periods, the conditions of external friction and very intense wear and tear are realized. Wear is a process whose origin and development are essentially related to boundary lubrication conditions. As the influence of the surface morphology effect cannot be ignored, the water-lubricated bearing lubrication mechanism is no longer pure fluid dynamic pressure lubrication or elastic flow lubrication. Surface roughness will directly affect the entire bearing's hydrodynamic characteristics, which in turn affect the bearing lubrication mechanism and lubrication performance. Aiming to solve the many problems within the micro-interface of water-lubricated bearings, both domestic and foreign scholars have carried out much research and made positive contributions. Wang et al. [1] established and investigated a hydrodynamic fluid-solid coupling model for water-lubricated polymer bearings with grooves, and explored the influence laws of factors such as the physical properties of the lining layer and the rotational speed on the surface characteristics. Zhang et al. [2] took biometric features into account in the design of surface structures from a biomimetic point of view and found that their surface structures were more helpful in increasing the load-carrying capacity and decreasing the friction force compared to geometric feature structures. Xie et al. [3,4] investigated the bearing micro-rough contact load ratio and lubrication state by modeling a water-lubricated bearing with micro-convex grooves, and further explored the relationship between the specific parameters and the contact ratio. Zhou et al. [5] processed circular grooves with different geometrical parameters on the surface of a titanium alloy by using a laser pitting technique and tested the friction characteristics of the titanium alloy on an experimental machine to analyze the effect of factors such as surface structural parameters on lubrication and tribological properties. In order to overcome the shortcomings of the traditional water-lubricated single-bushing bearings, Xie [6] proposed a novel double-liner bearing that enriched the lubrication theory system. Aiming to influence the misalignment effect on the performance of marine water-lubricated bearings, Xie [7] established a new lubrication model with a bidirectional misalignment effect, and the study proposed a new analysis method and provided important guidance for practical application. Surface texturing is the process of creating an array of pits or grooves with a specific size distribution on the friction surface by laser processing, ion etching, electro-deposition, abrasive flow jetting, etc. [8,9]. When relative motion is generated in a friction pair, the wedge effect of the pits generates additional hydrodynamic pressure between the oppositely moving surfaces, which improves the frictional properties of the lubricated contact surfaces. With advances in machining technology, surface texturing has been successfully applied to many products such as plain bearings, thrust bearings, mechanical face seals, cutting tools, piston rings, and disk storage devices. For example, Yang et al. [10] compared the tribological properties of surface textured and non-textured rubber samples under water lubrication, low speed, and overload conditions in order to control the wear of the propulsion-bearing rubber plate of an underwater vehicle, and the results showed that the frictional torque and coefficient of friction of the surface textured rubber samples were reduced by 15.5%, and the average hourly wear rate was reduced by 23.3%, which not only reduced friction but also improved the wear resistance. In 1966, Hamilton et al. [11] found for the first time that the irregular pits on the surface of rotary shaft end seals could generate additional fluid pressure and thus improve the surface load-carrying capacity. Anno et al. [12] further confirmed the lubrication improvement effect of end seals and thrust bearings.

After the discovery of the additional load-carrying property of surface texture, its cavitation effect attracted attention. In the dispersion zone of the texture unit, the pressure of the lubricant is lower than the gas saturation pressure or evaporation pressure, leading to the occurrence of cavitation. Hamilton et al. [11] observed the cavitation phenomenon at the pits through the transparent rotor test and proposed, for the first time, that the asymmetric pressure distribution caused by cavitation is the main reason for the texture load-carrying capacity. Qiu et al. [13,14] proposed a model for predicting the cavitation behavior in parallel textured friction pairs and found that the relative motion speed between the friction pairs has a significant effect on the cavitation phenomenon inside the texture. Zhang et al. [15] pointed out that the cavitation region changes with the number of textures and texture size, and when the operating conditions are stable, the morphology and the proportion of the cavitation region reach stability. Meng [16,17] found that the change in the area of the friction surface bubble induced by cavitation could increase the load-carrying capacity and decrease the surface friction, and that the change was affected by the distribution of the texture. Bai et al. [18] conducted an experimental study on the cavitation phenomenon in textured thrust bearings and found that the shape and area of the cavitation varied with the size of the texture. Li et al. [19] analyzed the loading mechanism of the texture and found that the non-isotropic superposition effect between the micro-dynamic pressure effect and the original dynamic pressure effect determines the loading capacity of the textured water-lubricated plain bearings. Fowell et al. [20] and Olver et al. [21], by investigating the tribological properties of textured bearings considering cavitation, suggested that the negative pressure in the cavitation region can draw more lubricant into the texture region, thus increasing the hydrodynamic pressure effect and load-carrying capacity. Dobrica et al. [22] found that cavitation has a great influence on the performance of parallel textured contact pairs by studying the effect of cavitation on the hydrodynamic properties of textured contact pairs. Wang et al. [23] found, through experimental studies, that under certain velocity and load conditions, surface texture induces cavitation in the lubricant and generates bubbles at the exit of the pits.

With the rapid development of techniques for solving the Navier-Stokes equations, the inertia effect of the texture was proposed by Arghir et al. [24]. By comparing the pressure generation effect inside the texture with different convective inertia, it is found that all the textures exhibit a net lift on the parallel wall as the convective inertia increases. Billy et al. [25] studied the influence of the inertia effect of lubricating medium on the load-carrying performance of the texture under high-speed conditions and found that with the enhancement of the texture inertia effect, the load-carrying capacity of the upper wall of the texture increased significantly. Sahlin et al. [26] carried out two-dimensional hydrodynamic analysis of a single texture and found that the load-carrying capacity of the texture is mainly caused by fluid inertia, and the load-carrying capacity of the texture increases with the increase in the Reynolds number. Feldman et al. [27] numerically analyzed the effect of the inertia effect in textured parallel surfaces and found that the inertia effect is negligible when the liquid film thickness is less than 3% of the texture diameter. Subsequently, Dobrica et al. [28,29] drew a completely opposite conclusion to the previous scholars and found that in general, the inertia had a negative effect and reduced the load-carrying capacity of the friction pair. For this reason, Cupillard et al. [30] stated that there is a special depth. Above this value, inertia negatively affects the load-carrying capacity, and below it, inertia positively affects the load-carrying capacity. These effects are amplified with an increasing Reynolds number for a given texture depth. Similar conclusions were obtained by Kraker et al. [31] and Syed et al. [32], who stated that the influence of inertial effects on the load-carrying capacity depends on the localized flow inside the texture. Wang et al. [33] analyzed the local lubrication mechanisms of textures in journal bearings from a microflow perspective while considering the interactions between textures and film formation in the entirety of the bearing. Khalil et al. [34] carried out a theoretical analysis and showed that inertial forces combined with centrifugal forces increase the pressure of the friction pair. Woloszynski et al. [35] found that the inertia effect is enhanced with the increase in Reynolds number.

In addition to these two effects, other effects have been gradually discovered and proposed to better explain the load-carrying mechanism of the texture. Tonde et al. [36–39] proposed two principles for generating pressure within the texture: the textured region provides a "virtual step" at the inlet; and the textured region acts as a resistance factor for lubricant escape, increasing the amount of lubricant available in the high-pressure zone. Subsequently, Etsion and Halperin [39] proposed the "delayed rupture effect" of the texture on the liquid film. In addition, Olver and Fowell [21] proposed the "entrainment and inlet suction" effect of the lubricant medium inside the texture, that is, the sliding wall generates a lower pressure than the external atmospheric pressure near the inlet of the texture, allowing more lubricant to be sucked into the friction pair. In addition to these roles, the formation of the lubrication film during sliding plays an important role in the frictional behavior of surface textures. Different lubrication states play different roles in the formation of lubricant film. Xu et al. [40] investigated the effect of the formation of lubricant film on the tribological properties of elliptical texture surfaces under fluid lubrication and boundary lubrication, respectively, and found that the surface friction coefficients of the surfaces with and without texture are similar under fluid lubrication, which suggests that it is the lubricant film adsorbed on the surfaces that dominates the tribological behaviors of the surfaces under fluid lubrication, rather than the textures. The new Navier-Stokes model proposed by Rom and Müller [41] is based on the modification of the Reynolds equations and provides more accurate results than the Reynolds and Stokes equations, which can be applied to deep surface texture and high-speed flow. However, there is still room for improvement in the setting of the boundary conditions so that the theoretical model can also be investigated in more depth by applying a simplified method to the unit problem. Surface micro-textures represent an important trend in tailoring friction and/or wear. In the future, synergistic effects with lubricant rheology and other surface modification approaches will come increasingly into focus [42].

The current research on the lubrication mechanism of the micro interface is relatively scattered, and the systematic research on the load-bearing mechanism of the micro-interface water-lubricated bearing still needs further development. Most of the studies on the micro-interface lubrication mechanism are based on parallel friction pairs. However, in practical applications such as plain bearings, the relative motion of the non-parallel surface is more common. Therefore, by establishing a non-parallel contact pair geometric model, computational fluid dynamics is used to calculate the inertia effect, cavitation effect, and inlet wedge effect at different degrees of parallelism and to analyze the effect of non-parallelism on the lubrication performance of the micro-cavities.

In summary, the current research on the micro-interfacial lubrication mechanism is relatively decentralized, and the systematic research on the micro-interfacial load-carrying mechanism still needs to be further developed. Most of the studies on the micro-interface lubrication mechanism are based on parallel friction pairs. However, in practical applications such as plain bearings, the relative motion of the non-parallel surface is more common. Therefore, to address the deficiencies in the studies of the above scholars, the analysis will focus on elaborating and explaining the influence of micro-cavities on the lubrication mechanism of the non-parallel micro interface. In this study, the influence of non-parallelism on the lubrication performance of microcavities is analyzed by establishing a geometrical model of non-parallel contact pairs and using computational fluid dynamics to calculate the inertia effect, cavitation effect, and inlet wedge effect under different parallelisms. This paper explores the change rule between non-parallelism and lubrication performance, grasps the evolution law of hydrodynamic properties with surface morphology, explains the inner mechanism of the law, and provides a theoretical basis for further development of optimization design of micro-interface morphology characteristics.

2. Numerical Methods

2.1. Physical Model and Boundary Conditions

The flow field inside the single micro-cavity is analyzed. The fluid distribution inside the micro-cavity is continuous. The N-S equation is satisfied. The following assumptions are used for convenience in the analysis and modeling:

(1) The volumetric force effects are neglected.
(2) The fluid inside the microcavity is incompressible, isobaric, and iso-viscous.
(3) The model is a two-dimensional model with a symmetric distribution in the z-direction.

The schematic diagram of a single micro-cavity in this study is shown in Figure 1. The upper wall slides from left to right with velocity v (as indicated by the arrows in the Figure 1), while the no-slip boundary condition is used. The inlet and outlet are periodic boundary conditions and the rest of the walls are fixed boundary conditions. The parameters of lubrication medium parameters and boundary conditions used are listed in Table 1. The subscript l denotes the liquid phase, v denotes the vapor phase, and ope denotes the operating environment.

To determine the suitable mesh number, a mesh independent study is carried out. As a result, the baseline model has approximately 70,000 grid elements.

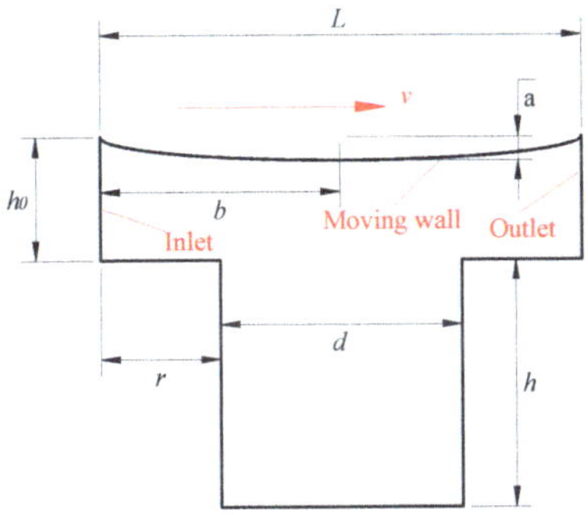

Figure 1. The single micro-cavity shape and boundary conditions (exaggerated non-parallelism).

Table 1. Parameters of geometry and boundary conditions.

Symbol	Parameter	Value
Textured geometry parameters		
L/mm	Length of the fluid domain	40
d/mm	Width of textured zone	20
h/mm	Depth of textured zone	20
h_0/mm	Liquid film thickness	10
r/mm	Inlet position	10
Lubrication medium parameters		
ρ_l/ kg/m^3	Density of liquid phase water	998.2
μ_l/Pa·s	Viscosity of liquid phase water	10^{-3}
ρ_v/kg/m^3	Density of vapor phase water	0.5542
μ_v/Pa·s	Viscosity of vapor phase water	1.34×10^{-5}
p_v/Pa	Saturated water vapor pressure	3540
v/(m·s^{-1})	sliding speed	50
P_{ope}/Pa	Environmental pressure	101,325

2.2. Governing Equation

In the present study, the lubricant is treated as an incompressible Newtonian fluid with a constant viscosity and density.

The degree of non-parallelism is denoted as θ, where $\theta = a/b$. In the equation, a is the distance of the thinnest part of the liquid film of the micro-cavities from the upper surface of the parallel friction pair; b is the distance from the pressure inlet to the centerline of a single micro-cavity. The upper surface of the friction pair is formed by a section of the elliptical arc. Keeping the value of b constant, θ is taken as 0.005, 0.0075, and 0.01 to build computational models for comparison with the parallel micro-cavity friction pair ($\theta = 0$). For the non-parallel contact pair models with different parallelisms, the lubricant film thickness equation is as follows:

$$H = h + h_0 \tag{1}$$

where H is the total oil film thickness; h_0 is the friction pair clearance; h is the depth of micro-cavity.

$$h_0 = \begin{cases} h_0(parallel\ contact\ pair) \\ h_0 - a(non - parallel\ contact\ pair) \end{cases} \tag{2}$$

where a is the decrease in value caused by non-parallelism.

The basic flow-governing equations including the continuity equation and the momentum conservation equation are given as follows:

$$\frac{\partial}{\partial t}(\rho_m) + \nabla \cdot \left(\rho_m \vec{v}_m\right) = 0 \tag{3}$$

$$\frac{\partial}{\partial t}(\rho_m v_m) + \nabla \cdot (\rho_m v_m v_m) = \nabla \cdot \left[\mu_m(\nabla v_m + \nabla v_m^T)\right] - \nabla p + S_v \tag{4}$$

$$v_m = \frac{\sum\limits_{k=1}^{n} \alpha_k \rho_k v_k}{\rho_m} \tag{5}$$

$$\rho_m = \sum_{k=1}^{n} \alpha_k \rho_k \tag{6}$$

where the ρ_m is the mixed phase density, v_m is the mass average velocity, and α_k is the volume fraction of phase k.

The "mixture" model is used to describe the multiphase flow of lubricant with cavitation. The Singhal [21] et al. "full cavitation model" can be used in the continuity equation and momentum equation of the mixture.

$$\frac{\partial}{\partial t}(\rho_m f_k) + \nabla(\rho_m v_m f_k) = \nabla(\mu_m \nabla f_k) + R_e - R_c \tag{7}$$

where the f is the mass fraction of the vapor phase, and R_e and R_c denote the rate of vapor production and dissolution, respectively. According to the different methods of R_e and R_c calculation, they can be classified into different cavitation models.

In this paper, the Zwart-Gerber-Belamri cavitation model [22] is used with the following expression:

$$\begin{cases} R_e = F_{vap} \frac{3\alpha_{nuc}(1-\alpha_v)\rho_v}{R_B} \sqrt{\frac{2}{3}\frac{(p_v - p)}{\rho_l}} & p \le p_v \\ R_c = F_{cond} \frac{3\alpha_v \rho_v}{R_B} \sqrt{\frac{2}{3}\frac{(p - p_v)}{\rho_l}} & p > p_v \end{cases} \tag{8}$$

where the α_{nuc} is the volume fraction of the nucleation point, generally taken as 5×10^{-4}. R_B is the radius of the bubble, generally taken as 10^{-6}. F_{vap} is the evaporation coefficient, and F_{cond} is the coefficient of condensation, generally taken as F_{vap} = 50, F_{cond} = 0.01. The subscripts l and v represent liquid and gaseous media, respectively.

3. Results and Discussion

3.1. Inertia Effect

In general, the Reynolds number, which is the main parameter for determining the state of fluid flow, is defined as:

$$R_e = \frac{\rho v l}{\mu} \tag{9}$$

where ρ is the fluid density, u is the dynamic viscosity, v is the linear velocity, and l is the feature length. For water-lubricated bearings, the Reynolds number equation can be organized as follows:

$$R_e = \frac{\rho R j \omega c}{\mu} \tag{10}$$

where R_j, c, and ω are the shaft diameter, bearing radius clearance, and shaft speed, respectively. Four Reynolds numbers of 0.1, 1, 10, and 100 are chosen to analyze the flow field characteristics inside the micro-cavity.

The Reynolds number is the vital factor influencing the inertia effect of micro-cavity pressure lubrication. Therefore, in order to investigate the inertia effect of the non-parallel friction pair, the effect of the Reynolds number on the pressure distribution of the friction pair is analyzed. The change in R_e is achieved by a change in linear speed. In Chapter 3, R_e denotes the Reynolds number unless otherwise specified. Figure 2 shows the contour of pressure distribution under different Reynolds numbers when the non-parallelism is 0.01. It can be seen from the figure that with the increase in Reynolds number, the maximum pressure in the friction pair increases, the area of the high-pressure region increases, and the area of the negative pressure region decreases. Due to the existence of non-parallelism, a convergence wedge is formed between the micro-cavities' moving wall and the pressure inlet, thus forming a new high-pressure zone.

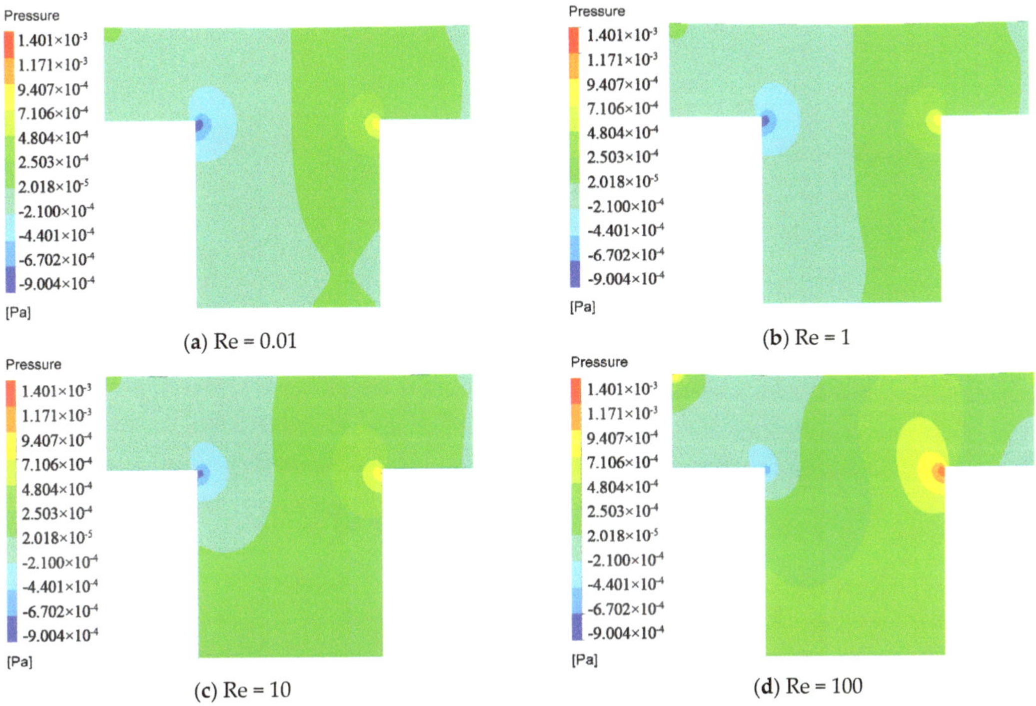

Figure 2. Pressure distribution at different R_e at $\theta = 0.01$.

The pressure distribution curve of the moving wall under different Reynolds numbers with the same non-parallelism is shown in Figure 3. It can be seen from the figure that the general trend of pressure distribution under the inertia effect of non-parallel friction pairs is consistent with that under the inertia effect of parallel friction pairs [33]. As the R_e increases, the pressure distribution curve moves upward, the proportion of the positive pressure area increases, and the load-carrying capacity increases. Compared with Figure 3b–d, it can be seen that with the increase in non-parallelism, the pressure rise inside the micro-cavity is limited.

In order to explore the influence of non-parallelism on lubrication performance under the same Reynolds number, the influence of non-parallelism on the pressure distribution curve of the moving wall of the friction pair under the same Reynolds number is shown in Figure 4. Under the same Reynolds number, the wedge effect caused by the non-parallelism makes the pressure at the inlet of the texture rise, thus increasing the load-carrying capacity of the friction pair. At the same time, the existence of non-parallelism will prevent the pressure distribution curve from moving up, reducing the load-carrying capacity of the friction pair. In addition, the increase in Reynolds number further limits the pressure rise in the micro-cavity. Figure 5 shows the velocity vector diagram in the micro-cavity with different Reynolds numbers when the non-parallelism is 0.1. It can be seen from the diagram that under the same non-parallelism, the larger the Reynolds number, the greater the downward velocity of the fluid in the micro-cavity, resulting in an enhanced plugging effect. Thus, the increase in non-parallelism leads to a stronger limiting effect on the pressure rise inside the micro-cavity.

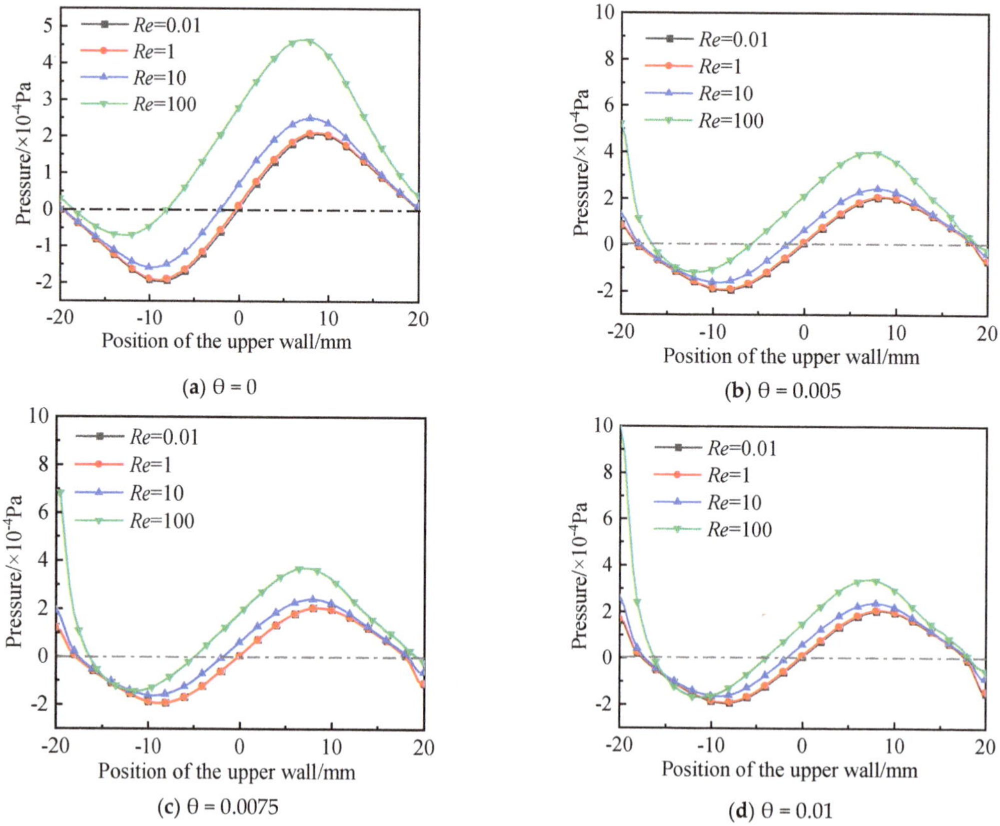

Figure 3. Pressure distribution under different R_e.

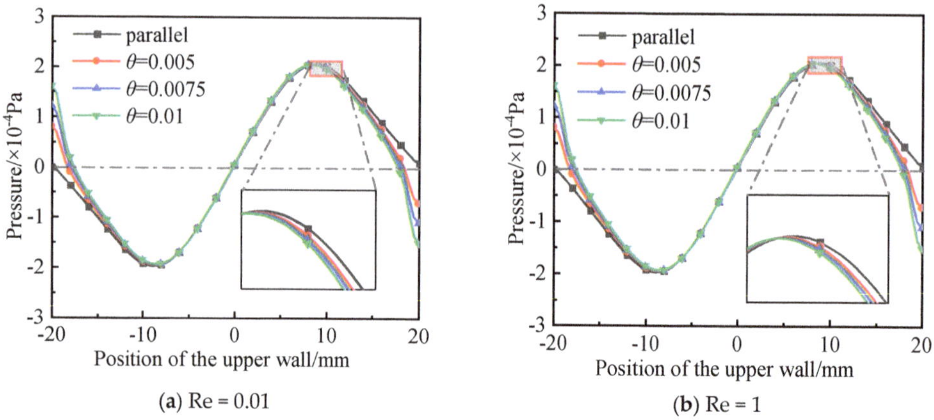

Figure 4. *Cont.*

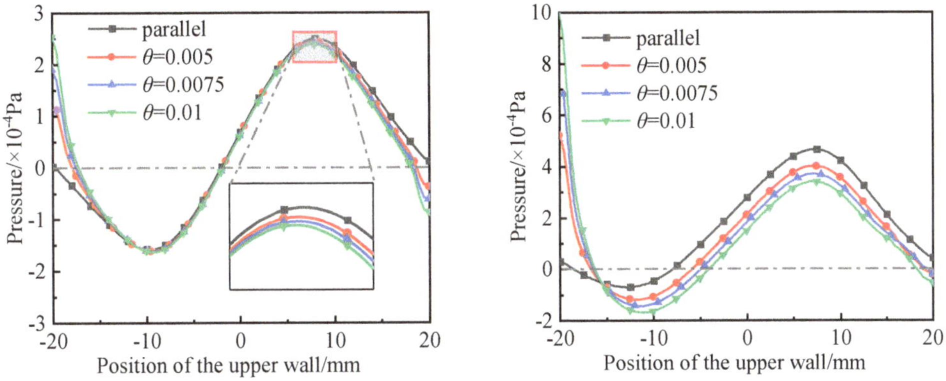

(**c**) Re = 10 (**d**) Re = 100

Figure 4. Influence of non-parallelism on the pressure distribution of the moving wall at the same R_e.

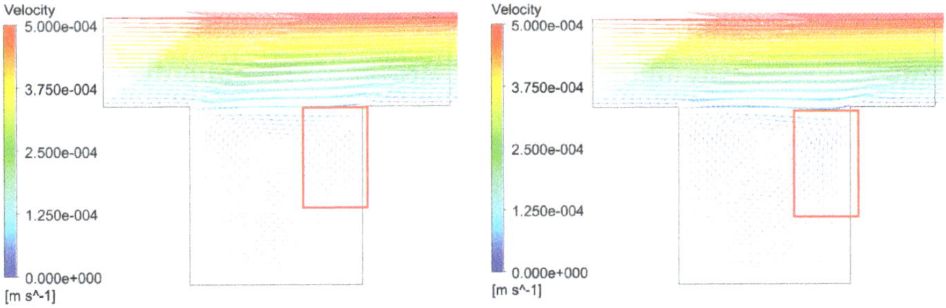

(**a**) Re = 0.01 (**b**) Re = 100

Figure 5. Velocity vector diagram at different R_e.

In summary, the influence of the non-parallelism on the inertia effect of the friction pair depends on the wedge effect and the limit on the maximum pressure. The influence of non-parallelism on the load-carrying capacity of the friction pair at different R_e is shown in Figure 6. It can be seen that at R_e of 0.01, and 1, the wedge effect caused by the non-parallelism plays a dominant role, and the load-carrying capacity increases with the non-parallelism. However, at Reynolds numbers of 10 and 100, the load-carrying capacity decreases with the increase in non-parallelism, which is due to the fact that the presence of non-parallelism restricts the pressure rise, making the negative effect dominant.

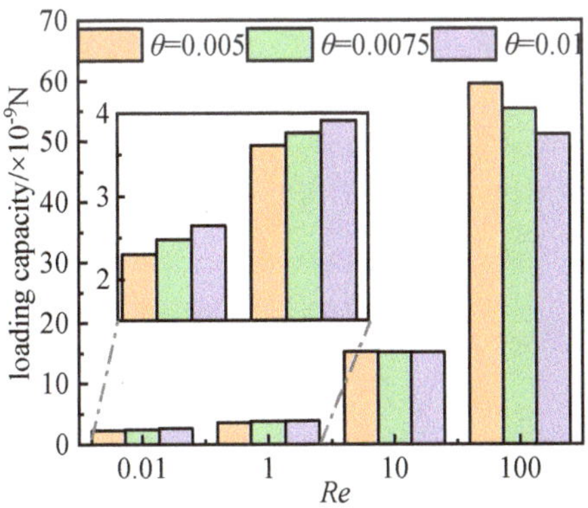

Figure 6. The load-carrying capacity of a non-parallel friction pair.

3.2. *Cavitation Effect*

In order to explore the cavitation effect under the same non-parallelism, the influence of the cavitation effect on the pressure distribution is analyzed. Figure 7 shows the pressure distribution contour of the friction pair under different cavitation pressures when the non-parallelism is 0.01. It can be seen from the figure that the cavitation effect causes the minimum negative pressure of the friction pair to increase, while the negative pressure proportion decreases, and the positive pressure proportion decreases slightly with the cavitation pressure. Figure 8 compares the influence of the cavitation effect on the pressure distribution curve of the moving wall of the micro-cavity between parallel friction pairs and non-parallel pairs with a degree of 0.005 and 0.01. From the figure, the general trend of the pressure distribution of the non-parallel friction pairs under the cavitation effect is consistent with that of the parallel friction pair, and the pressure reduction at the inlet of the micro-cavity region is limited.

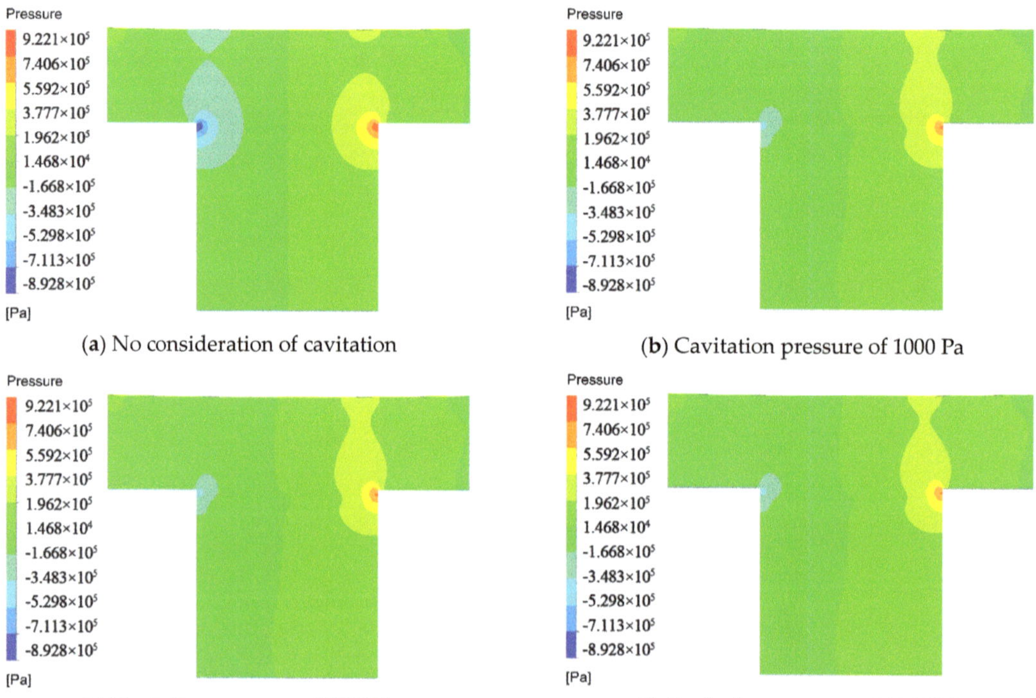

(**a**) No consideration of cavitation

(**b**) Cavitation pressure of 1000 Pa

(**c**) Cavitation pressure of 3540 Pa

(**d**) Cavitation pressure of 10,000 Pa

Figure 7. Pressure distribution under different cavitation pressures at $\theta = 0.01$.

Sun et al. [43] investigated the cavitation phenomenon of oil films in dynamically loaded journal bearings using high-speed photography and pressure measurements simultaneously, and the experimental results indicated that the pressure in the cavitation region was close to 0.5 MPa. According to the experimental results of Braun et al. [44] and Etsion et al. [45], the pressure in the cavitation region is close to the atmospheric pressure. It indicates that the cavitation pressure is uncertain for actual working conditions. In order to explore the influence of non-parallelism on the lubrication condition of the friction pair under the same cavitation pressure, the pressure distribution curves of the moving wall of the friction pair with different non-parallelisms are given in Figure 9 for the cavitation pressures of 3540 Pa and 10,000 Pa. The wedge effect caused by the non-parallelism significantly increases the pressure distribution at the pressure inlet of the friction pair of the non-parallel micro-cavity. After considering the cavitation effect, the maximum pressure in the high-pressure region caused by the micro-cavity has no obvious change with the increase of non-parallelism.

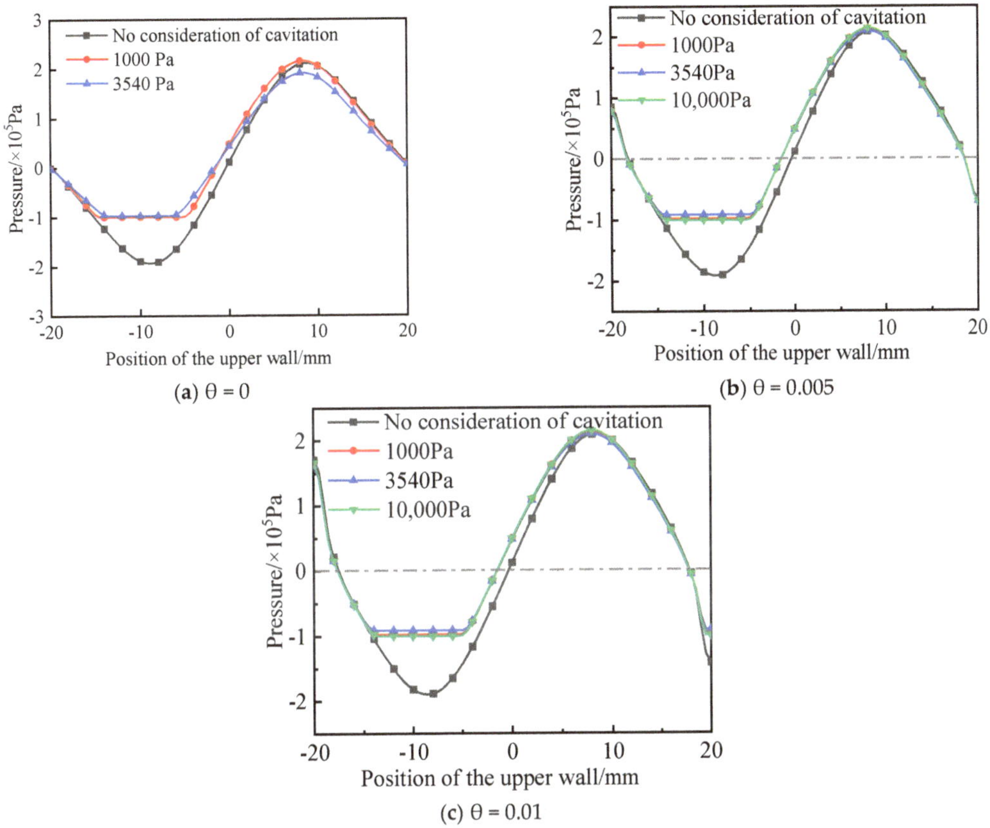

Figure 8. Cavitation effect under the same non-parallelism.

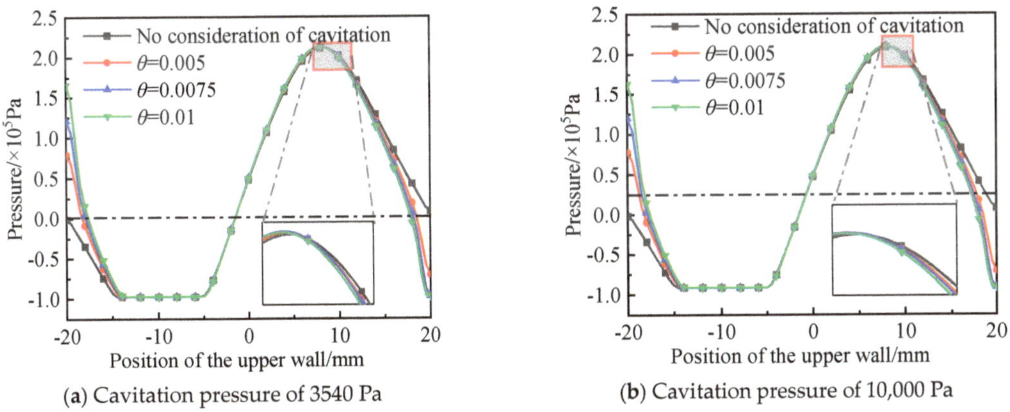

Figure 9. Influence of non-parallelism on the pressure distribution of the moving wall under the same cavitation pressure.

The percentage contribution of cavitation is defined as the "difference in load-carrying capacity before and after considering cavitation"/"load-carrying capacity when considering cavitation", and the remaining part is the percentage contribution of other effects (for example, inertia effects and the inlet wedge effect). Figure 10 shows the contribution percentage of the cavitation effect and other effects to the load-carrying capacity under different non-parallelisms. It can be seen from the figure that, under the same non-parallelism, when the R_e is small, the negative pressure has not reached the cavitation pressure, and other effects other than the cavitation effect account for 100%. With the R_e, cavitation increases gradually, and the percentage of contribution to load-carrying capacity increases gradually.

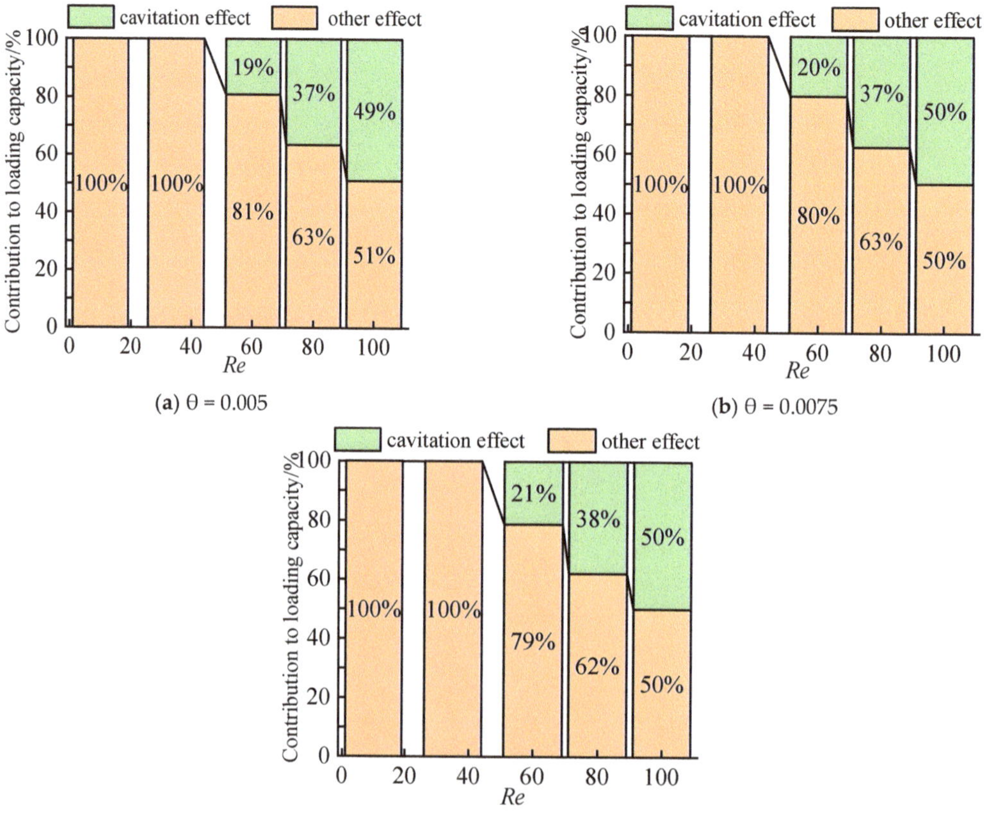

Figure 10. Contribution of the cavitation effect and other effects to the load-carrying capacity.

3.3. *Inlet Wedge Effect*

In order to explore the wedge effect of the inlet of the friction pair with the same non-parallelism, r was defined as the distance between the micro-cavity inlet and the pressure inlet boundary conditions, and it was taken as 0 mm, 5 mm, 10 mm, 15 mm, and 20 mm, respectively. Figure 11 shows the pressure distribution of the friction pair at different r when the non-parallelism is 0.01. As r increases, the overall pressure distribution in the non-parallel friction pairs decreases, while the proportion of the high-pressure zone decreases, the proportion of the negative pressure zone increases, and the load-carrying capacity of non-parallel friction pairs decreases. Figure 12 shows the pressure distribution curve of the moving wall surface under the inlet wedge effect when the degree of non-parallelism is 0.005 and 0.01. It can be seen that the moving wall pressure moves to the right and down with the increase at the inlet position. In addition, the area of the positive pressure area gradually decreases and the friction load-carrying capacity decreases, all of which are influenced by the inlet wedge effect.

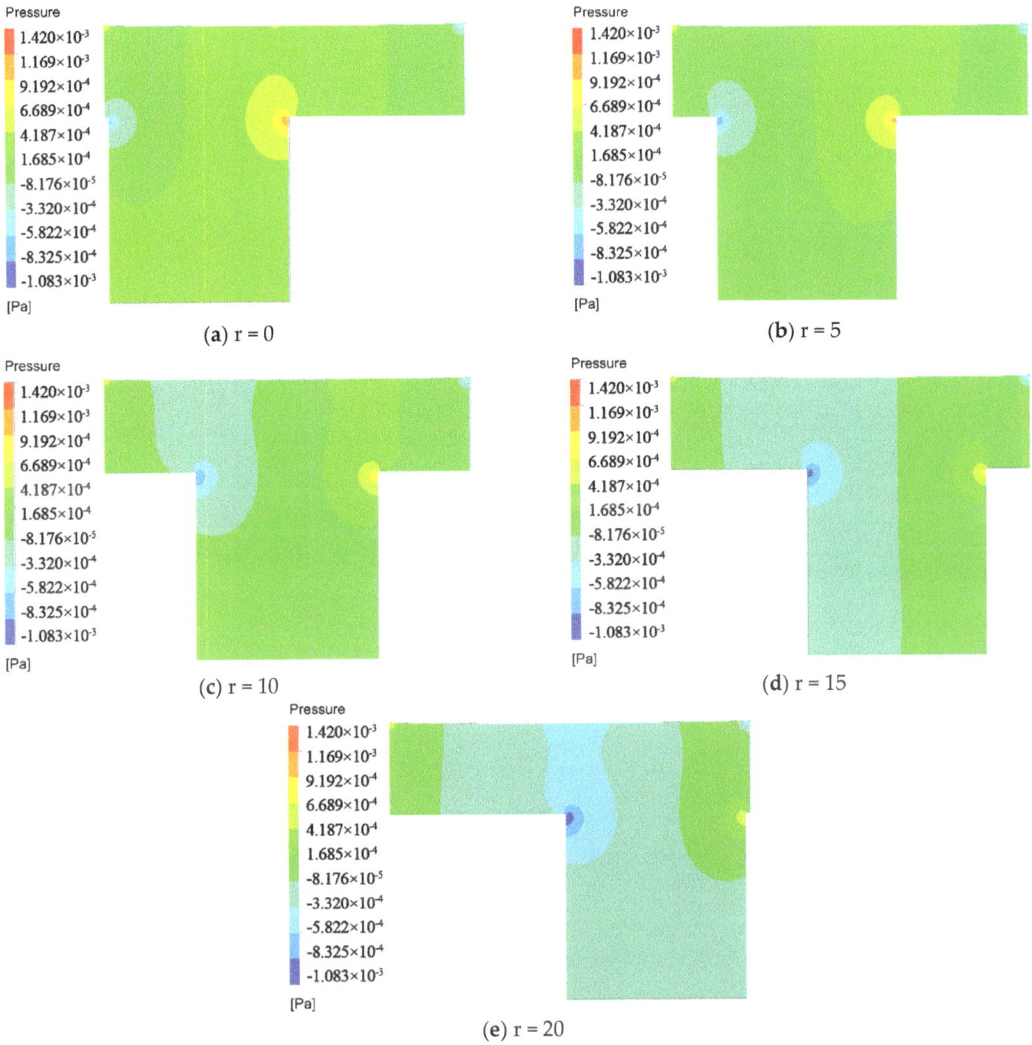

Figure 11. The pressure distribution at different inlet positions at $\theta = 0.01$.

In order to explore the influence of non-parallelism on the lubrication condition of friction pairs at the same inlet position, the determination of non-parallelism on the pressure distribution curve of the moving wall at the same inlet distance is shown in Figure 13. The wedge effect caused by non-parallelism significantly increases the pressure distribution at the pressure inlet, thus increasing the load-carrying capacity of friction pairs. As can be seen from Figure 13a–c, when the inlet distance is less than or equal to 10 mm, the existence of non-parallelism reduces the proportion of the negative pressure zone inside the micro-cavity and increases the proportion of the positive pressure zone, which improves the load-carrying capacity of the texture friction pair. In addition, the shorter the inlet distance, the more pronounced the pressure increase. As can be seen from Figure 13d,e, when the inlet distance is greater than 10 mm, the change trend is reversed. In addition, the larger the inlet distance is, the more obvious the reduction of non-parallelism on the pressure inside the micro-cavity is. Figure 14 shows the variation trend of load-carrying capacity with non-parallelism at different inlet positions. When the inlet distance is less than or equal to 10, the load-carrying capacity of non-parallel friction pairs increases with the increase in non-parallelism.

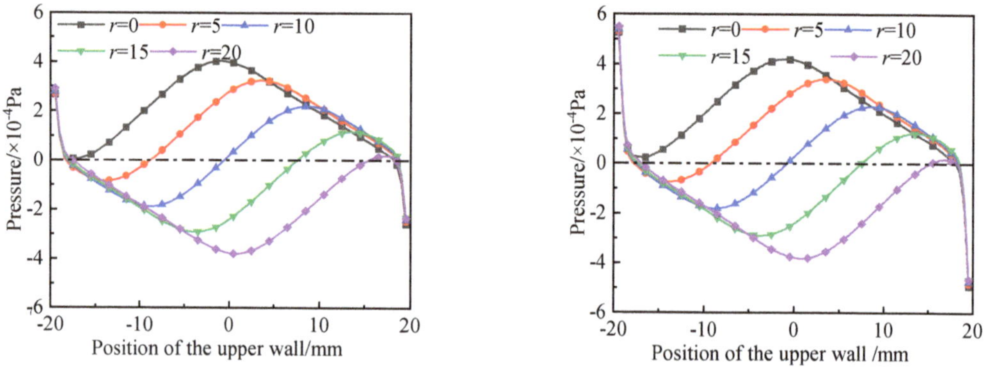

(**a**) θ = 0.005 (**b**) θ = 0.01

Figure 12. The same inlet distance under the inlet suction effect.

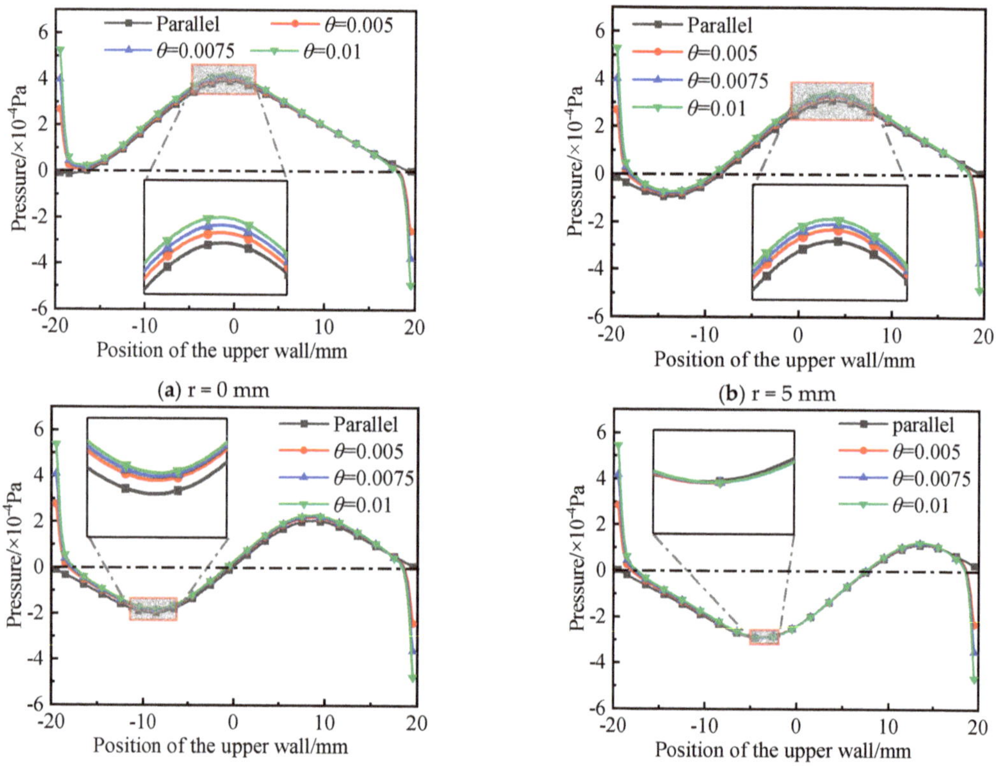

(**a**) r = 0 mm (**b**) r = 5 mm (**c**) r = 10 mm (**d**) r = 15 mm

Figure 13. *Cont.*

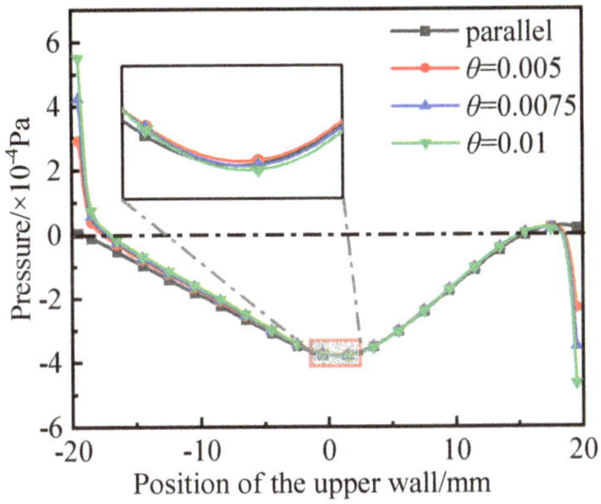

(e) r = 20 mm

Figure 13. Influence of non-parallelism on the pressure distribution of the moving wall at the same inlet position.

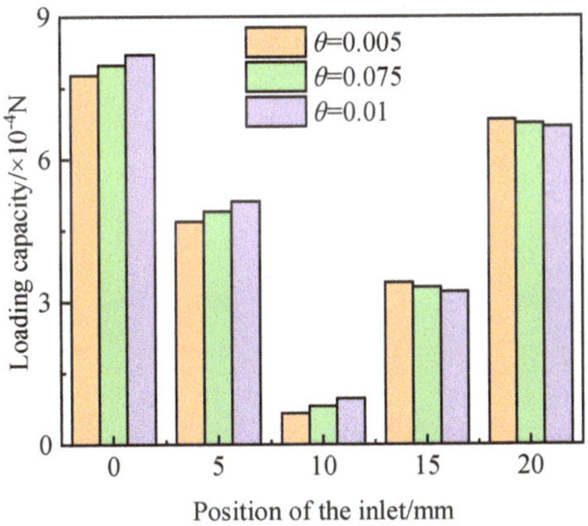

Figure 14. Load-carrying capacity at different inlet positions.

4. Conclusions

The textural friction pair models with different non-parallelisms are established, and the inertia effect, cavitation effect, and inlet wedge effect were analyzed, respectively. The following conclusions are drawn:

(1) The wedge effect caused by the non-parallelism itself causes the pressure at the inlet of the micro-cavity to rise, thus increasing the load-carrying capacity of the friction pair.

(2) The existence of non-parallelism limits the rise of the highest pressure of the inertia effect in the micro-cavity, and reduces the load-carrying capacity of the friction pair.

(3) After considering the cavitation effect, the maximum pressure in the high-pressure zone caused by the micro-cavity does not change significantly, and the load-carrying capacity of the contact pair increases with the non-parallelism.

(4) The presence of non-parallelism decreases the proportion of the negative pressure zone and increases the proportion of the positive pressure zone inside the micro-cavity, thus increasing the load-carrying capacity of the friction pair.

The research results and conclusions lay the theoretical foundation for further research and propose scale standards for water-lubricated bearings.

Author Contributions: Conceptualization, L.S. and Z.L. (Zhaozeng Liu); Validation, J.S., Y.W. and Z.L. (Zhaozeng Liu); Investigation, Z.L. (Zhen Li); Writing—review & editing, T.J. and Y.W. All authors have read and agreed to the published version of the manuscript.

Funding: This research is supported by the National Natural Science Foundation of China (No. 52176050), the General Program of Natural Science Foundation of Shandong Province (ZR2020ME174).

Data Availability Statement: Data are contained within the article.

Conflicts of Interest: The authors declare that the research was conducted in the absence of any commercial or financial relationships that could be construed as a potential conflict of interest.

References

1. Wang, J.; Wang, J.; Tian, J.B.; Wang, G.W.; Yue, Z.H. Finite element analysis of load-carrying property of surface texture water-lubricated polymer bearings. *Lubr. Eng.* **2021**, *46*, 90–96.
2. Zhang, P.Y.; Zhang, Y.H.; Hua, X.J. Lubrication design and development analysis of micro-texturing surfaces. *Surf. Technol.* **2021**, *50*, 14–32.
3. Xie, Z.L.; Zhu, W.D. Theoretical and experimental exploration on the micro asperity contact load ratios and lubrication regimes transition for water-lubricated stern tube bearing. *Tribol. Int.* **2021**, *164*, 107105. [CrossRef]
4. Xie, Z.L.; Shen, N.W.; Zhu, W.D.; Tian, W.; Hao, L. Theoretical and experimental investigation on the influences of misalignment on the lubrication performances and lubrication regimes transition of water lubricated bearing. *Mech. Syst. Signal Process.* **2021**, *149*, 107211. [CrossRef]
5. Zhou, X.X.; Pang, X.J.; Yue, S.W.; Zhang, Y.; Zhang, H.; Yu, B. Tribological properties of combination of surface texture and ionic liquids. *Tribology* **2021**, *41*, 995–1003.
6. Xie, Z.L.; Jian, J.; Stanislaw, W. The fluid-structure interaction lubrication performances of a novel bearing: Experimental and numerical study. *Tribol. Int.* **2023**, *179*, 108151. [CrossRef]
7. Xie, Z.L.; Jian, J.; Zhao, B.; Zhang, J.; Xu, F. Theoretical and experimental research on the effect of bi-directional misalignment on the static and dynamic characteristics of a novel bearing. *Mech. Syst. Signal Process.* **2024**, *208*, 111041. [CrossRef]
8. Gachot, C.; Rosenkranz, A.; Hsu, S.M.; Costa, H.L. A critical assessment of surface texturing for friction and wear improvement. *Wear* **2017**, *372*, 21–41. [CrossRef]
9. Pettersson, U.; Jacobson, S. Influence of surface texture on boundary lubricated sliding contacts. *Tribol. Int.* **2003**, *36*, 857–864. [CrossRef]
10. Yang, J.; Liu, Z.L.; Cheng, Q.C.; Liu, X.K.; Deng, T.Y. The effect of wear on the frictional vibration suppression of water-lubricated rubber slat with/without surface texture. *Wear* **2019**, *426–427*, 1304–1317. [CrossRef]
11. Hamilton, D.B.; Walowit, J.A.; Allen, M. A theory of lubrication by micro irregularities. *J. Basic Eng.* **1966**, *88*, 177–185. [CrossRef]
12. Annao, J.N.; Walowit, J.A.; Allen, C.M. Load support and leakage from micro asperity-lubricated face seals. *J. Lubr. Technol.* **1969**, *91*, 726–731. [CrossRef]
13. Qiu, Y.; Khonsari, M.M. Experimental investigation of tribological performance of laser textured stainless steel rings. *Tribol. Int.* **2011**, *44*, 635–644. [CrossRef]
14. Qiu, Y.; Khonsari, M.M. On the prediction of cavitation in dimples using a mass-conservative algorithm. *J. Tribol.* **2009**, *131*, 041702. [CrossRef]
15. Zhang, J.Y.; Meng, Y.G. Direct observation of cavitation phenomenon and hydrodynamic lubrication analysis of textured surfaces. *Tribol. Lett.* **2012**, *46*, 147–158. [CrossRef]
16. Meng, F.M.; Zhang, L. Effect of cavitation on tribological performances for textured surfaces. *Adv. Mater. Res.* **2012**, *472–475*, 391–394. [CrossRef]
17. Meng, F.M. On influence of cavitation in lubricant upon tribological performances of textured surfaces. *Opt. Laser Technol.* **2013**, *48*, 422–431. [CrossRef]
18. Bai, L.Q.; Meng, Y.G.; Zhang, V.R. Experimental study on transient behavior of cavitation phenomenon in textured thrust bearings. *Tribol. Lett.* **2016**, *63*, 27. [CrossRef]
19. Li, Q.; Wang, Y.J.; Zhang, S. Investigation on hydrodynamic super position loading mechanism and micro-hydrodynamic effect of textured water-lubricated beating. *Surf. Technol.* **2019**, *48*, 180–187.
20. Fowell, M.; Olver, A.V.; Gosman, A.D. Entrainment and inlet suction: Two mechanisms of hydrodynamic lubrication in textured bearings. *J. Tribol.* **2006**, *129*, 336–347. [CrossRef]
21. Olver, A.V.; Fowell, M.T.; Spikes, H.A. 'Inlet suction', a load support mechanism in non-convergent, pocketed, hydrodynamic bearings. *Proc. Inst. Mech. Eng. Part J J. Eng. Tribol.* **2006**, *220*, 105–108. [CrossRef]
22. Dobrica, M.B.; Fillon, M.; Pascovici, M.D. Optimizing surface texture for hydrodynamic lubricated contacts using a mass-conserving numerical approach. *Proc. Inst. Mech. Eng. Part J J. Eng. Tribol.* **2010**, *224*, 737–750. [CrossRef]
23. Wang, J.; Yan, Z.; Fang, X. Observation and experimental investigation on cavitation effect of friction pair surface texture. *Lubr. Sci.* **2020**, *32*, 404–414. [CrossRef]

24. Arghir, M.; Roucou, N.; Helene, M. Theoretical analysis of the incompressible laminar flow in a macro-roughness cell. *J. Tribol.* **2003**, *125*, 309–318. [CrossRef]
25. Billy, F.; Arghir, M.; Pineau, G. Navier-stokes analysis of a regular two-dimensional roughness pattern under turbulent flow regime. *J. Tribol.* **2006**, *128*, 122–130.
26. Sahlin, F.; Glavatskih, S.; Almqvist, T.R. Two dimensional CFD-analysis of micro-patterned surfaces in hydrodynamic lubrication. *J. Tribol.* **2005**, *127*, 96–102. [CrossRef]
27. Feldman, Y.; Kligerman, Y.; Etsion, I.; Haber, S. The validity of the reynolds equation in modeling hydrostatic effects in gas lubricated textured parallel surfaces. *J. Tribol.* **2006**, *128*, 345–350. [CrossRef]
28. Dobrica, M.B.; Fillon, M. About the validity of Reynolds equation and inertia effects in textured sliders of infinite width. *Proc. Inst. Mech. Eng. J. Eng. Tribol.* **2009**, *223*, 69–78. [CrossRef]
29. Dobrica, M.B.; Fillon, M. Thermo hydrodynamic behavior of a slider pocket bearing. *J. Tribol.* **2006**, *128*, 312–318. [CrossRef]
30. Cupillard, S.; Glavatskih, S.; Cervantes, M.J. Inertia effects in textured hydrodynamic contacts. *J. Proc. Inst. Mech. Eng. J. Eng. Tribol.* **2010**, *224*, 751–756. [CrossRef]
31. Kraker, A.D.; Ostayen, R.A.J.V.; Rixen, D.J. Development of a texture averaged Reynolds equation. *Tribol. Int.* **2010**, *43*, 2100–2109. [CrossRef]
32. Syed, I.; Sarangi, M. Hydrodynamic lubrication with deterministic micro textures considering fluid inertia effect. *Tribol. Int.* **2014**, *69*, 30–38. [CrossRef]
33. Wang, Y.; Jacobs, G.; König, F.; Zhang, S. Investigation of Microflow Effects in Textures on Hydrodynamic Performance of Journal Bearings Using CFD Simulations. *Lubricants* **2023**, *11*, 20. [CrossRef]
34. Khalil, M.F.; Kassab, S.Z. Ismail A S. Influence of inertia forces on the performance of turbulent externally pressurized bearings. *J. Tribol. Int.* **1992**, *25*, 17–25. [CrossRef]
35. Woloszynski, T.; Podsiadlo, P.; Stachowiak, G.W. Evaluation of inertia effect in finite hydrodynamic bearings with surface texturing using spectral element solver. *J. Tribol. Int.* **2015**, *91*, 170–176. [CrossRef]
36. Tonder, K. Dynamics of rough slider bearings: Effects of one-sided roughness/waviness. *Tribol. Int.* **1996**, *29*, 117–122. [CrossRef]
37. Tonder, K. Hydrodynamic effects of tailored inlet roughnesses: Extended theory. *Tribol. Int.* **2004**, *37*, 137–142. [CrossRef]
38. Tonder, K. Inlet roughness tribodevices: Dynamic coefficients and leakage. *Tribol. Int.* **2001**, *34*, 847–852. [CrossRef]
39. Etsion, I.; Halperin, G. A laser surface textured hydrostatic mechanical seal. *Tribol. Trans.* **2002**, *45*, 430–434. [CrossRef]
40. Xu, Y.F.; Zheng, Q.; Abuflaha, R. Influence of dimple shape on tribofilm formation and tribological properties of textured surfaces under full and starved lubrication. *Tribol. Int.* **2019**, *136*, 267–275. [CrossRef]
41. Rom, M.; Müller, S. A new model for textured surface lubrication based on a modified Reynolds equation including inertia effects. *Tribol. Int.* **2019**, *133*, 55–66. [CrossRef]
42. Marian, M.; Almqvist, A.; Rosenkranz, A.; Fillon, M. Numerical micro-texture optimization for lubricated contacts—A critical discussion. *Friction* **2022**, *10*, 1772–1809. [CrossRef]
43. Sun, D.C.; Brewe, D.E. A high-speed photography study of cavitation in a dynamically loaded journal bearing. *ASME J. Tribol.* **1991**, *113*, 287–292. [CrossRef]
44. Braun, M.J.; Hannon, W.M. Cavitation formation and modelling for fluid film bearings: A review. *Proc. Inst. Mech. Eng. J. Eng. Tribol.* **2010**, *224*, 839–863. [CrossRef]
45. Etsion, I.; Ludwig, L.P. Observation of pressure variation in the cavitation region of submerged journal bearings. *ASME J. Lubr. Technol.* **1982**, *104*, 157–163. [CrossRef]

lubricants

MDPI

Article

Investigation of Cavitated Flow in Water-Lubricated Bearings Considering Surface Roughness, Thermal, and Elastic Effects

Dimitris Charamis and Pantelis G. Nikolakopoulos *

Machine Design Laboratory, Department of Mechanical Engineering and Aeronautics, University of Patras, 26504 Patras, Greece; up1054503@ac.upatras.gr

* Correspondence: pnikolakop@upatras.gr

Abstract: Contrary to conventional journal bearings, which operate using oil-based substances, water-lubricated bearings (WLBs) utilize water and, thus, constitute a more environmentally responsible solution. The shipping industry, among others, as already been introduced to this technology with a lot of commercial ships using water-lubricated stern tube systems; in other cases, hydropower plants manage to keep up with the strict environmental regulations by implementing the use of WLBs in water turbines. However, there are a lot of challenges when it comes to transitioning from conventional bearings to water-based ones. Such challenges are caused by the low viscosity of water and lead to phenomena of high complexity. Such phenomena are related but not limited to cavitation and turbulent flow due to the interaction between the lubricating water and bearing surface. In this study, a numerical method will be used to simulate the fluid film and bearing geometries in order to perform a thermo-elastohydrodynamic (TEHD) analysis. The dynamic characteristics of the bearing will be calculated and the results will be discussed. The novelty of the study is evident in but not limited to the determination of the elastic deformation of a WLB during operation, as well as the effect of surface roughness, cavitation, and thermal effects on bearing characteristics.

Keywords: water-lubricated bearings; marine applications; green tribology; CFD; surface roughness

Citation: Charamis, D.; Nikolakopoulos, P.G. Investigation of Cavitated Flow in Water-Lubricated Bearings Considering Surface Roughness, Thermal, and Elastic Effects. *Lubricants* **2024**, *12*, 107. https://doi.org/10.3390/lubricants12040107

Received: 31 January 2024
Revised: 15 March 2024
Accepted: 17 March 2024
Published: 24 March 2024

1. Introduction

Journal bearings constitute one of the most basic machine elements used in rotational machines. Their function is based on the theory of hydrodynamic lubrication, which has been studied by several scientists and researchers like Yukio Hori [1] and describes the ideal type of lubrication, during which a thick enough (full) film of lubricant is formed between the bearing and the machine shaft so that the two elements do not come into direct contact.

Nowadays, there is an extraordinarily high annual demand for lubricants all around the world. Taking into consideration the fact that the need for responsible and environmentally friendly consumption has never been higher, it is apparent that a demand of this magnitude is likely to prove problematic. Several new technologies are being developed so that the consumption of lubricants can be mitigated. When it comes to journal bearings, one of these options is the use of water-lubricated journal bearings. Ref. [2] elaborated on the importance of implementing the use of water-lubricated bearings and conducted a variety of experiments to better understand the behavior of such bearings.

Extensive research into journal bearings has shown the detrimental effect of turbulence and cavitation in operation. Especially for the case of WLBs, the low viscosity easily results in the inception of cavitation or in turbulent flow, even more so when roughness or complex geometries are in effect. This results in challenges regarding the tribological design of WLBs. The knowledge about lubrication in journal bearings has progressed greatly over the last years and is something that has been studied thoroughly by many researchers. Zhou et al. [3] used the twofold secant superliner iteration method to calculate the equilibrium position of a plain oil-based journal bearing with great accuracy, by solving Reynolds' equation numerically.

Another issue that is of concern when it comes to the design and production of all types of bearings, water-lubricated ones included, is the bearing roughness. Many studies have been conducted on the drawbacks that a rough surface has on the lubrication regime, while some researchers even investigated certain conditions under which a surface roughness arrangement could be beneficial. Tauviqirrahman et al. [4,5] studied the possibility of utilizing a heterogenous rough/smooth surface profile, which would provide superior tribological properties compared to the equivalent homogenous smooth or rough bearing.

Zhang et al. [6] also studied WLBs and performed a CFD analysis to determinate the stiffness coefficient of a bearing during cavitated steady flow. Similarly, Zheng et al. [7] proposed an improved Schnerr-Sauer cavitation model based on the three-dimensional element of NACA66.

On the other hand, Tamboli et al. [8] conducted experimental investigations of WLBs and concluded that, like in the case of conventional bearings, the load carrying capacity of WLBs was highly dependent on the selection of the bearing surface finishing, radial clearances, and the water feed rate.

The contribution of Xie et al. [9,10] was also notable. Xie investigated the effect of novel surface texturing in a water-lubricated bearing and supported his findings with both computational and experimental findings. Among other findings, it was shown that the position of said texturing could locally increase the load carrying capacity of a bearing, while also reducing the coefficient of friction, but ill-placed texturing could also affect the continuity of the flow and, thus, of the pressure profile. In other studies, Xie [10] also proposed a model to calculate the dynamic coefficients in WLBs.

In this study, an effort will be made to better understand how the use of water as a lubricant affects the static and dynamic properties of a bearing, while the effect of bearing surface roughness will also be investigated. A CFD (computational fluid dynamics), as well as a FSI (fluid-structure interaction) approach will be chosen, in order to extract the results needed for the tribological design of water-lubricated bearings.

The innovations of the study are based on the utilization of a dynamic mesh method that will allow the transient solution of the bearings for a given external load. By using this method, the accurate calculation of the bearing elastic deformation and thermal expansion is possible. The change in the water viscosity will be taken into consideration and the cavitation phenomenon will be studied. Furthermore, given the importance that grooves have in WLBs (for providing a generous supply of lubricating water), the study will be based on a grooved journal bearing, which was used by Brito et al. [11,12] in both computational and experimental investigations of how heat transfer affects bearing operation. Last but not least, great progress is made in the investigation of surface roughness in WLBs, where for several different roughness heights the static and dynamic characteristics are determined and discussed.

2. Materials and Methods

In this study a simulation will be modeled for journal bearings lubricated using either conventional lubricants or water. For these analyses, the computational fluid dynamics (CFD) software FLUENT will be used, which solves the Navier-Stokes equations numerically. In order to add some case-specific functions to the analyses, a few UDF (user defined function) algorithms will be coded and compiled. More specifically, an algorithm will be used to find the equilibrium position of the journal and another one will be compiled containing Vogel's equation to account for the change in water (liquid) viscosity with temperature. The resulting analysis will constitute a flexible, accurate, and efficient method for the tribological design of any kind of journal bearing, since the same algorithms will be able to be used in any different set of geometric, physical, and operating parameters (as long as the relevant models are correctly set) to produce high quality results. The material chosen for the WLB is a polymer made by the company Thordon(Burlington, ON, Canada) and has the commercial name COMPAC(Valencia, Spanish). Initial water properties are used for a temperature of 15 °C, while a table of water properties is inserted to account for vaporization pressure at different temperatures.

2.1. Governing Equations

One of the most challenging parts of the analysis of journal bearings is solving the flow equations that apply to the lubricant. In this study, the Navier-Stokes equations are solved numerically to produce accurate results. The Navier-Stokes equations are employed for the solution of a wide spectrum of problems and refer to the following laws.

The law of conservation of mass:

$$\nabla \cdot \overline{u} = 0 \tag{1}$$

where

$$\nabla \cdot u = \partial u \partial x + \partial v \partial y + \partial w \partial z. \tag{2}$$

The conservation of momentum:

$$p\frac{D\overline{u}}{Dt} = -\nabla p + \mu \nabla^2 \overline{u} + p\overline{F}. \tag{3}$$

The law of conservation of energy, which is an expression of the first law of thermodynamics:

$$\rho\frac{De}{Dt} = \nabla \cdot (k\nabla T) + \frac{DP}{Dt} + \Phi\,. \tag{4}$$

Reynolds' equation [13] is also used very frequently in journal bearings and is also the one that is used in the study conducted by Zhou et al. [3], which is used for the validation of this model. Reynolds is considered to be one of the pioneers of tribology who inspired by Tower's experiments [14,15] conducted his own theoretical investigation to develop his equation. In Reynold's equation certain assumptions are proposed, in order to simplify the more general and complex equations (Navier-Stokes), which describe fluid flow.

2.2. Cavitation Model

When it comes to analyzing the flow in water-lubricated bearings, certain effects must be taken into consideration. One of those effects is cavitation. The drop in pressure in the fluid below the saturation pressure results in the creation of bubbles or cavities. Bubbles rupture at areas of elevated pressure within the liquid, resulting in the release of a substantial force of impact. This impact force can lead to severe repercussions on both the journal and bearing surfaces. Given the considerable detrimental impact of cavitation effects within the system, this investigation also encompasses the gaseous component. Among the three common cavitation models, the Zwart-Gerber-Belamri model stands out due to its robust coupling solution, rapid convergence, and suitability for applications in rotating machinery.

Since the cavitation phenomenon is present in water-lubricated bearings, the Zwart-Gerber-Belamri cavitation model is applied, according to which the mass transfer rates are given as follows:

$$\dot{m}_a^- = -C_v\, r_{nuc}\, \frac{3\rho_v}{R_B}\sqrt{\frac{2}{3\rho_i}\frac{1}{|P-P_v|}}\min(P-P_v,0), \tag{5}$$

$$\dot{m}_a^+ = C_c\, \frac{3\rho_v}{R_B}\sqrt{\frac{2}{3\rho_i}\frac{1}{|P-P_v|}}\max(P-P_v,0) \tag{6}$$

where r_{nuc} is the nucleation site volume fraction. Like the SS (Schnerr-Sauer) model, it is based on the simplified Rayleigh-Plesset equation but assumes a constant nucleation site volume. Thus, both the nucleation site volume fraction, r_{nuc}, and nucleation site radius, RB, are model constants. According to Zwart-Gerber-Belamri, these constants are $r_{nuc} = 5.0 \times 10^{-4}$ and $RB = 1.0 \times 10^{-6}$ m by default.

The vaporization pressure (Pv) is defined using the saturation pressure–temperature table of water [16], which is implemented to account for temperature changes.

2.3. Roughness Modelling

For the modelling of a rough bearing surface, the modified-for-roughness law of the wall is applied. While most analyses assume a hydrodynamically smooth ($K_s^+ \leq 2.25$) bearing, most real cases either have or at some point obtain characteristics of rough surfaces. Whether there are asperities that exist on the bearing surface during installation that need to be taken into consideration during operation (in hours) or whether there are surfaces that are roughened as a result of wear, it is important for this case to be investigated in the designing process. While the effect of asperities might seem insignificant, this is not the case. Contrary to smooth bearings, where, in most cases, the flow is laminar, the varying Reynolds number (Re), in this case, tends to exceed the critical Reynolds number (Re_c) resulting in turbulent flow phenomena. For those cases, the flow is studied through the use of the Reynolds-averaged Navier–Stokes equations (RANS). In reality, the rough surfaces are mostly nonuniform and the asperities are of the random roughness constant and height. In this study, a uniform sand-grain roughness model is used similar to the one used by Bhatt et al. [17]. The law of the wall for mean velocity modified for roughness has the following form:

$$\frac{u_p\, u^*}{\tau_w/\rho} = \frac{1}{k} \ln\left(E \frac{\rho\, u^* y_p}{\mu} \right) - \Delta B \tag{7}$$

where $u^* = C_\mu^{1/4}\, \kappa^{1/2}$ and

$$\Delta B = \frac{1}{k} \ln f_r \tag{8}$$

where f_r is a roughness function dictated by the kind and magnitude of the rough surface. For the purposes of this study, the roughness constant C_s is set to 0.5, while the roughness height K_s will take values between 0 and 100 μm.

2.4. Bearing Bush Structure Model

The structure of the bearing bush is modelled in ANSYS and solved in the transient structural analysis component. The transient dynamic equilibrium equation is based on the following linear structure:

$$[M]\{\ddot{u}\} + [C]\{\dot{u}\} + [K]\{u\} = \{F^a\} \tag{9}$$

where $[M]$ is the structural mass matrix;

$[C]$ is the structural damping matrix;

$[K]$ is the structural stiffness matrix;

$\{u\}$ is the nodal displacement vector and {Fa} is the applied load vector.

The structural matrixes are calculated for the geometry using the algorithms in [18], which utilize the properties of the corresponding element stiffness and damping matrixes, respectively. In this particular case, a hexahedral finite element is used.

Using finite difference expansions in the time interval Δt, the Newmark method is implemented, which forms Equation (10):

$$(a_0[M] + a_1[C] + [K])\{u_{n+1}\} = \{F^a\} + [M](a_0\{u_n\} + a_2\{\dot{u}_n\} + a_3\{\ddot{u}_n\}) + [C]\ (a_1\{u_n\} + a_4\{\dot{u}_n\} + a_5\{\ddot{u}_n\}) \tag{10}$$

where α_0–α_5 are Newmark integration parameters.

Lastly, the full solution method is chosen for higher accuracy. The full solution method solves equation 10 numerically, using the Newton-Raphson method.

2.5. Numerical Scheme

The geometry of the journal bearing is created in the integrated software called DesignModeler and the mesh was created by using hexahedral elements. This choice was made for their low requirements, regarding processing power and the accuracy they provide. Thorough analysis of grid sensitivity has been conducted as part of the study. Multiple sets of grid sensitivity calculations were executed. An optimum balance between the processing time and calculation accuracy is considered best. The quality of the mesh at the liquid film's inlet and outlet plays a crucial role in determining the overall accuracy of the calculations. When dividing the liquid film thickness direction into ten layers, it was observed that the grid exhibited a relatively high quality. At this time, the number of elements reaches 69,600 and the orthogonal quality is 0.99996 with a skewness of 5.9095×10^{-3}. It is essential to avoid having an excessive number of layers throughout the film's thickness. This is because an excessively high number of layers leads to a significant increase in the overall aspect ratio of the grid, making it prone to distortion and, consequently, lowering grid quality. Hence, achieving the right balance between grid quality, element quantity, layer count, and computation time becomes crucial. This balance optimization aims to enhance both calculation accuracy and efficiency. After conducting numerous trial calculations and thorough investigations, we have determined that a layer count of 15 across the film length achieves the optimal equilibrium point.

In the case of the FSI (fluid–structure interaction) analysis there is a variety of things to be investigated. The first part is the analysis of the flow of the lubricant. The second part is the heat transfer in the bush and between the bush and the lubricant. The last part is the resulting stress and strain on the bearing as well as its deformation. All of those can be fully analyzed using the following components to produce a thermo-elastohydrodynamic analysis: CFD (computational fluid dynamics), FLUENT, mechanical, transient thermal, and transient structural. For the case of the rough bearing, the transition shear stress transport (SST) k-ω formulation is utilized. The model utilizes k-ω near the wall, while it switches to a k-ε type in the free stream. For the FSI case, laminar flow is modelled using the numerical solutions of the Navier–Stokes equations, which are described in the Materials and Methods section.

The utilization of the commercial software in addition to the UDF algorithms provides notable flexibility and calculation efficiency. In the present model, a grooved bearing is analyzed using the Reynolds-averaged Navier–Stokes (RANS) equations in the transient state. The two different models handle equation sets of high complexity, which, among others, include lubricant pressure–temperature–viscosity models, varying water vaporization pressure based on tabular data, thermal effects, cavitation, and either turbulent flow inception or elastic deformation of the bush. The inclusion of the above in the model allows the analysis to capture certain subtleties, such as properties throughout the thickness of the lubricant. The model can also effectively provide vital information like the extent of cavitation in the flow, which, in this case, is limited to a reasonable extent, since the vaporization pressure, heat transfer, and water viscosity are all variables; they are dependent on each other and their values tend to prevent the cavitation effect from increasing uncontrollably.

For all calculation criteria, the following convergence criteria are used. The sum of the forces acting on the journal (external load and hydrodynamic forces) should be less than 0.01 N. For the solution of the Navier–Stokes equations, the scaled residuals of continuity, momentum, and energy equations should be less than 10^{-4}. The static characteristics calculated for the bearing are the pressure, temperature, viscosity, and vapor volume fraction distributions. For the case of turbulent flow, the turbulent kinetic energy and energy dissipation rate are also calculated.

The solution procedure executed in the study is described by the flow chart depicted in Figure 1:

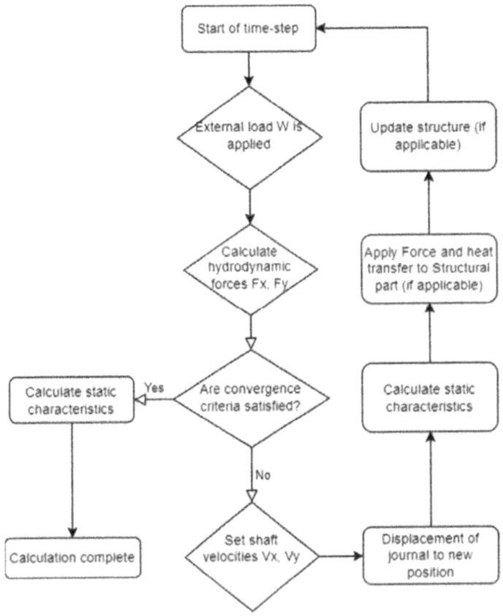

Figure 1. Numerical model solution procedure flowchart.

3. Model Verification

It should be noted that for the purposes of this study, a hydrodynamic lubrication regime is assumed. During this steady operating condition, a generous quantity of lubricant forms a thick full film to support the external load. Researchers like Failawati et al. [19] studied the change of bearing dynamic characteristics based on different lubrication regimes, while others like de Kraker et al. [20] studied the stribeck curve for the case of WLBs. For conventional bearings such diagrams can be constructed with the help of tables, as described by Shigley et al. [21] and in other machine element designing theory literature.

For the validation of the model, three different studies are used. The study conducted by Zhang et al. [6] is used to verify the validity of the cavitation model results. The equilibrium position results are compared to the results calculated in the study of Zhou et al. [3] and the experimental validation is based on the results measured by Brito et al. [11,12]. The accuracy of the results of this study will be examined in comparison to the method published by Zhou [3]. In this case, the following parameters are set as shown in Table 1.

Table 1. Parameters for journal bearing.

Parameters	Geometrical Parameters			Physical Parameters		
Symbol	cr (mm)	L (m)	Rb (m)	W (N)	u_j (rpm)	μ (pa s)
Value	0.6	0.12	0.1	1920	3000	0.04

The equilibrium position results are displayed in Table 2. After comparing the results it is clear that the UDF solution appears to be almost the same as the one calculated by Zhou. In fact, since FLUENT works by solving the Navier–Stokes equations numerically, while Zhou utilized Reynolds' equation, it is safe to say that the solution extracted in the present study is in fact closer to the actual case.

Table 2. Comparison of equilibrium position results.

Results	Current Study	Zhou [3]
Eccentricity ratio ε	0.378	0.372
Attitude angle θ	69.71°	67.42°

The simulation is repeated for L/D ranging from 0.5 to 1.0. In Figure 2, the equilibrium locus is calculated and compared to the one calculated by Zhou [3].

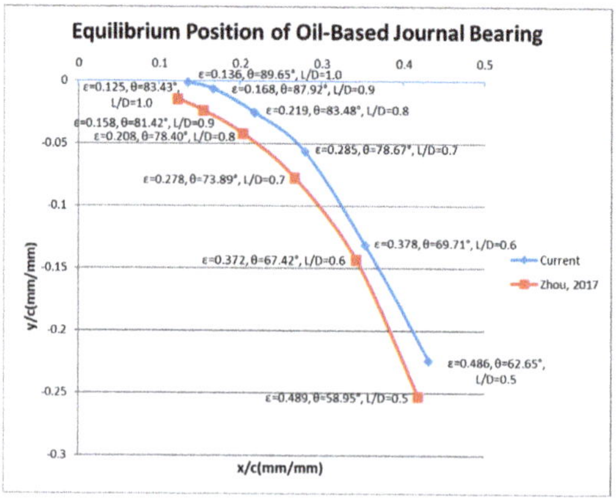

Figure 2. Influence of bearing length (L) on the equilibrium position [3].

The equilibrium locus is also extracted for the case of a water-lubricated bearing. It is evident by observing Figure 3 that the water-lubricated bearing cannot support the same loads as easily due to the low viscosity of water.

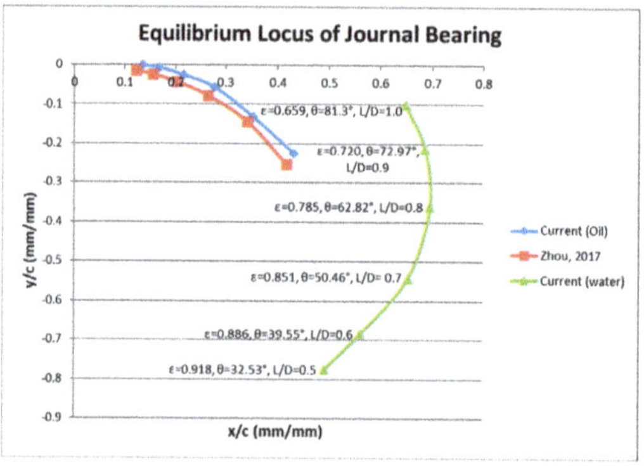

Figure 3. Influence of bearing length (L) on the equilibrium position of water-lubricated bearings [3].

For the validation of the roughness modelling, the study made by Tauviqirrahman et al. [4] was recreated. The choice of roughness size in this particular study (and studies referenced) is unusually high yet purposefully chosen for the purposes of this study—to show with clarity and emphasize the effect of roughness on the bearing performance indicators. More specifically, a heterogenous rough surface of 25 μm roughness height was modelled with an angular velocity of 2000 rpm and an eccentricity ratio of 0.8; the below results were calculated.

Both the pressure and vapor volume fraction distributions are almost identical in contour shape as well as maximum magnitude. More specifically, Tauviqirrahman et al. [4] calculated a maximum pressure of 380 kPa and maximum vapor volume fraction of 0.832, whereas the corresponding values in this study are 376 kPa and 0.952, respectively, which is depicted in Figure 4.

It is evident, after observing Figure 5, that the cavitation effect is most intense after the pressure decrease, which, of course, is in agreement with what is known about the phenomenon.

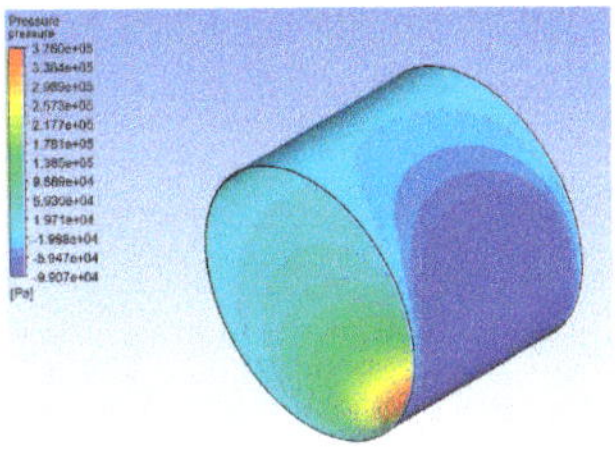

Figure 4. Pressure distribution for roughness model validation.

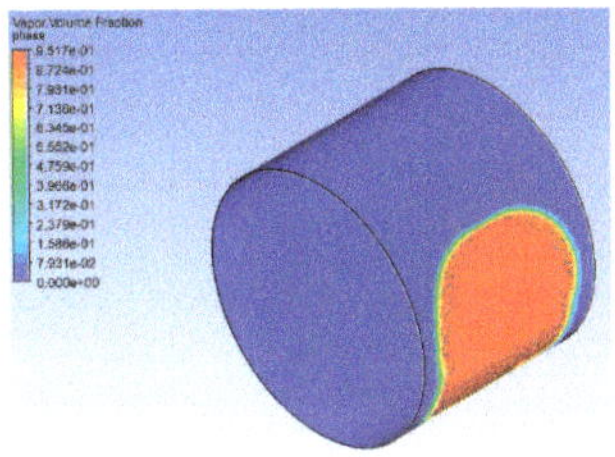

Figure 5. Vapor volume fraction distribution for roughness model validation.

4. Results

4.1. Rough Grooved Journal Bearing

In order to have a point of reference, the geometric parameters used in the rough rigid bearing analysis will be the same ones used in the case of the elastic FSI bearing. More specifically, the geometric parameters used are listed in Table 3.

Table 3. Geometric parameters of grooved journal bearing.

Geometric Parameters of Grooved Journal Bearing [11,12]	
Length (mm)	80
Bush inner diameter (mm)	100
Diametral clearance (μm)	171
Groove length (mm)	80
Circumferential extension of grooves (mm)	16
Groove number	0.5

The groove number is equal to the groove radius divided by the radial clearance of the bearing. Geometric characteristics like grooves can have several benefits to the operation of journal bearings either by providing a greater supply of lubricant or by increasing certain performance indicators like the load carrying capacity. Bombos et al. [22] investigated multi-step bearings, while Gong et al. [23] studied water lubricated micro groove bearings.

It is noted that, in this study, the assumption is made that no contact between the shaft and bearing or asperities takes place. So, to ensure hydrodynamic lubrication regardless of roughness height, the bearing is modelled in a way that the final inner diameter is always 100 mm (including the roughness height).

The grid is created as per the methodology described in the numerical scheme chapter.

This bearing has been studied extensively both in simulations and experimental investigations by Brito et al. [11,12]. The grooved bearing is subjected to an external load of W = 2 kN and an angular velocity of u_J = 4000 rpm. The lubricant is set to water and a uniform roughness ranging from 0 to 100 μm is applied. The numerical scheme is modeled using the RANS equations, thus taking into consideration the occurrence of turbulence, which is especially expected in the cases where roughness is present.

In the end, the following results regarding the friction coefficient were observed. The friction coefficient is derived from the equation

$$F_f = f\,W \tag{11}$$

where W is the external load applied to the bearing and F_f is the force of friction and equal to

$$F_f = \frac{M_f}{R_b} \tag{12}$$

where R_b is the bearing radius and M_f is the moment of friction acting on the bearing surface, calculated using the algorithm during the solution process.

The results appear to be in line with what would be expected based on logic and with the research published on the subject. As seen in Figure 6, the values of the friction coefficient appear to steadily increase when the roughness height rises, which is expected.

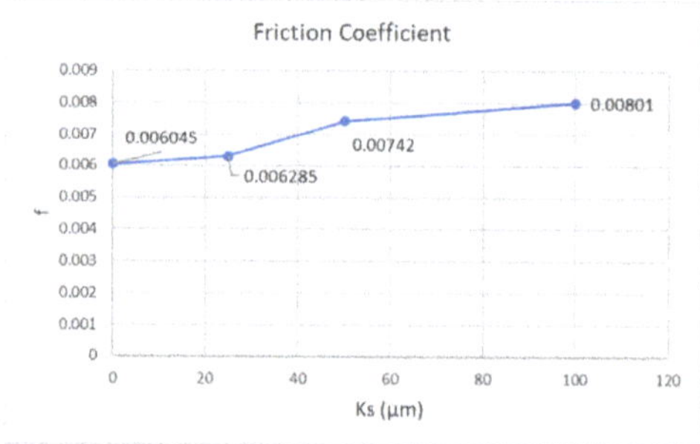

Figure 6. Friction coefficient for various roughness values.

Figure 7 exhibits the pressure distribution for each one of the cases described above. There is a lot of interest in the fact that, while the eccentricity values increase, the maximum pressure values observed appear to decrease with every roughness increase.

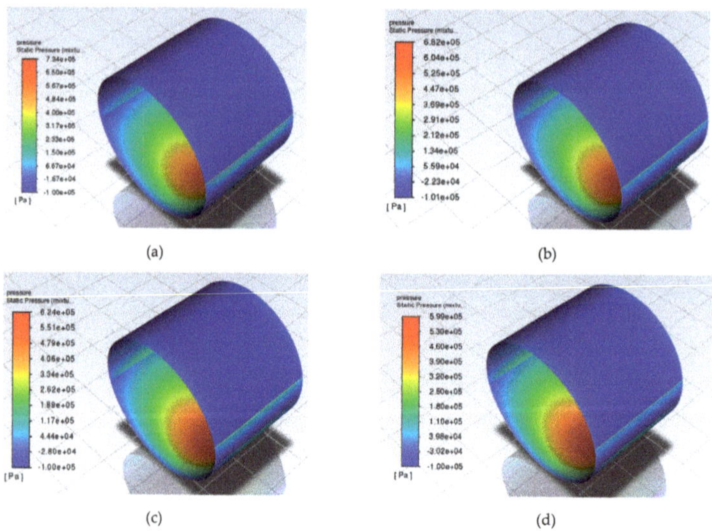

Figure 7. Static pressure distribution for various roughness values. (**a**) Smooth Bearing. (**b**) 25 μm Roughness Bearing. (**c**) 50 μm Roughness Bearing. (**d**) 100 μm Roughness Bearing.

Tauviqirrahman et al. [4] investigated and showed in their study that the addition of a well-chosen heterogeneous rough surface can be advantageous to the tribological performance of a bearing. In this case, the same load is distributed to a larger area, since the pockets between the asperities allow a larger quantity of lubricant to be in contact with the structure. Therefore, the maximum local pressure is not as high. The change in eccentricity ratio, friction coefficient and maximum pressure with respect to roughness height is calculated and shown in Table 4 as a point of reference.

Table 4. Rate of change of dynamic characteristics of bearing with roughness height change.

Roughness Height/ Change Ratio	**K_s = 0 (Point of Reference)**	**K_s = 25 μm**	**K_s = 50 μm**	**K_s = 100 μm**
Eccentricity ratio	0.459	+3.05%	+2.40%	+4.14%
Friction coefficient	0.006045	+3.97%	+22.75%	+32.51%
Maximum Pressure	0.734 MPa	−7.08%	−14.99%	−18.39%

The change in the friction coefficient is as significant as expected due to the rough texturing. Even in the case of a uniform roughness surface, it is evident that the roughness height of the bearing surface is something that should be investigated very carefully during the designing and production procedure.

Figure 8 shows not only that the turbulent kinetic energy increases with respect to the roughness height but also that the turbulence is most intense at the start and end of the grooves, where the change in geometry makes the flow unsteady.

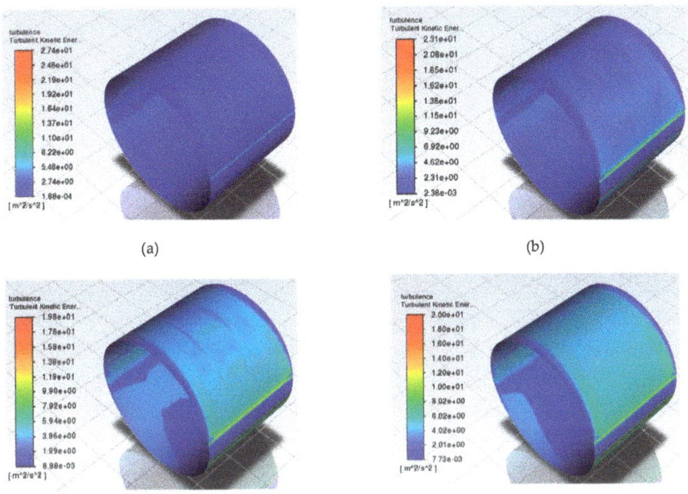

(a) (b) (c) (d)

Figure 8. Turbulent kinetic energy k distribution for various roughness values. (**a**) Smooth Bearing. (**b**) 25 μm Roughness Bearing. (**c**) 50 μm Roughness Bearing. (**d**) 100 μm Roughness Bearing.

It is important to note that the turbulent kinetic energy is greater and, thus, depicted for the outer wall of the fluid film (the bearing surface). This is expected as the geometry of the bearing surface contains the roughened texturing as well as the two axial grooves, all of which enhance the phenomenon. On the contrary, the turbulent kinetic energy is not as significant in the journal boundary layer. Figure 9 depicts the above and corresponds to the turbulent kinetic energy distribution in the inner (journal) and outer (bearing) wall, respectively. The case of 100 μm roughness is used.

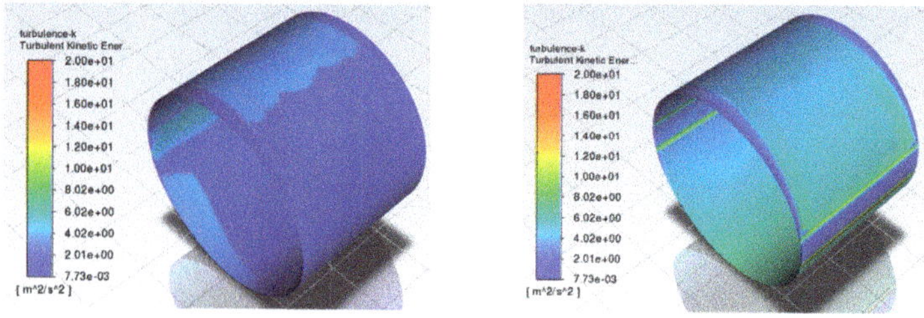

100 μm Roughness Bearing (inner wall) 100 μm Roughness Bearing (outer wall)

Figure 9. Turbulent kinetic energy distribution in journal and bearing boundary layers.

The vapor volume fraction distribution is also depicted in Figure 10. The cavitation effect appears to be most prominent in the area after the upstream groove, where the pressure drops and the change in geometry makes the phenomenon much more likely to occur.

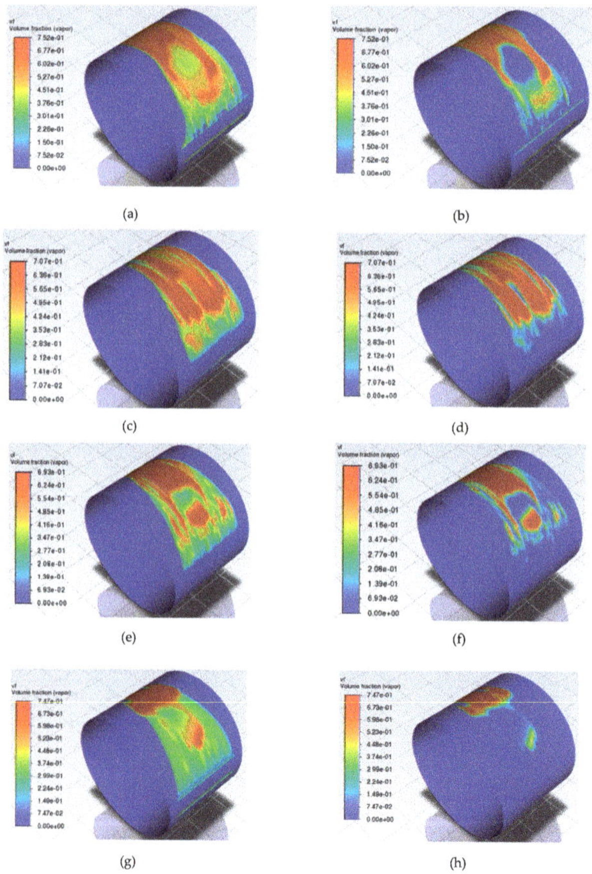

Figure 10. Vapor volume fraction distribution for various roughness values. (**a**) Smooth Bearing (inner wall). (**b**) Smooth Bearing (outer wall). (**c**) 25 μm Roughness Bearing (inner wall). (**d**) 25 μm Roughness Bearing (outer wall). (**e**) 50 μm Roughness Bearing (inner wall). (**f**) 50 μm Roughness Bearing (outer wall). (**g**) 100 μm Roughness Bearing (inner wall). (**h**) 100 μm Roughness Bearing (outer wall).

For the case of 100 μm roughness height, Figures 11 and 12 depict the turbulent dissipation rate (e) and dynamic pressure distributions, respectively. After observing the results, a connection between the grooves, turbulent effects, and cavitation is perceptible.

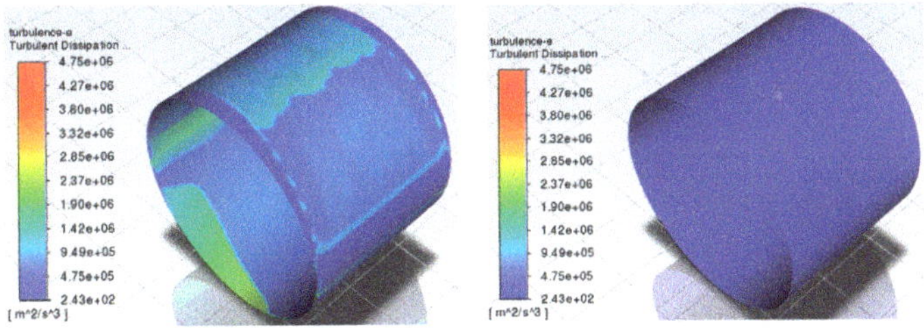

100 μm Roughness Bearing (inner wall) 100 μm Roughness Bearing (outer wall)

Figure 11. Turbulent dissipation rate distribution in journal and bearing boundary layers.

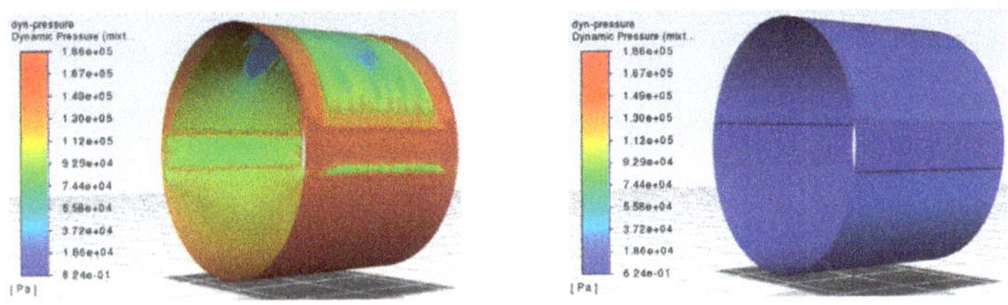

100 μm Roughness Bearing (inner wall) 100 μm Roughness Bearing (outer wall)

Figure 12. Dynamic pressure distribution in journal and bearing boundary layers.

4.2. Fluid-Structure Interaction

In this section, the effect of bearing elastic deformation and thermal expansion will be investigated. Contrary to the previous section, the flow will be assumed to be laminar and the bearing surface will be smooth.

The grooved journal bearing model used is the one used by Brito [11,12]. The bush will be modeled as described by Brito and its material will be set to the widely used Thordon water-lubricated bearing material called COMPAC [24]. Although such polymers exhibit high deformation rate under load, they are often chosen for their tribological properties. Ren et al. [25] and Serdjuchenko et al. [26] also studied the behavior of non-metallic elastic WLBs. Other researchers that have provided insight on the operation of polymer-based water-lubricated bearings are Liu et al. [27], who studied WLBs operating in high angular velocities and Litwin et al. [28–30], who conducted experiments for several different types of materials, in order to compare the most widely used WLB materials and features, such us lubricating grooves and surface roughness. Contrary to Brito's experiments, a load of smaller magnitude is used due to the lower viscosity of water, which makes it impossible for the water-lubricated bearing to have the same load capacity as the equivalent oil-based one. Therefore, a load of 2 kN is used instead of 10 kN. For a better comparison of the two bearings, oil-lubricated metal and water-lubricated polymer bearings, respectively, and, for verification purposes, Brito's case is recreated (for a load of 10 kN).

Regarding the bearing geometric and physical parameters, the following model is implemented.

The bush has an inner diameter of 100 mm and the same groove characteristics as in the CFD case above. The external diameter is 200 mm and the mesh of the model is designed using hexahedral elements with 80 divisions in the axial direction, 30 in the radial direction with inflation set at the inner wall of the bush (where the FSI occurs), and 360 in the circumferential (one division per 1°). The bush material properties are provided in

Table 5. Brito et al. [11,12] utilized a bush material of X22CrNi17, whereas the material COMPAC is used for the WLB case.

Table 5. Bearing bush structure properties.

No.	Property Designation	Symbol	COMPAC	X22CrNi17
1	Density (kg/m^3)	ρ	1160	7700
2	Elastic Modulus (GPa)	E	0.44	220
3	Ultimate Strength (MPa)	S_u	37.5	980
4	Poisson's Ratio	v	0.45	0.3
5	Thermal Conductivity (W/m*K)	w	0.25	16
6	Specific Heat (J/kg*K)	c_p	1500	460
7	Coefficient of Thermal Expansion Strain (10^{-6}/K)	α_{th}	151	16

For the oil-lubricated bearing (OLB), the equivalent rigid (CFD) bearing is the one used in the experimental validation of the model.

The results calculated in comparison to the values measured by Brito et al. [11] are displayed in Table 6.

Table 6. Comparison of grooved journal bearing model to Brito's experimental results.

Result	Present Work	Experimental [12]
Maximum Pressure (MPa)	2.76	2.55
Maximum temperature (°C)	82	69
Eccentricity ratio	0.507	0.560
Attitude angle (°)	60.3	45.5

Considering the fact that the bearing structure was not modelled in the rigid bearing, the bearing surface was modelled as an isothermal wall. Still, the results in this analysis can be considered quite accurate and no unreasonable discrepancies are noted.

For the TEHD analysis of the oil-based, lubricated X22CrNi17 bearing, the eccentricity ratio is ε = 0.494803 and an attitude angle of θ = 61.81° is used. The maximum pressure calculated is Pmax = 2.72 MPa, the moment of friction is Mf = 4.37 N m, and the maximum temperature in the oil film is Tmax = 73 °C, which is only 4 °C higher than the experimental value of Tmax = 69 °C. The maximum total deformation of the bush is DLmax = 3.5×10^{-7} m = 0.35 µm, the maximum equivalent (Von–Mises) stress is σmax = 2.22 MPa, and the maximum equivalent strain is 1.009×10^{-5}.

Figures 13 and 14 display the pressure and temperature distribution in the oil film, respectively. The thermal effect of the grooves is once again perceptible, as a drop in temperature is evident.

On the other hand, the equivalent WLB rigid (CFD) bearing solution is given as an equilibrium position of ε = 0.63125 and θ = 63.01°. The moment of friction is M_f = 0.73 N m and the maximum pressure is P_{max} = 797 kPa

For the TEHD analysis of the water-lubricated COMPAC bearing, the eccentricity ratio is ε = 0.768136 and an attitude angle of θ = 56.17° is used. The maximum pressure calculated is P_{max} = 725 kPa, the moment of friction is M_f = 0.66 N m, and the maximum temperature in the oil film is T_{max} = 23 °C. The maximum total deformation of the bush is DL_{max} = 5.07×10^{-5} m = 50.7 µm, the maximum equivalent (von–Mises) stress is σ_{max} = 915 kPa, and the maximum equivalent strain is ε_{max} = 0.0015. Figures 15 and 16 show the pressure and temperature distribution, respectively. It is important to note how the effect of cavitation is apparent in the temperature distribution above the upstream groove. Figure 17 also depicts the elastic deformation of the bush during operation.

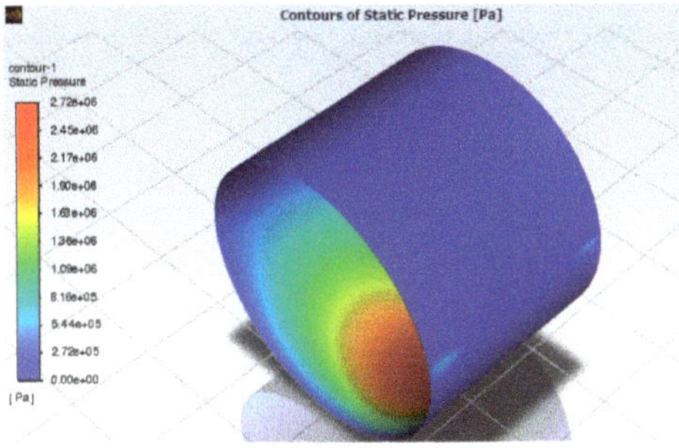

Figure 13. Pressure distribution of oil-based grooved journal bearing in FSI analysis.

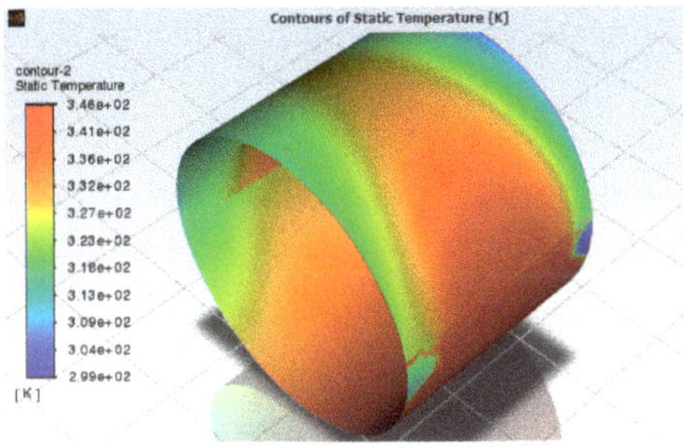

Figure 14. Temperature distribution of oil-based grooved journal bearing in FSI analysis.

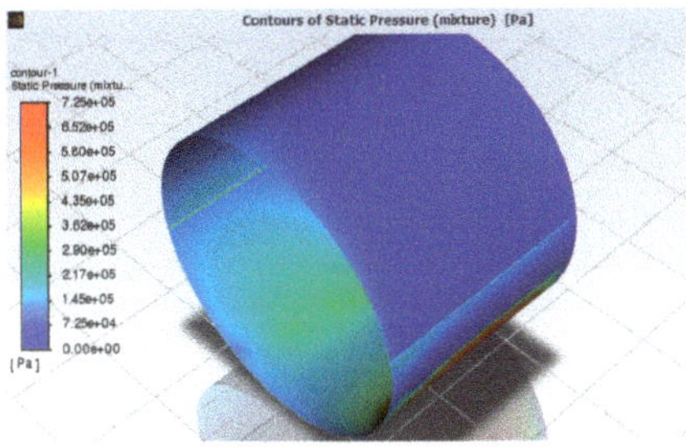

Figure 15. Pressure distribution of grooved water-lubricated journal bearing in FSI analysis.

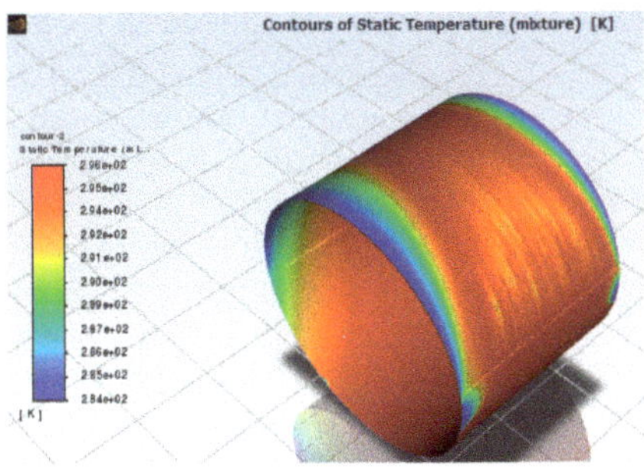

Figure 16. Temperature distribution of grooved water-lubricated journal bearing in FSI analysis.

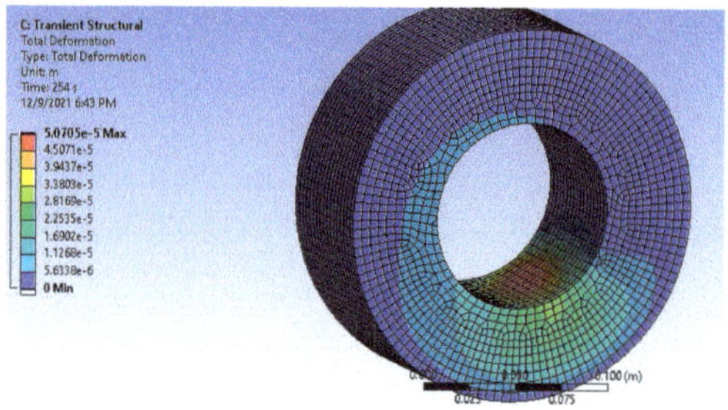

Figure 17. Total deformation distribution in water-lubricated bush in FSI analysis.

5. Discussion

In the CFD cases, with respect to different values of surface roughness, the following results are observed.

The change in geometry is especially important in the cases where the RANS equations are implemented. Here, the phenomenon of turbulence appears.

The turbulent kinetic energy is at its maximum in the water–bush boundary layer, while it is very low in the water–journal boundary layer.

The intensity of the cavitation phenomenon is high in both the inner and outer layer of the water film. Moreover, the research findings point to the phenomenon being more prevalent in the inner layer of the water film, especially for higher roughness values. This can be associated with two factors in the flow:

1. One of them is the effect of turbulence. As shown in Figure 9, the turbulent kinetic energy k is higher in the outer layer of the film, while it is minimized towards the inner layer. The logical conclusion is that the turbulent kinetic energy dissipates into heat through viscous forces and eddy production, which would make the flow erratic, cause pressure fluctuations, and lead to cavitation inception. Indeed, the results depicted in the turbulent dissipation rate (ε) distribution validate this assumption. These can be seen in Figure 11 for the case of a 100 μm rough bearing, where the turbulent dissipation rate ε distribution is shown for both the inner and outer layer of the water film.
2. The other factor is the rotation of the shaft, which, coupled with the eddy production and dissipation, leads to drops in pressure and, thus, enhances the cavitation phenomenon. This assumption can be verified by investigating the dynamic pressure in the water flow, which represents the decrease in the static pressure and increase in the dynamic one, due to the velocity of the fluid. Figure 12 exhibits how the dynamic pressure is much higher in the inner layer of the water film, while it is almost non-existent in the outer one, which is of course expected as the bearing wall is stationary.

Taking the above into consideration, a kind of pattern forms. The geometry of the bearing (high surface roughness and axial grooves) coupled with the low viscosity of the water (liquid) leads to an increase in turbulent kinetic energy near the bearing surface. Energy dissipation is the outcome of turbulence. The turbulent kinetic energy leads to eddy production. The eddies interact with each other at the initial phase of dissipation, producing larger eddies. As the flow comes to rest, the eddies break down into smaller ones until such patterns are no longer present. This procedure is much more prominent in the inner wall (water–journal boundary layer), which is clearly shown in Figure 11, where the turbulent dissipation rate is much higher in the inner wall. The smooth surface and high velocity of the journal (dynamic pressure) promote a laminar flow, which explains why the dissipation rate (e) is so high at this area.

On the other hand, after observing the thermo-elastohydrodynamic analyses, it is noted that the solution of the WLB FSI (fluid–structure interaction) or TEHD analysis was affected to a high degree by the bearing bush properties. More importantly, the drop in the maximum pressure is of high interest, as the same external load can be supported more evenly. This is because the deformation forms micro pockets on the bearing surface, where the pressure can be distributed over a larger area. On the other hand, in the case of the equivalent OLB, it was observed that, although the FSI solution was closer to the experimental results measured by Brito [11], the difference was not as significant as in the case of the WLB. For that reason, there is clearly a need to be able to properly include the bearing structure into the tribological design of WLBs, since the materials from which they are made (mostly polymers) exhibit much greater elastic deformation and thermal expansion compared to their conventional bearing counterparts, which are made of metal materials. As a result, emitting the bearing deformation from the study for certain load ranges can lead to results, which, even for small external forces, might lead to large errors and uncertainty.

6. Conclusions

Several conclusions can be reached by considering the results. The addition of Vogel's equation as well as tabular data of vaporization pressure and saturation temperature make the simulation of the flow more accurate. The larger bearing surface roughness values proved to have a negative effect on the performance of the water-lubricated bearing, since, for the same load applied (2 kN), the friction coefficient became much greater, a characteristic that is strongly associated with a loss of efficiency, a rise in temperature, and wear. Moreover, the addition of higher roughness was also proven to increase the turbulent kinetic energy k, which intensified the cavitation phenomenon. On the other hand, it is known from lubrication theory that a completely smooth surface is not ideal, since it does not allow for the formation of a thick enough oil film; consequently, a sufficient quantity of lubricant cannot become entrapped in the surface asperities and establish a stable oil wedge. It is, therefore, necessary that extra attention is given to the processing of the bearing surface during the manufacturing procedure. In the case of the FSI simulation, higher accuracy compared to the rigid (CFD) counterpart was achieved, since in the TEHD case the bush geometry and material were also modeled and, therefore, allowed thermal boundary conditions to be applied and accounted for the elastic deformation of the bush due to pressure and temperature change. The latter was proven to be of great importance in the case of water-lubricated bearings, where the bush material led to significant deformation, thus affecting the results calculated to a greater magnitude. Suggestions for further research are the study of misalignment in water-lubricated journal bearings, the use of the FSI method for studying different bush materials, and the in-depth investigation of the cavitation phenomenon in water-lubricated journal bearings. The phenomenon of mixed lubrication is also a subject of interest, while the study of bearing roughness running through the use of an FSI method would also have value for future research.

Author Contributions: Conceptualization, D.C.; methodology, D.C.; software, D.C.; validation, D.C.; formal analysis, D.C.; investigation, D.C.; resources, D.C.; data curation, D.C.; writing—original draft preparation, D.C.; writing—review and editing, D.C. and P.G.N.; visualization, D.C.; supervision, P.G.N.; project administration; P.G.N. All authors have read and agreed to the published version of the manuscript.

Funding: This research received no external funding.

Data Availability Statement: The raw data supporting the conclusions of this article will be made available by the authors on request.

Conflicts of Interest: The authors declare no conflict of interest.

Nomenclature

Symbol	Definition	Unit
L	Length of bearing	m
D	Inner diameter of bearing	m
R_b	Radius of bearing	m
R_j	Radius of journal	m
e	Eccentricity	m
ε	Eccentricity ratio	-
cr	Radial clearance	m
θ	Attitude angle	°
Φ	Bearing angle	°
P	Pressure	Pa
uf	Angular velocity of journal	rpm
μ	Dynamic viscosity	Pa s
Cp	Thermal capacity	J/(kg K)
W	Journal load	N
Mf	Moment of friction	N m
Ff	Friction force	N
f	Coefficient of friction	-
g	Gravitational acceleration	m/s^2
K_s	Roughness height	m
C_s	Roughness constant	-
E	Modulus of elasticity	Pa
S_u	Ultimate strength	Pa
v	Poisson's ratio	-
w	Thermal conductivity	W/m K
a_{th}	Coefficient of thermal expansion strain	10^{-6}/K
σ_{max}	Maximum equivalent stress	Pa
ε_{max}	Maximum equivalent strain	-
DL_{max}	Maximum deformation	m
T	Temperature	K

References

1. Hori, Y. *Hydrodynamic Lubrication*; Springer: Tokyo, Japan, 2006.
2. Nosonovsky, M.; Bhushan, B. *Green Tribology: Biomimetics, Energy Conservation and Sustainability*; Springer: Berlin/Heidelberg, Germany, 2012.
3. Zhou, W.; Wei, X.; Wang, L.; Wu, G. A superlinear iteration method for calculation of finite length journal bearing's static equilibrium position. *R. Soc. Open Sci.* **2017**, *4*, 161059. [CrossRef]
4. Tauviqirrahman, M.; Jamari, J.; Wicaksono, A.A.; Muchammad, M.; Susilowati, S.; Ngatilah, Y.; Pujiastuti, C. CFD Analysis of Journal Bearing with a Heterogeneous Rough/Smooth Surface. *Lubricants* **2021**, *9*, 88. [CrossRef]
5. Tauviqirrahman, M.; Paryanto, P.; Indrawan, H.; Cahyo, N.; Simaremare, A.; Aisyah, S. Optimal Design of Lubricated Journal Bearing Under Surface Roughness Arrangement. In Proceedings of the 2019 International Conference on Technologies and Policies in Electric Power & Energy, Yogyakarta, Indonesia, 21–22 October 2019; pp. 1–5. [CrossRef]
6. Zhang, X.; Yin, Z.; Gao, G.; Li, Z. Determination of stiffness coefficients of hydrodynamic water-lubricated plain journal bearings. *Tribol. Int.* **2015**, *85*, 37–47. [CrossRef]

7. Zheng, X.B.; Liu, L.L.; Guo, P.C.; Hong, F.; Luo, X.Q. Improved Schnerr-Sauer cavitation model for unsteady cavitating flow on NACA66. *IOP Conf. Ser. Earth Environ. Sci.* **2018**, *163*, 012020. [CrossRef]
8. Tamboli, K.; Athre, K. Experimental Investigations on Water Lubricated Hydrodynamic Bearing. *Procedia Technol.* **2016**, *23*, 68–75. [CrossRef]
9. Xie, Z.; Li, J.; Tian, Y.; Du, P.; Zhao, B.; Xu, F. Theoretical and experimental study on influences of surface texture on lubrication performance of a novel bearing. *Tribol. Int.* **2024**, *193*, 109351. [CrossRef]
10. Xie, Z.; Jiao, J.; Zhao, B.; Zhang, J.; Xu, F. Theoretical and experimental research on the effect of bi-directional misalignment on the static and dynamic characteristics of a novel bearing. *Mech. Syst. Signal Process.* **2024**, *208*, 111041. [CrossRef]
11. Brito, F.P.; Bouyer, J.; Miranda, A.S.; Fillon, M. Experimental Investigation of the Influence of Supply Temperature and Supply Pressure on the Performance of a Two Axial Groove Hydrodynamic Journal Bearing. *J. Tribol.* **2007**, *129*, 98–105. [CrossRef]
12. Brito, F.P.; Miranda, A.S.; Fillon, M. Analysis of the effect of grooves in single and twin axial groove journal bearings under varying load direction. *Tribol. Int.* **2016**, *103*, 609–619. [CrossRef]
13. Reynolds, O. On the Theory of Lubrication and its Application to Mr. BEAUCHAMP TOWER'S Experiments, including an Experimental Determination of the Viscosity of Olive Oil. *Phil. Trans. R. Soc.* **1886**, *177*, 157–167.
14. Tower, B. First Report on Friction-Experiments (Friction of Lubricated Bearings). *Proc. Inst. Mech. Eng.* **1883**, *34*, 632–659. [CrossRef]
15. Tower, B. Second Report on Friction-Experiments (Experiments on the Oil Pressure in a Bearing). *Proc. Inst. Mech. Eng.* **1885**, *22*, 58–70. [CrossRef]
16. Çengel, Y.A. *Thermodynamics: An Engineering Approach*; McGraw-Hill Higher Education: Boston, MA, USA, 2008.
17. Bhatt, C.P.; McClain, S.T. Assessment of Uncertainty in Equivalent Sand-Grain Roughness Methods. In Proceedings of the ASME 2007 International Mechanical Engineering Congress and Exposition, Seattle, WA, USA, 11–15 November 2007; Volume 8, pp. 719–728. [CrossRef]
18. Kohnke, P. *Ansys Theory Reference, 001242*, 11th ed.; Ansys Inc.: Canonsburg, PA, USA.
19. Failawati, V.; Yamin, M.; Poernomosari, S.; Naik, A.R. CFD Analysis of the Eccentricity Ratio on Journal Bearing due to Differences in Lubrication Type. *J. Nov. Eng. Sci. Technol.* **2022**, *1*, 51–56. [CrossRef]
20. de Kraker, A.; van Ostayen, R.A.J.; Rixen, D.J. Calculation of Stribeck curves for (water) lubricated journal bearings. *Tribol. Int.* **2007**, *40*, 459–469. [CrossRef]
21. Shigley, J.; Mischke, C.; Brown, T. *Standard Handbook of Machine Design*, 3rd ed.; Mc Graw-Hill: Boston, MA, USA, 2014.
22. Bombos, D.; Nikolakopoulos, P. Tribological design of a multistep journal bearing. *Simul. Model. Pract. Theory* **2016**, *68*, 18–32. [CrossRef]
23. Gong, J.; Jin, Y.; Liu, Z.; Jiang, H.; Xiao, M. Study on influencing factors of lubrication performance of water-lubricated micro-groove bearing. *Tribol. Int.* **2019**, *129*, 390–397. [CrossRef]
24. *Thordon Engineering Manual Version E2006.1*; Thordon, ON, Canada, 2006; pp. 2–12.
25. Ren, G. Hypo-elastohydrodynamic lubrication of journal bearings with deformable surface. *Tribol. Int.* **2022**, *175*, 107787. [CrossRef]
26. Serdjuchenko, A.; Ursolov, A.; Batrak, Y. Asymptotic estimation of the fluid film pressure in non-metallic water-lubricated staved stern tube bearings. *Tribol. Int.* **2022**, *175*, 107798. [CrossRef]
27. Liu, G.; Li, M. Experimental study on the lubrication characteristics of water-lubricated rubber bearings at high rotating speeds. *Tribol. Int.* **2021**, *157*, 106868. [CrossRef]
28. Litwin, W. Properties comparison of rubber and three layer PTFE-NBR-bronze water lubricated bearings with lubricating grooves along entire bush circumference based on experimental tests. *Tribol. Int.* **2015**, *90*, 404–411. [CrossRef]
29. Litwin, W.; Dymarski, C. Experimental research on water-lubricated marine stern tube bearings in conditions of improper lubrication and cooling causing rapid bush wear. *Tribol. Int.* **2016**, *95*, 449–455. [CrossRef]
30. Litwin, W. Influence of Surface Roughness Topography on Properties of Water-Lubricated Polymer Bearings: Experimental Research. *Tribol. Trans.* **2011**, *54*, 351–361. [CrossRef]

MDPI AG
Grosspeteranlage 5
4052 Basel
Switzerland
Tel.: +41 61 683 77 34

Lubricants Editorial Office
E-mail: lubricants@mdpi.com
www.mdpi.com/journal/lubricants

www.ingramcontent.com/pod-product-compliance
Lightning Source LLC
Chambersburg PA
CBHW061449210926
51850CB00008B/1665
* 9 7 8 3 7 2 5 8 2 1 2 5 9 *